ECODEVELOPMENT
Glocal Perspectives

ECODEVELOPMENT
Glocal Perspectives

Essays in Honour of
R. B. SINGH

Edited by
SUBHASH ANAND
Associate Professor
Department of Geography
Swami Shraddhanand College
University of Delhi, Delhi
India

RESEARCH INDIA PRESS
New Delhi (INDIA)

ISBN : 978-81-89131-46-3

First Edition: 2011

Cataloging in Publication Data—DK
Courtesy: D.K. Agencies (P) Ltd. <docinfo@dkagencies.com>

Ecodevelopment : glocal perspectives : essays in honour of R.B. Singh / edited by Subhash Anand. — 1st ed.
p. cm.
Includes bibliographical references and index.
ISBN 9788189131463

1. Economic development—Environmental aspects—India. 2. Sustainable development—India. I. Singh, R. B. (Ram Babu), 1955- II. Anand, Subhash, 1972-

DDC 338.9270954 22

Typesetting by
G. R. Sharma
9968281826

Published by :
RESEARCH INDIA PRESS
E-6/34, Sangam Vihar
New Delhi-110062
Phone: 011-26047013, (M) 9818085794
E-mail: researchindiapress@yahoo.co.in

Contents

Section II
DEVELOPMENT

Foreword

This book is a fitting dedication to Dr. R.B. Singh in that it provides an excellent overview of a series of critical environment and development issues at the forefront of Indian society. Dr. Singh has devoted much of his career for researching such problems and to training students in ways to face the policy challenges that are presenting themselves in relation to such important issues as ecodevelopment, climate variability and waste management. Over the twenty-two chapters of the book, the various authors have captured the essences of environment and development often through detailed case studies, many of which have broader applicability to the country as a whole. Taken as a whole, the collection represents a valuable addition to our understanding of the state of the Indian environment and its management.

We were particularly pleased to see that the book contained a number of chapters on issues related to climate change issues and community level involvement in resource and environmental management. These topics are not just important in there own right, but also represent two areas where Dr. R.B. Singh has placed considerable effort and made a significant contribution over the years. Dr. Singh has a number of published papers on the issue of climate change and it is a topic he feels personally very dedicated to. In our research projects with Dr. Singh he has also shown his strong passion for community involvement. He not only assisted us in taking a community-based approach to our research, but also ensured that local people and leaders participated in workshops and dissemination activities related to the work that we were doing.

This book is an important addition to the library of any researcher doing work on ecodevelopment in India and to any student studying the environment.

Dr. James S. Gardner
Professor Emeritus
Victoria, Canada

Dr. A. John Sinclair
Professor, University of Manitoba
Winnipeg, Canada

Preface

The present volume on ecodevelopment is a modest attempt to honour Prof. R.B. Singh, who is teaching in the Department of Geography, Delhi School of Economics, University of Delhi, Delhi since last more than twenty five years. A number of experts have contributed the papers for this volume. I received papers on different dimensions of environment and development at Glocal Scale (Global to Local). Ecodevelopment is a multi-disciplinary and multi-stakeholders strategy. Basically, ecodevelopment aims at a combination of development and attainment of equilibrium with the environment for sustainability. The linkages between environment and development have been discussed from a variety of perspectives like social, demographic, economic, political, technological and administrative besides others. At a time when the world is fast reckoning the impeding horrors of global warming, it is equally important to take cognisance of the fact that environmental degradation also drastically offsets improvements achieved by economic prosperity. The primary purpose for economic development is to improve the life of the people by providing them employment and thereby enabling them for better shelter, food, healthcare and education. Economic development can be achieved through industrialisation and agriculture development. The economic development must be accompanied by inclusive and sustainable growth. As human development reaches a climax, people all over the world realize the significance of conservation of the environment and the need for a sustainable development.

It has been argued that the present rates are unsustainable, and hence, if economic growth has to be sustained in the long run, the only way to do it is to ensure that there is no depletion and degradation of the natural resource base. All developmental activities, irrespective of their scale, magnitude and nature have environmental effects short as well as long terms. The shifting of developmental paradigms from Classical Environmental Determinism to Marxist gave new shape to the utilisation of land and labour so as to optimize the spatial returns in terms of development. Class inequities, geographical and political limitations greatly compound India's environmental situation. But with greater sensitivity towards basic needs, one hope India's future development will be harmonized and reconciled with environmental contradiction.

This book is an outcome of the valuable contributions made by eminent scientists, teachers and scholars who have been striving to develop alternative strategies solutions and models for ecodevelopment through research, screening, monitoring and experiments varying from regional to global level. This volume is an attempt to respond to the environmental challenge due to developmental activities. We must appreciate that our human society, like nature itself, is dynamic, not independent, and that we must live within its bounds. We have the potential to alter our environment globally but we must exercise this power with discretion. The book is divided into two sections, first section Environment consists fourteen papers on various environmental issues while second section Development having eight papers on different developmental perspective from local level to global level. This volume will be helpful and useful for research scholars, environmentalists, planners, social scientists, academicians of others connected in the field of environment and development. I do hope that this book can provide atleast context for the debate that must take place to ensure a sustainable future for independent global society. I am sincerely thankful to all contributors who extended their contributions to make this volume worth presentation. I would like to take this opportunity to acknowledge my deepest gratitude to my all colleagues, friends, relatives and students for their good cooperation, encouragement and moral support. I am greatly thankful to Shouraseni Sen Roy for writing a biographical sketch of Prof. R.B. Singh. I am truly indebted to my all family members particulary my parents. I would like to pay a true tribute to my father Late Sh. Bhoop Singh who left this world 12 years ago, he is always source of inspiration for me. I wish to express my sincere thanks and gratitude to my wife Mrs. Premwati for her full cooperation, affectionate encouragement all the times. She showed extreme patience and stood by at each and every stage. My thanks are also due to Research India Press, New Delhi for bringing out this volume on time.

20/02/2011 **Subhash Anand**
Delhi

List of Contributors

Ambika, Department of Chemistry, Hansraj College, University of Delhi, Delhi

Anju Dabas, Department of Geography, Kanya Mahavidalaya, Sonipat, Haryana.

Ashwini Kumar, Department of Geography, M.D. University, Rohatak, Haryana.

Binu Sangwan, Department of Geography, M.D. University, Rohtak, Haryana.

B.S. Yadav, Department of Economics, DNPG College, CCS University, Meerut, U.P.

Fazlurehman, Institute of Geography, Urban and Regional Planning, University of Peshawar Khyber Pakhtoonkhwa, Pakistan.

Iffat Tabassum, Institute of Geography, Urban and Regional Planning, University of Peshawar Khyber Pakhtoonkhwa, Pakistan.

Inder Jeet, Department of Geography, M.D. University, Rohtak, Haryana.

Jagbir Singh, Department of Geography, Swami Shraddhanand College, University of Delhi, Delhi.

Jitender Dwivedi, Gorakhpur Environmental Action Group, Gorakhpur, U.P.

Kaushal Kumar Sharma, Department of Geography, Kirorimal College, University of Delhi, Delhi

Kiran Dabas, Department of Geography, Swami Shraddhanand College, University of Delhi, Delhi.

K.K. Singh, Gorakhpur Environmental Action Group, Gorakhpur, U.P.

Kuldeep Singh, Department of Economics, Kurukshetra University, Kurukshetra, Haryana.

M. Mumtaz, Institute of Geography, Urban and Regional Planning, University of Peshawar Khyber Pakhtoonkhwa, Pakistan.

Monika Ahlawat, Department of Geography, Dr. Bhim Rao Ambedkar College, University of Delhi, Delhi.

Nitin Punit, Department of Geography, University of Delhi, Delhi.

Poonam Kumaria, Department of Geography, Miranda House, University of Delhi, Delhi.

Pradeep Pratap Singh, Department of Chemistry, Swami Shraddhanand College, University of Delhi, Delhi.

Rakhi Parijat, Department of Geography, Miranda House, University of Delhi, Delhi.

Randhir Singh Sangwan, Department of Geography, M.D. University, Rohtak.

Sachi Rana, Department of Management Studies, IAMR College of Engineering (UP Technical University), Meerut, U.P.

Savita Kaushal, Department of EMIS, National University of Educational Planning and Administration, New Delhi.

Shalini Sheel, Department of PGDM, CMD, Modinagar, Ghaziabad, U.P.

Shouraseni Sen Roy, Department of Geography and Regional Studies, University of Miami, Coral Gables, Florida-33124, USA.

S. K. Bandooni, Department of Geography, SBS(E) College, University of Delhi, Delhi.

Sneh Sangwan, Department of Geography, B.P.S. Mahila Vishwavidyalya, Sonipat, Haryana.

Subhash Anand, Department of Geography, Swami Shraddhanand College, University of Delhi, Delhi.

Surender Mor, Department of Business Administration, Ch. Devi Lal University, Sirsa, Haryana.

Suresh Kumar, Department of African Studies, University of Delhi, Delhi.

Swarnima Singh, Department of Geography, Kirorimal College, University of Delhi, Delhi.

Vaneeta Chandana, Department of Geography, SBS(E) College, University of Delhi, Delhi.

Victor R. Savage, Department of Geography, National University of Singapore, Singapore.

R. B. SINGH: Biographical Profile and Academic Achievements

SHOURASENI SEN ROY
Department of Geography and Regional Studies, University of Miami, USA

Professor R. B. Singh is one of the most renowned geographers of India specifically in the field of environmental geography and GIS. As a result of his well-established national reputation, coupled with strong research and collaborative skills, several international scholars have chosen him as the key resource person to work on research projects in India. This is clearly evident from the multiple international collaborations he has successfully completed, and still ongoing throughout his career. Through the rest of this article I will make an attempt to summarize his career and contributions as a role model to many young aspiring Geographers and scholars in general.

Prof. Singh was born on February 3rd, 1955 in Bihar, India. As a son of Freedom Fighter he excelled at all levels as a brilliant and hardworking student. He completed his Masters with First Class in 1977 at the age of 22 from one of the most elite universities in India, Banaras Hindu University (BHU), followed by PhD after four years in 1981 from the same university. While studying in BHU, he also received a First Class Diploma in Statistics that greatly helped him in his research in the future. He also received his training in GIS during its early stages of development through the U.N.

Research Training in GIS Technology from the UNEP/UNITAR, Geneva and EPFL Laussane. The rest of this article is an attempt to summarize all his academic achievements as he served in different roles in various capacities in the academic world.

For the past almost three decades he has devoted most of his time into research and advancement of teaching in various capacities at various levels of higher education. He started his career as a Pool Officer in BHU for two years followed by his initial appointment in University of Delhi as a Lecturer for three years. He was later appointed as UGC Research Scientist-B for eight and half years two months, followed by his promotion to UGC Research Scientist C (Professor). He served in this position for almost six years. He was promoted to Associate Professor in the Department of Geography, University of Delhi. In short, he has been constantly recognized for his academic achievements as reflected by regular promotions throughout his academic career.

His achievements can be summarized in four main roles – teacher, mentor, collaborator, and administrator. Usually in academics one is very good in one of the above mentioned roles or maximum two. However, Prof. Singh excelled in all of the roles and never shied away from taking on new challenges and roles throughout his career serving as a great role model for aspiring young academics. In 1988 the UNESCO/International Social Science Council (Paris) awarded him Research and Study Grants Award in Social and Human Sciences.

As a teacher/professor he taught twelve different courses in the Department of Geography. Given the wide applicability of his expertise to other disciplines, he taught four courses in other departments in the University of Delhi. He has been teaching a wide variety of courses ranging from fundamentals in environment and ecology, GIS and remote sensing, to advanced topics in environmental security, environmental impact assessment, to regional planning and development. His classes have always filled up given wide relevance of the topics covered in his classes. There is hardly any student doing their Masters in Geography at the University of Delhi, who has not taken at least one of his environmental or GIS courses, with a majority of them deciding to pursue higher degrees in the field of environmental geography. This is evident from the number of graduate students that have completed their degrees with him (25 PhD students and 59 MPhil students), as well as the number of students who are currently enrolled under his guidance (5 PhD students and 12 MPhil students). All of his former students are pursuing successful careers in various academic institutions across the world, as well as environmental research organizations.

His professional contributions can be assessed from the large list of research work that has been published throughout his academic career in the form of peer reviewed journal articles and book chapters in renowned international publication outlets, edited volumes, as well as popular and general review write-ups published in various local magazines, newspapers, and reports. Prof. Singh's research addresses critical issues in the field of regional development, land and water management, land use and land cover change, disaster management, and remote sensing and GIS technology. He has very successfully propagated the findings of his research to the wider community by organizing regular workshops, conferences, as well as press releases at regular intervals. The results of his research have been published in the form of 37 research volumes/books, 169 research papers published in national and international journals (i.e. Agriculture, Ecosystem and Environment, Hydrological Processes, Mountain Research and Development, Journal of Mountain Science, Advances in Earth Science, Advances in Limnology, IAHS Pub., Geographica, Asian Geographer, Acta Hydrotechnica, Indonesian Jl. of Geography, Journal of Geography Education, AOGS Pub., Tourism Recreation Research, Annals of NAGI, Transactions of IIG, The Geographer, The Indian Geographical Jl., Physical Geography).

His research contributions can be mainly grouped into two critical areas of the Indian subcontinent including the Delhi Metropolitan Region and the Indian Himalayas. He has completed successfully several collaborative research projects funded by various International organizations. A few of the successfully completed research projects are listed below:

1. Co-Investigator: Shastri Applied Research Project (SHARP) on Role of Public, Private and Civil Sectors in Sustainable Environmental Management collaborated with University Of Manitoba and the

University of Winnipeg, Winnipeg, Canada (2003-2005).

2. Project Director: ICSSR-Indo-Dutch Program on Alternative in Dev. (IDPAD) on Environmental Implications and its Socio-Economic Implications in Rural-Urban Fringe of Delhi -University of Delhi and Univ. of Groningen, The Netherlands. (1997-2002).
3. Chief Co-investigator-CIDA-SICI: Partnership Project-II on Urban Development and Environmental Impacts in Mountain Context, University of Delhi and University of Manitoba, Canada. (1998-2002)
4. Project Director, DFID: Research Project on Enhancing Food Chain Integrity Pollution Impact on Vegetable System in Peri-Urban Areas, Collaboration with Imperial College, London, UK (2000-2002).
5. Chief Co-investigator-CIDA-SICI: Partnership Project-I on Sustainable Development of Mountain Environment in India and Canada, University of Delhi & University of Manitoba, Canada, (1994-1997).
6. Project Director: Ministry of Agriculture Project for Preparation of Perspective Plan for Land Resources in N. Zone, India (1994).

Dr. Singh has contributed significantly to the academic activities of the International Geographical Union. Since 1984 IGU Paris Congress, he has attended all Main Congresses and most IGU Regional Conferences and has been actively participating in the academic programmes of various IGU Commissions and Working Groups. His previous assignments as Full Member in the IGU Commissions/Working Groups/Study Groups include Mountain Geoecology and Resource Management (1988-92), Development Issues in Marginal Lands (1992-1996), Land Use and Cover Change (1996-2004) and Biogeography and Biodiversity (2004-2008). Presently, he is Vice Chair of the IGU Commission on Biogeography and Biodiversity (2008-2012). As a geographer, he strongly believes in demonstrating competence, experience, interdisciplinary orientation and professionalism in the use of geographical knowledge and skills. In 1991, he had also organized an interdisciplinary IGU Seminar at Delhi in which 4 IGU Commissions/Groups participated with the partial support provided by the IGU. As Secretary General of the National Association of Geographers, he has organized various sessions in the national Congresses for promoting IGU Initiative Culture and Civilization for Human Development. As Head of the Indian Delegation to IGU Tunis (2008), he also contributed in the IGU Finance Committee during General Assembly. He is committed to maintain professional standards in the field of geography through ongoing academic research and knowledge dissemination activities.

He has actively participated in conferences worldwide by presenting results from his research at various national and international platforms with the support of IGU, UNESCO, UNEP, UNU, UNCRD, UN-IDNDR, WMO-WCRP, ICIMOD, NASDA, UNITAR, IGBP, IAHS, GWSP and IHDP held in USA, Canada, Mexico, Japan, Australia, France, Finland, Denmark, Spain, U.K., Netherlands, Norway, Germany, Switzerland, Russia, Poland, Czech Rep., Mongolia, Malaysia, Thailand, Egypt, China, Taiwan, Tunisia, Sweden, Israel, South Korea, Ireland, S. Africa, Brazil, Singapore, Sri Lanka, Indonesia, Nepal and Bhutan. He has been recognized by various well known organizations for his professional contributions by being nominated to the IUGG-IGU Joint National committee of the Indian National Science Academy, member of one of the groups of NCR 2020 Perspective Plan. He has also served as the Secretary General of the National Association of Geographers India (NAGI, which is the leading Geographers association in the Indian subcontinent) and is currently serving as the Managing Editor of NAGI. Under nine years of General Secretaryship of Dr. Singh, National Association of Geographers, India (NAGI) achieved new horizons and heights by enlarging membership and financial base. Some of his individual published works and national and international academic achievements are listed below:

DR. R. B. SINGH: ACADEMIC ACHIEVMENTS AT INTERNATIONAL AND NATIONAL LEVEL

International Award/Fellowship:

1. UNESCO/ISSC, Paris Research and Study Grant Award in Social & Human Sciences, 1988.

2. UNEP-UNITAR Fellowship for Training in GIS Technology in the Field of Environment, at EPFL-Laussane and GRID-Geneva, October 1988 to April 1989.
3. Guest Fellowship for Asian Scholar, Nordic Institute of Asian Studies, Copenhagen, Denmark during Sept.-Oct. 1998.
4. Several Travel Fellowships/ Support from UNEP, UNITAR,UNU, UNCRD, WCRP, IAHS, IGU, NASDA, INSA, UGC, SICI, MAIRS and Univ. of Delhi etc. for participating and presenting papers, Chairing session and discussing research projects in different countries.

International / National Academic Recognition:

1. Nominated Member-IUGG-IGU Joint Nat. Committee, Indian National Science Academy (2008-12).
2. Vice-Chair-IGU Commission on Biogeography and Biodiversity (2008-12)
3. Managing Editor- Nat. Asso. Of Geographers, India , NAGI (2010-13)
4. Elected Secretary General-Nat. Asso. Of Geographers India (2002-2010); Joint Secretary in 2001
5. Leader, INSA National Delegation to Tunis International Geography Congress, 2008, Tunisia.
6. Nominated Steering Comm. Members of Various Commissions - Int. Geographical Union (IGU) Commissions (1988-2008).
7. Dy. Coordinator, UGC-DRS Programme, Department of Geography, University of Delhi.
8. Nominated Special Representative-World Asso. Of Soil and Water Conservation, since 2007.
9. Associate Faculty, Global Earth System Governance Project of ESSP, The Netherlands since 2008.
10. Nominated Member Curriculum Development for Geoinformatics jointly nominated by UGC and Department of Space, Govt. of India.
11. Regional Promoter-South Asia, IGU Initiative Culture, Civilisations and Human Dev. since 2006.

12. Member-Editorial Advisory Board, Indian Jl. Of Development Research and Social Action.
13. Member, INSA National Delegation to Glasgow International Geography Congress, 2004
14. Elected Secretary, Hydrology Section, Asia-Ocenia Geoscience Society, Singapore, 2004-2007.
15. Nominated Member –Study Group for Preparation of NCR Plan-2021, Govt. of India.
16. Nominated Member-Board of Studies, TERI School of Advanced Studies, New Delhi since 2003.
17. Nominated Member, UGC Xth Plan Visiting Team to University of Pune (2002).
18. South Asian Rep. in Managing Board of Commonwealth Geographical Bureau (1992-2000).
19. Nominated Member, ICAR Programme Impl. Comm. on Land Use/Cover Change and Food Security in Indo-Gangetic Plain.
20. Expert - UGC-CEC, Glossary of Geography, Comm. of Sc. & Tech. Terminology, Government of India, NRC for Evn. Education, NCERT.
21. Editor, Asso. of Geographical Studies, Delhi University-2002-2007.
22. Chairman - NAGI -Comm. on Env. Monitoring and GIS Since 1989-2000.
23. Judge for Delhi and National Level Fairs for the INTEL Science Talent Discovery Fair, 2001.
24. Elected Recorder - Earth System Science Section of Indian Science Congress Association (1996-98).

R.B.SINGH: LIST OF PUBLICATIONS, RESEARCH SUPERVISION

A. Published Books/Atlas

1. Environmental Change and Biodiversity, (Jointly with Suraj Mal), Rawat Publication, Jaipur, 2009.

2. Medicinal Plants and Sustainable Livelihood, (Jointly with M.Panwar and S.K.Sharma), R.K.Books, New Delhi, Pages 342, 2010.
3. The World Atlas-Earth Concise (Jointly with Chief Consultant Charles F.Gritzner and several other Contributors), Millennium House Ltd., NSW, Australia, pages 576, 2008.
4. Natural Hazards and Disaster Management (jointly with Balaka Dey), CBSE, Delhi, pages 45, 2006. *www.cbse.nic.in/natural%20hazards%20&%20disaster%20management.pdf (Internet Pub.)*
5. Risk Assessment and Vulnerability Analysis, IGNOU PG Diploma in Disaster Management- MPA-003, New Delhi., pages 355, 2005.
6. Urban Development and Anthropogenic Climate Change-Experience in Indian Metropolitan Cities, (jointly with B.Kumar), Manak Pub., New Delhi, pages 216, 2003.
7. Practical Work in Geography, Part 2, Text Book for Class XII, NCERT, New Delhi, pages 58, 2003.
8. Climate Variability, Extreme Events and Agricultural Productivity in Mountain Regions, (jointly with S.Sen Roy), Oxford & IBH Pub., New Delhi, pages 232, 2002.
9. Practical Work in Geography, Part 1, Text Book for Class XI, NCERT, New Delhi, pages 65, 2002.
10. Predicting Global Change Impacts on Mountain Hydrology and Ecology: Integrated Catchment Hydrology/ Altitudinal Gradient Studies, (Together with others), IGBP Report 43, Royal Swedish Academy of Sciences, Stockholm, Pages 61, 1997.
11. Space Technology for Disaster Monitoring and Mitigation in India, INCEDE, Inst. Of Industrial Science, University of Tokyo, Pages 58, 1994.
12. Geography of Rural Development, New Delhi, Inter India Publisher, 1986.
13. Changing Frontiers of Indian Village Ecology, (jointly with R. P.B. Singh), Varanasi, Banaras Hindu University-NGSI, 1981.

B. Edited Research Volumes:

14. Glocal Environmental Education, (Jointly with Y.Himiyama, F. Kanda and J. Hindson), Rawat Publications, Jaipur, Pages 246, 2010.
15. Biogeography and Biodiversity, Rawat Publications, Jaipur, Pages 340, 2009.
16. Land Use-Reflection on Spatial Informatics, Agriculture and Development, (Jointly with M.M.Jha), Concept Publishing Company, New Delhi, Pages 318, 2008
17. Spatial Information Technology for Natural Resources Management, Ed. (Jointly with N.Mohammad and A. Datta), Concept Publishing Company, New Delhi, Pages 180, 2007.
18. Advances of Geosciences: Vol. 4: Hydrological Sciences (HS), Ed., World Scientific Publishing Co., Singapore, Pages 295, 2006.
19. Natural Hazards and Disaster Management, Ed. R.B.Singh, Rawat Pub., Jaipur, Pages 352, 2006.
20. Sustainable Urban Development, Ed., Concept Publishing Company, New Delhi, Pages 431, 2006.
21. Human Dimensions of Sustainable Development, Ed. Rawat Pub., Jaipur, pages 279, 2002
22. Urban Sustainability in the context of Global Change, Ed., Science Pub., Inc., Enfield (NH), USA and Oxford & IBH Pub., New Delhi, Pages 282, 2001.
23. Land Use and Cover Change, jointly edited with J. Fox and Y. Himiyama, Science Pub., Inc., Enfield (NH), USA and Oxford & IBH Pub., New Delhi, pages 299, 2001.
24. Disaster Management, ed., Rawat Pub., Jaipur, pages 333, 2000.
25. Space Technology for Sustainable Development, ed. jointly with Shunji Murai, Oxford & IBH Publisher, New Delhi, 1998 , pages, 368.
26. Sustainable Development of Mountain Environment of India and Canada, ed., New Delhi, Oxford & IBH Pub., 1998, Pages 345.

27. Ecological Techniques and Approaches to Vulnerable Environment, Ed. Science Pub., Inc., Enfield (NH), USA And New Delhi, Oxford & IBH Pub., pages 255, 1998.
28. Development Issues in Marginal Regions, Ed. jointly with Roser Majoral, New Delhi, Oxford & IBH Pub., 1996, Pages 310.
29. Environmental Law - Issues and Responses, Ed. jointly with S. Misra, New Delhi, Concept Pub., 1996, Page 303.
30. Disasters, Environment and Development, New Delhi, Oxford & IBH Pub., Pages 600, 1996.
31. Research in Geography, Vol. II Disaster and Environment - Monitoring and Forecasting, New Delhi, APH Pub., 1996, Pages 357.
32. Research in Geography, Vol I - Land Use Changes and Sustainable Development, New Delhi, APH Pub.,1996, Pages, 335.
33. Sustainable Reconstruction of Highland and Headwater Regions, Ed., jointly with M.J. Haigh, New Delhi, Oxford & IBH Pub. 1995, Pages 681.
34. Global Environmental Change-Perspectives on Remote Sensing and GIS, Ed., New Delhi, Oxford & IBH Pub., 1995 Pages 321.
35. Environmental Monitoring - Application of Remote Sensing and GIS, Ed., Hong Kong: Geocarto Int. Centre, Pages 192, 1991.
36. Dynamics of Mountain Geosystems, ed., New Delhi: Ashish Pub., 1991, Pages 375,
37. Environmental Geography, Contribution to Indian Geography Vol. XI, New Delhi, Heritage Pub., 1990, Pages 352 XV.

C. Research Papers Published in Foreign Countries: Foreign Journal/Peer Reviewed Publications

1. Community Response to Mountain Tourism-A Case in Bhyundar Valley, Indian Himalaya, Journal of Mountain Science, Vol.6, No.4, December, pp.394-404, 2009.
2. Integrated Water Resources Management in the Vulnerable Indian Environment, IAHS Red Book 330, pp. 187-93, 2009.

3. Sustainable Forestry in Mega-cities of India for Mitigating Carbon Sequestration: Case Study of Delhi, (Jointly with A.Goel), Advances in Earth Science, Vol.21, No.2, Feb., pp.144-150, 2006.
4. Climate Variability and Drought in Rajasthan, (Jointly with A.Goel), in Advances of Geosciences: Vol. 4: Hydrological Sciences (HS), Ed., World Scientific Publishing Co., Singapore, pp. 57-68, 2006.
5. Green Tourism in Mountain Regions-reducing Vulnerability and promoting People and place Centric Development in the Himalayas, (Jointly with D.K.Mishra), Journal of Mountain Science, Vol. 1, No. 1, pp. 57-64, 2004.
6. Accelerated Tourism Development and its Impacts in Kullu-Manali, H.P., India (jointly with J.Gardner, J.Sinclair, F.Berkes), In Tourism Recreation Research, Vol. 27 (3), pp. 9-20, 2002.
7. Sustainable Management of Water Resources in an Agriculturally Developed Region of the Indo-Gangetic Ecosystem, India, Pub. In FRIEND 2002-Regional Hydrology:Bridging the Gap Between Research and Practice, IAHS Red Book 274, pp. 315-21, 2002.
8. Impact of Land-use change on groundwater in the Punjab-Haryana Plains, India, Pub. In Impact of Human Activity on Groundwater Dynamics, IAHS Red Book 269, pp. 117-122, 2001.
9. Sustainable Regional Water Management in the Yamuna River Basin: Case Study of Delhi Region, Pub. In Regional Management of Water Resources, IAHS Red Book 268, pp. 25-32, 2001.
10. Environmental Consequences of Agricultural Development: A Case study from the Green Revolution State of Haryana, India, In Agriculture, Ecosystem and Environment, 1637 (2000), 1-7.
11. Land Use and Cover Change and Comprehensive Food Security in the Indo-Gangetic Plains: Challenges and Responses, Journal of Geography Education, Vol. 43, pp. 27-31, 2000.
12. Urban Impact on Groundwater Quality in the Delhi Region, in Proceedings of IAHS Symposium 5, Birmingham, Ed.

J.B.Ellis, IAHS Press, IAHS Red Book 259, pp. 227-234, 1999.

13. Local Knowledge in the Assessment of Resource Sustainability: Case Studies in Himachal Pradesh, India and British Columbia. Canada, Mountain Research and Development, vol.18, No.1, 1998, pp. 35-49.
14. Land Use/Cover Changes, Extreme Events and Ecohydrological Responses in the Himalayan Region, in Hydrological Processes, 12, 1998.
15. Monitoring Hydrological Parameters and Sustainable Water Resource Development in the Mahanadi River Basin, (Jointly with R. P. Verma, Proceedings of 1997 FRIEND Conf., Acta hydrotechnica, 15/19 (1997) Ed. M. Mikos, Univ. of Ljubljana, pp.159-162.
16. Water Resources Monitoring for Environmental Sustainability under increasing uncertainty in the Himalayan Region, Proceedings of IAHS Rabat Symposium, Sustainability of Water Resources Under Increasing Uncertainty, IAHS Red Book 240, 1997, pp. 479-485.
17. Environmental Monitoring of Particulate Matter and Pollution in Major Rivers of India, in Particulate Matter in Rivers and Estuaries, Hamburg: Max Planck Inst. pp. 246-251, 1994 also published in Advances in Limnology, Vol. 47, 1996.
18. Spatial Distribution of Rural Settlement System: Micro-Level Indian Case Study, Asian Geographer, Vol. 1, No. 2, Hong Kong, 1982, pp. 51-56.
19. The role of Education in Rural Development : Spatial and Structural Strategy in Rural India, The Indonasian Jl. of Geography, Vol. 10, No. 39, 1980, pp.45-53 also published in Human Resource Development, Vol. Ed. A. Kumar, Anmol Pub., New Delhi, 1991
20. Ecological Perspectives of Rural Development in North Indian Plain: A Case of Siwan District, (With Rana P.B. Singh *et al.*) Geographica, Kaula Lampur, Vol. 13, 1979, pp. 49-57.

D. Chapters in Foreign Books and Proceedings

21. Land Use Environment and Development in the Uttarakhand Himalaya, India, in Volume Towards Sustainable Land Use in Asia, Ed. Y. Himiyama, SLUAS Science Report 2010, Hokkaido University of Education, Asahikwa, pp.47-64, 2010.
22. Land Use/Cover Change, Environment and Climate Change in Delhi Metropolitan Region: Towards Promoting Sustainable City, In Proceedings of the Int. Symp. On Sustainable Urban Environment 2007, Tokyo Metropolitan University, Tokyo, pp. 73-79. 2007.
23. Impact of Floods on Biodiversity in the Brahmputra Valley, in Proceedings of the International Conference on Monitoring, Prediction and Mitigation of Water-Related Disasters, (Jointly with M.Bortamuly) Eds Takara, K et al., Kyoto University, Kyoto, Japan, pp. 599-606, 2005.
24. Forest-Cover Mapping of Uttaranchal State, India, In Understanding Land-Use and Land-Cover Change in Global and Regional Context (Jointly with P. Mishra) Eds Milanova, E. *et al.* Science Publisher, Inc Enfield (NH), USA. Pp 249-258, 2005.
25. Water and Environment Related Disasters and Vulnerability in Indian Mega Cities: Case of Delhi Metropolitan Region (Jointly with Anju Singh), in Proceedings of the International Conference on Monitoring, Prediction and Mitigation of Water-Related Disasters, Eds Takara, K et al., Kyoto University, Kyoto, Japan, pp. 691-606, 2005.
26. Water Resources Sustainability in the Context of Increasing Disasters in the Himalaya-Gangetic Region, In. Proceedings of the 2nd Asia Pacific Association of Hydrology and Water Resources Conference, Vol. 11, AOGS, Singapore, pp. 566-74., 2004. Impact of Climate Change on Water Resources Sustainability in the Himalayan-Gangetic Region, in Annals of the National Association of Geographers, India, Vol XXIV, No 1, pp.32-39, 2004.
27. Remote Sensing Based Land Use and Cover Change Monitoring as an Hydrological Indicator of Sustainability

in the Himalayan Ungauged Basin (Jointly with P. Mishra), In Sustainable Water Resources Management in the Changing Environment of the Monsoon Region, Vol. 1 Eds. S. Herath *et al.*, UNU and Nat. Water Resources Secretariat, Sri Lanka, Colombo, pp.425-31, 2004.

28. Assessment of Wastewater Generation, Management Pattern and its Re-Use Potential in Urban India, IAHS Pub. 285, International Association of Hydrological Sciences, Wallingford, pp. 19-22, 2004.
29. Monitoring, Mapping and Mitigation of Flood Disaster in India Using Remote sensing and GIS: A Case Study,(jointly with Dilip Kumar), Pub. In Monitoring, Prediction and Mitigation of Disasters by Satellite Remote Sensing, Eds. K.Takara *et al.*, DPRI, Kyoto University, pp. 53- 59, 2004.
30. Local Knowledge and community participation in water management for poverty alleviation: towards integrating water science and ethics in India, Proceedings of the Asia-Pacific Asso. of Hydrology and Water Resources, Eds. K. Takara and Kojiima, Disaster Prevention Inst., Univ. of Kyoto, 2003.
31. Tourism Development and Environmental Implications for the Indian Frontier Region: A Study of Himachal Himalaya, Pub.in Tourism in Frontier Areas, Eds. Shaul Krakover and Yehuda Gradus, Lexington Books, Lanham, Maryland, pp.177-94, 2002.
32. Towards Promoting PUB Monitoring Network and Predicting Erosion Vulnerability in the ungauged Basin of the Indian Himalaya using Remote Sensing and GIS, Communications of IAHS-PUB Workshop, Eds. P.Hubert et al., Brasilia, Brazil, 2002.
33. Environmental Degradation and its impact on Land Use/ Land Cover Changes in Agriculturally Developed Regions of India, Proceedings of the IGU-LUCC International Conference, Eds. I.Blicik et al., Charles University, Prague, pp. 67-74, 2002.
34. Environmental Implications of the Tala Hydro Electric project, Bhutan, Jointly with Smriti Kumari, Pub.in Wasserwirtschaft IM Wandel, Ed. A. Hoffmann, Kasseler

Wasserbau-Mitteilungen, Number 12, Herkules Verlag Kassel, pp. 165-174, 2001.

35. Hazard-induced Hydrological problems in Himalayan Watershed-A study of an Indian Hot Spot., Pub. In Hydrological Challenges in Transboundary Water Resources Management, Unesco-IHP/OHP Secretariat, Koblenz, pp. 75-80, 2001.
36. Ecological Impact of Land Use Change in Delhi Ridge: Anthropogenic Stress and Spatial Realities, (Jointly with Rakhi Parijat), Pub. In Urban Sustainability in the Context of Global Change, Ed. R.B.Singh, Science Pub., Inc., Enfield (NH), USA, pp.189-201, 2001.
37. Sources of Air Pollution Emissions in Selected Metropolitan Cities of India, (Jointly with B.Kumar), Pub. In Urban Sustainability in the Context of Global Change, Ed. R.B.Singh, Science Pub., Inc., Enfield (NH), USA, pp.29-48, 2001.
38. Large -scale Dams as Drivers of Land Use/cover Change in the Tropical Deciduous Forests-A Case -Study of Sardar Sarovar Project (SSP), India., (Jointly with A. Mathur), Pub. In Land Use and Cover Change, Eds. R.B.Singh, J.Fox and Y. Himiyama, Science Pub., Inc., Enfield (NH), USA, pp.139-153, 2001.
39. Diversification of Rural Agricultural Land Due to Urban Intrusion-A Case -Study of the Metropolitan Periphery of Delhi, India, (Jointly with Jag Mohan), Pub. In Land Use and Cover Change, Eds. R.B.Singh, J.Fox and Y. Himiyama, Science Pub., Inc., Enfield (NH), USA, pp.155-168, 2001.
40. Urbanization in India's metropolis Delhi: Problem and Perspectives, In Resolving Urban Environmental and Spatial Conflicts, eds. Donald Miller & Gert De Roo, Geo Press / Geo Pers Pubs. Groningen, pp. 27-37, 2000.
41. The Depletion of Common Lands in the Rural-Urban Fringe of Delhi: A Case Study of Unsustainable Development, Pub. Dimensions of Sustainable Rural Systems, Ed. I.R. Bowler *et. al.*, Groningen: Netherlands Geographical Studies, pp. 105-14., 1998.
42. Impact of Agriculture and Land Use/Cover Change on Soil and Water Resources and Ecosystem Sustainability: A Case

Study of Punjab, Sustaining Rural Systems in the context of Global Change, Ed. R. Epps, Armidale: Univ. of New England, 1998, pp.335-42.

43. Impact of Tourism on Ecohydrology in the Headwater Region of the Beas, Himachal Pradesh,(jointly with B.W.Pandey), Ecohydrology of High Mountain Areas, Eds. S.R.Chalise et al., Kathmandu: ICIMOD, 1998, pp. 519-525.

44. Environmental Change study using Land use/Cover Data in LIS Framework in India, Information Bases for Land Use/Cover Change Research, Proceedings IGU SG LUCC, Eds. Y. Himiyama and L. Crissman, 1997, Asahikawa : Hokaido Univ. of Education. pp. 98-104.

45. Environmental, Economic and Social Sustainability through Human Development Initiatives: Case study of an Indian Marginal Region-Upper Kullu Valley, Proceedings of IGU SG Meeting Dev. Issues in Marginal Regions, Ed. G.Jones *et al.*, 1997, Glasgow: Univ. of Strathclyde.

46. Challenges and Crisis of Sustainability and Opportunities for Resource Management in Indian Himalaya, Proceedings of NASDA/UNCRD/ICIMOD seminar on Space Informatics for Sustainable Development, Nagoya, UNCRD, Proceeding series No.21, 1997 pp. 51-60.

47. Monitoring Tropical Geosystem, Underground Topography and Potential Water Resources for Sustainable Development Using Geoinformatics-Indian Case Study, in Proceedings of the 12th Int. Congress of Speleology, Ed. Pierre-Yves Jeannin, La Chaux-de-Fonds, pp. 369-372. 1997.

48. Sustainability of a Mountain Watershed in the Himachal Pradesh Himalaya: Background and Overview, (Jointly with F.Berkes et al.), Sustainability of Mountain Environment in India and Canada, Eds. F.Berkes and J.S. Gardner, Univ.of Manitoba, Winnipeg, 1997, pp. 19-48.

49. Risk from Natural Hazards in the Manali Area, (Jointly with J.S.Gardner and B.W. Pandey), in Sustainability of Mountain Environments in India and Canada, eds. F.Berkes and J.S.Gardner, Winnipeg, Univ. of Manitoba, pp. 237-254, 1997.

50. Environmental Monitoring and Natural Resource Management in Indian Arid Region using Space Informatics, Proceedings of the International Symp. on the Role of Remote Sensing for the Environmental Issues in Arid and Semi-Arid Regions, Eds. S. Shindo and A. Kondoh, CEReS, Chiba Univ., pp. 9-21, 1997.
51. Snow Melting Variation and Regional Environmental Change - A Micro-Level Study of Himalayan Rivers, Proceedings of CGB Symp. on Global Change and the Commonwealth,ed. Yue-man Yeung, Hong Kong: the Chinese Univ. of Hong Kong, pp.433-45, 1996.
52. Remote Sensing and GIS Based Land Use Information System for Natural Resource Monitoring and Management: The Indian Experience in Proceedings of Int. Workshop held at Ullan Bataar, Nagoya, UNCRD, pp.99-106, 1996.
53. Landslide Hazard in Indian Himalaya and Canadian Rockies: A ComparativeAnalysis, Proceedings of the Eighth Int.Conf.on Lanslides.Eds. J.Chacon et al. Rotterdam: A.A Balkema, pp.63-69,1996.
54. Recent Earthquake in Maharashtra, India : Uncertainty, Existing Reality, and Task Ahead in Disaster Management in Metropolitan Areas for the 21st Century, Proceedings of the IDNDR Aichi/Nagoya Int.conf., UNCRD, Nagoya also in Research in Geography, Disasters and Enviroment : Monitoring and Forecasting, eds. R.B. Singh, New Delhi, APH Pub., Vol. II, pp. 83-89, 1996.
55. Land Use, Development and Environmental Criticality in Marginal Regions: A Case Study of Himachal Pradesh, in Marginality and Development Issues in Marginal Regions, Ed. Chang-Yi David Chang, Taipei, Nat. Taiwan Univ., 1994. pp. 19-40; Also publisherd in Contemporary Approaches to Indian Geography, Ed. Rais Akhtar, pp. 109-140, 1997.
56. Desert Region in India: Resource Management Issues and Research Strategy, "Towards Solving the Global Desertification Problems, eds. T. Miyazaki and A. Tsunekawa, Tsukuba: NIES pp. 1-18, 1994. Also pub. Environmental Strategies, Ed. R. M. Lodha, 1993.

57. Monitoring Soil Loss and Sediment Deposition in Indian Perspective, Erosion and Sediment Transport Monitoring Programmes in River Basins, Ed. Jim Bogen, NWREA, Oslo, 1992. pp. 138-42.
58. Remote Sensing Data Base and GIS Projects for Monitoring National Environmental Change in India in Environmental Change and GIS, Editor I. Ota, Proc. of INSEG -1991,Vol. 2. 344-51.
59. Role of Geographical Monitoring and Forecasting in Ecosystem Modelling and Management: The Indian Experience Editor - Y. Aruga, Proceedings of the UNESCO/ MAB Seminar on Future Research Trends in MAB, Tokyo: MAB, 1991, pp. 209-252. Also pub. in Global Environmental Change, ed. R. B. Singh, *op.cit.*, pp. 1-21, 1995.
60. Domestic Energy Use in Delhi Metropolitan Fringe: A Study of Rural and Urban Villages, Proc. of 10th Int. Symp. on Asian Studies, Hong Kong; ARS, 1989, pp. 1003-9.
61. The state of environmental and Resource Management in Drylands of North India, in Resource Management in Drylands, eds. W. Mecalien *et al.* Stuttgarter (Germany) : Stuttgarter Geographische Studien, 1985, pp. 127-40 also published in Indian Desert : Resources and perpectives Development ed. T. S. Chauhan, Jaipur, Printwell: pp. 254-62, 1995.
62. Spatial Organisation of Service Centres: case of Jaunpur District Proc, of IGU Symp. on Rural Development, Liege (Belgium), ed. Ch. Christians, 1983, pp. 321-35.
63. Ecological Approach of Irrigation System : Potentiality Constrains and Management in Traditional Soceity, Problems of the Management of irrigated Land in Areas of Traditional and Modern Cultivation, ed. H.G. Mensching, Proc. of IGU Working Group Stmp., Cairo-ElMinia (Egypt), Humberg (Germany), 1983, pp 123-32.
64. Transformation of Agriculture for Rural Development : the Indian Micro Level Experience, Changing Scene of Rural Developing Countries eds. Yuihama et al, 24th IGU Tokyo pub., Okyohama : Geography Department, 1980, pp.145-53.

65. Rural Spatial Taxonomy for Integrate Rural Development: the Indian Experience, Changing Scene of Rural Habitat in Developing Countries, eds. S. Yuihama et al., 24th IGC Tokyo Pub., Okyohama : Geography Department, 1980, pp. 129-44.

E. RESEARCH PAPERS PUBLISHED IN INDIAN JOURNALS

66. Migration and Agrarian Changes in Mountain Regions: A Case Study of the Pindar Basin of Uttarakhand Himalaya, India, (Jointly with V.P.Sati), Annals of the Nat. Ass. Of Geographers, India (NAGI), Vol. 29, No. 1, June, pp. 20-33, 2009.
67. Remote Sensing and Geographical Information System, Pub. In Geography in India-Selected Themes, Ed. L.S.Bhat, Pearson, Delhi, pp.203-257, 2009. Also see Remote Sensing Research in India: Trends and Prospects in Trans. Inst. Indian Geographers, Vol. 15, No. 1, 1993, pp. 71-83 also pub. in Third World Impact, June-Aug. 1993.
68. Mitigating Natural Disasters for Achieving Inclusive Development and Human Security in India, In Journal of Social Science, Vol. 10, No.2, July-December, pp 258-72, 2008.
69. Current Status of Municipal Solid Waste Management Practices in Delhi, (Jointly with Subhash Anand), Annals of the Nat. Ass. Of Geographers, India (NAGI), Vol. 28, No. 1, June, pp. 36-50. 2008.
70. Estimation of Soil Erosion by Using RUSLE and GIS for a Himalayan Sub-Watershed in India, (Jointly with Mehtab Singh, Suresh Kumar and J.Prasad), Annals of the Nat. Ass. Of Geographers, India (NAGI), Vol. 27, No. 2, December, pp. 1-11, 2007.
71. Hydroinformatics for Integrated Watershed Management-Combining Science and Community Participation, Indian National Geographer, Vol.21, No. 1 & 2, 2006. 59-66.
72. Biodiversity in the High Altitude Lakes of Ladakh: Threats and Opportunities, (With A.Anand), Annals of NAGI, Vol. 26, No.2, December, pp. 10-17, 2006.

73. Geographical Mosaics of Culture, Environment and Human Development, Annals of NAGI, Vol. 25, No.2, December 2005, pp.20-32.
74. Prediction in Ungauged Basins (PUB): Hydrological Challenges for India, Annals of NAGI, Vol. 22, No.1, June 2002, pp. 89-98. Also Pub. In PUB as Disaster Mitigation Tool in the Himalayan River Basin, Proceeding of the World Congress on Natural Disaster Mitigation, Vol. 2, The institution of Engineers (India), New Delhi, pp.256-64, 2004.
75. Mapping Environmental Stress in the Context of Land Use Changes and urban Sprawl in Delhi Metropolitan Region (jointly with B.Kumar and Anju Singh), The Geographer, Vol. 49, Number 2, July, pp. 44-60, 2002
76. Management of Water Related Disasters in the Context of the Himalayan Mountain Region of India, Annals of the National Asso. Of Geographers, India (NAGI), Vol. 21, No. 2, December 2001, pp. 41-46.
77. Levels of Atmospheric Quality in Mumbai, Calcutta and Delhi, (jointly with B. Kumar), The Indian Geographical Jl., Vol. 74, No.2, December, pp. 112-25, 1999.
78. Rainfall Variability and its Relation with climate and Ecosystem: A Case of Western India, (jointly with U.Kumar), Indian Jl. of Landscape System and Ecological Studies, Vol. 22, No.1, June, pp. 29-35, 1999.
79. Prediction and Detection of Anthropogenic Climate Change: Case of Delhi (jointly with B.Kumar), Transactions, Institute of Indian Geographers, Vol.21, No.1 ,Jan., pp. 61-75.1999.
80. Human Dimensions of Environmental Change: Geographical Perespective, In Bhartiya Samajik Chintan, Vol. XXII Nos. 1-4, Jan.-Dec., pp. 55-60, 1999.
81. Urbanization, Environmental Crisis and Urban Planning in India, The Geographer, Vol. XLIV, No. 2., July 1997, pp. 5-21.
82. Environmental Regeneration of Wasteland through Agroforestry in Rajasthan (jointly with G.S.Chauhan), The Geographer, Vol.XLIV, No. 1, Jan. 1997. 79.

83. Hydrological Monitoring for Integrated Watershed Management in India, Indian Cartographer, Vol. 17, pp. 260-64, 1997.
84. Forest and Forestry in India: Spatial Nature, Uses & Issues of Sustainability, National Geographical Jl. Of India,Vol. 4, March 1995, pp. 55-67.
85. Spatial Variability of Anthropogenic Influences on Atmospheric Chemistry in Indian Mega-Cities (jointly with Bhuwan Kumar) in Geographical Review of India, Vol. 55, No.1, 1993, pp. 22-33.
86. Environmental and Human Response to Hydrogeomorphology: Micro- Level Experience of Ganga Basin, (Jointly with B. W. Pandey) in Journal of Scientific Research, Vol. 43, (i&ii) 1993. pp. 153- 167.
87. Remote Sensing and GIS as Decision Support System in the Himalaya, Annals, NAGI, Vol.XII No. 1 & 2, 1992, pp.103-11. Also pub. In Application of GIS Technology for Environmental Monitoring and Forecasting in India, in Research Methodology in Geography, Eds. H.N.Misra and V.P.Singh, New Delhi: Rawat Pub., 1998, pp.255-71.
88. Application of Digital Image Processing and GIS Technology for Land Use Mapping and Monitoring Land Degradation Hazards in U.P. Himalaya, Indian Cartographer, Vol. IX-I, 1989, pp. 432-42. Also pub. In Environmental Monitoring, Editor R.B. Singh Geocarto Int. Centre, Hong Kong, 169-78, 1991.
89. Determinants of Energy Use - A study of Rural Delhi, (with Afzal Raza Khan), Indian Nat. Geographer, Vol. 2 (1& 2), 1987, pp. 69-78.
90. Energy Potentialities and Related Rural Programmes in India : An Assessment of Emerging Trends, Nat. Geog. Jl. Of India, Vol. J31, pt. 2, 1985, pp. 147-50.
91. Micro-Level Development Planning of Mohamadpur Gram Panchayat, (with S.H. Ansari) The Jl. of Scientific Research, B.H.U., Vol. 33(2), 1983, pp.145-56.
92. The role of Rural Industralization Towards Eradication of Rural Poverty : Case Study of Siwan District, Rural System, Vol. I Part 3, Varanasi, 1983, pp. 225-36. Also in Manpower

Planning in India, ed. A. Kumar, Criteria Pub., New Delhi, 1989, pp.141-58.

93. Demographic Structure and Topology of Population Resource, A Case Study of Siwan District (Bihar), Indian Geographical Studies, Res. Bull. 16, 1981, Patna, pp. 33-40.
94. Regional Development through Social Planning : A Micro - level Study from India, (with S.L.Kayastha) Indian Jl. Of Regional Science, Vol. XIII, No. 1 , 1981, pp. 28-45.
95. Emerging Dynamics of Integrated Rural development (with S.L. Kayastha) Nat. Geog. Jl. of India, Vol 26, Pt. 3-4, pp. 28-38, 1980.

F. Chapters in Indian Books and Proceedings

96. Designing and Shaping Global to Local Environmental Education: The Indian Experience, Pub. In Glocal Environmental Education, Eds. Y.Himiyama et al., Rawat Publications, Jaipur, pp.62-80, 2010.
97. Forest Resources, Biogeography and Biodiversity: Indian Experience, Pub. In Biogeography and Biodiversity, Ed. R.B.Singh, Rawat Publications, Jaipur, pp.1-21, 340, 2009.
98. Remote Sensing and GIS Based Modelling Approach for Land Capability Classification (Case Study of East Dehradun Valley) (Jointly with Mehtab Singh, Suresh Kumar and P.M.Kala), Pub.In Land Use, Biodiversity and Climate Change, Eds. R.B.Singh et al., MMC College, Marthandam, pp.13-26, 2009.
99. Landslides Hazard and its Influences on Biodiversity in Nanda Devi Biosphere Reserve, (Jointly with Suraj Mal) Pub. In Biogeography and Biodiversity, Ed. R.B.Singh, Rawat Publications, Jaipur, pp.1-21, 340, 2009.
100. Geographical Analysis of Water logging Induced Land Degradation in Karnal District of Haryana, (with Mehtab Singh), In. Perspectives in Resource Management in Developing Countries, Vol. 3, Ed. B.Thakur, Concept Pub., New Delhi, pp. 196-211, 2008.
101. Local Governance in Disaster Management through Vulnerability Analysis and Mitigation, Pub. In Disaster

Management in India-Perspective, Issues and Strategies, Eds. N. Rai and A.K.Singh, New Royal Book, Lucknow, pp. 93-107, 2007.

102. Challenges of Flood Disaster Management: A Case Study of NOIDA (with Swarnima Singh), In Disaster Management: Future Challenges and Opportunities, Ed. Jagbir Singh, I. K. Int. Pub. House, New Delhi pp.15-30, 2007.

103. Temporal Transport Hazard Dynamics: A Case Study of Delhi (with Swarnima Singh), In Disaster Management: Future Challenges and Opportunities, Ed. Jagbir Singh, I.K.Int. Pub.House, pp.95-112, 2007.

104. Integrated Watershed Management for Livelihood Security:A Case Study of Uppalgaon Watershed, Uttaranchal Himalaya, (With Nawal Prakash) Pub. In Use and Misuse of Land and Water Resources, Ed. A.L.Singh, B. R. Pub., Delhi, pp.113-127, 2006.

105. Introduction to Natural Hazards and Disasters: Conceptual Framework, Vulnerability, Spatial Pattern, Impacts and Mitigation Strategy: Indian Experience, In Natural Hazards and Disaster Management, Ed. R.B.Singh, Rawat Pub., Jaipur, pp 3-39, 2006.

106. Tsunami Disaster in India: Warning, Responses, Impacts and Mitigation, In Natural Hazards and Disaster Management, Ed. R.B.Singh, Rawat Pub., Jaipur, pp. 104-119, 2006.

107. Urban Development and Environmental Impacts in a Mountain Context, (With J.S.Gardner) In Sustainable Urban Development, Concept Publishing Company, New Delhi, pp. 67-86, 2006.

108. Slums, Environment and Development in Metropolitan Cities of India, (With D.K.Mishra), In Sustainable Urban Development, Concept Publishing Company, New Delhi, pp. 261-71, 2006.

109. Brick Kiln Industry as Drivers of Land Use Change and Related Land Degradation in Rural-Urban Fringe of Delhi, (With Jag Mohan), In Sustainable Urban Development, Concept Publishing Company, New Delhi, pp. 279-88, 2006.

110. Geoinformatics for Sustainable Development-A case Study of Ferozpur Jhirka Block of Gurgaon District in Haryana (Jointly with Dilip Kumar), In Geoinformatics Applications for Sustainable Development, Eds. A.K.Singh et al., ISRS (Delhi Chapter) and IARI, New Delhi, pp.126-33, 2004.

111. Agricultural Transformation and Rural Sustainability in a Himalayan Watershed: Case Study of Kullu Valley (Jointly with D.K.Mishra), In Resource Conservation and Food Security, Ed. T. Singh, Concept.Pub., New Delhi, pp. 345-55, 2004.

112. Impact of Climate Change on Mountain Ecosystems, (Jointly with S.Sen Roy), In Assessment of Climate Change in India, Eds. S.K.Das and Prakash Rao, WWF, New Delhi, 221-30, 2004.

113. Remote Sensing and GIS for Resource Management in India, Perspectives in Resource Management in Developing Countries, Vol. 1-Resource Management: Theory and Techniques, Ed. B.Thakur, Concept Pub. Company, New Delhi, pp. 485-99, 2003.

114. Urban Environmental Data Base, Planning and Policies, (Jointly with B.Kumar), In Management of Environmental Pollution, Eds. Anju Kohli et al., Book Enclave, Jaipur, pp. 252-69, 2003.

115. Himalayan Environment, Land Use Cover Changes and Highland Lowland Interactive System with Indus-Ganga -Brahmputra Plains, Pub. In Land Use-Historical Perspectives-Focus on Indo-Gangetic Plains, Eds. Y.P.Abrol *et al.*, Allied Pub., New Delhi, pp.475-98, 2002.

116. Environmental Impact and Response Analysis of Urban and Tourism Infrastructure Development in the Himachal Himalaya, In. Recent Advances in Geomorphology, Quaternary Geology and Environmental Geosciences: Indian Case Studies, Eds. S.K.Tandon and B.Thakur, Manisha Pub., New Delhi, pp. 385-94., 2002.

117. Land Use Planning in Resettlement and Rehabilitation Areas in Highlands, In Human Dimensions of Sustainable Development, Ed. R.B.Singh, Rawat Pub., Jaipur, pp.254-66, 2002.

118. Landslide Hazard in Himachal Himalaya, in Geoenvironmental Hazards, Eds. V. Subramanian and Jayakumar,UNESCO-ENVIS, New Delhi, pp.71-84, 2000.

119. Human Induced Hazards and Development in a part of Upper Yamuna Watershed, Garhwal Himalaya (jointly with Sarita Ghai) in Contributions to Environmental Geoscience, Eds. A.M.Pathan and S.S. Thigale, Aravali Books Int., New Delhi, pp. 151-160. 2000.

120. Environmental Impact Assessment and Management in Highlands and Headwater Regions, In Environmental Impacts Assessment of Water Resources Projects, Ed. R.S.Goel, Oxford & IBH. Pub., New Delhi, pp. 343-356, 2000.

121. River Pollution as Environmental Disaster in Yamuna at Delhi: Perspectives for 21st Century, (Jointly with D.K. Mishra), In Disaster Management, ed. R.B. Singh, Rawat Pub., Jaipur, pp. 319-330, 2000.

122. Governmental Response to Flood Disaster Preparedness, Relief and Mitigation in Haryana, (Jointly with A.K. Shrivastava), In Disaster Management, ed. R.B. Singh, Rawat Pub., Jaipur, pp. 308-318, 2000.

123. Man-made and Natural Geo-Hazards in Kullu Valley, (Jointly with Dhananjay Tingal), In Disaster Management, ed. R.B. Singh, Rawat Pub., Jaipur, pp. 259-286, 2000

124. Remote Sensing and GIS for Disaster Management, In Disaster Management, ed. R.B. Singh, Rawat Pub., Jaipur, pp. 11-24, 2000.

125. Dynamics of Village Common Land and its Sustainable Management in the Himalayan Upper Beas Basin (jointly with M.Babar Ali), In The Himalayan Environment, Eds. S.K. Dash and J. Bahadur, New Age International (P) LTD-Pub., New Delhi, pp. 246-57, 1999.

126. Geographic Information System for Environmental Monitoring and Management, In Proceedings of National Seminar on Environmental Management in Hydro Electric Projects, New Delhi, 1999, Ministry of Environment & Forests, Govt. of India, Organised by Institute for Resource Management and Economic Development, pp. 119-128.,

1999. Also Pub. In Geographic Information System as a Tool of Environmental Decision Making, Ed. B. Misra, Proceedings of SCADA Conference . New Delhi: Mittal Pub.,1998.

127. Delhi Metropolis: Housing and Quality of Life, Pub. in Million Cities of India, Volume 1, Ed. R.P. Misra and K. Misra, New Delhi: Sustainable Development Foundation, pp. 196-228, 1998

128. Status of Environmental Quality and Decision-making in India: Past, Present and Future Trends, Pub. in Techniques and Approaches to Vulnerable Environment, Ed. R.B. Singh, New Delhi: Oxford and IBH Pub., pp. 350-68, 1998.

129. Dam and Water Resource Development: Myth or Reality - Case of Hirakud Dam, Pub. in Ecological Techniques and Approaches to Vulnerable Environment, Ed. R.B. Singh, New Delhi: Oxford and IBH Pub., pp. 223-42, 1998.

130. Earth Science Data and Information Technology Network Development in India, in Space Informatics for Sustainable Development, Eds. R.B.Singh and S.Murai, New Delhi: Oxford & IBH Pub., pp. 47-62. 1998.

131. Understanding High Mountain Land Use Towards Sustainable Environmental in Jammu and Kashmir, in Headwaters-Water Resources and Soil Conservation, Eds. M.J.Haigh et al., New Delhi: Oxford & IBH Pub. 1998, pp. 409-422.

132. Headwater Control: Matters Arising (Jointly with M.J.Haigh and J.Krecek),in Headwaters- Water Resources and Soil conservation, Eds. M.J.Haigh et al., New Delhi: Oxford & IBH Pub., 1998, pp. 3-24.

133. Genensis and Impacts of A Landslide-The Case Study of Kharadi-Khanera Slope Section, U.P. Himalayas(Jointly with Sarita Ghai)in Sustainable Development of Mountain Environment in India and Canada, Ed. R.B.Singh, New Delhi: Oxford & IBH Pub., 1998, pp. 298-308.

134. Forest and Foretry: Degradation and Renewal in Arrow Lake Forest District, British Columbia, Canada (Jointly with B.W.Pandey) in Sustainable Development of Mountain Environment in India and Canada, Ed.

R.B.Singh, New Delhi: Oxford & IBH Pub., 1998, pp. 242-256.

135. Environmental Hazards and Mitigation Measures in Upper Arrow Lake Region,British Columbia, Canada (Jointly with B.W.Pandey) in Sustainable Development of Mountain Environment in India and Canada, Ed. R.B.Singh, New Delhi: Oxford & IBH Pub., 1998, pp. 217-241.

136. Population, Urbanization and Tourism in Kullu District (Jointly with B.W. Pandey and Tripta Singh) in Sustainable Development of Mountain Environment in India and Canada, Ed. R.B.Singh, New Delhi: Oxford & IBH Pub., 1998, pp. 153-171.

137. Village Common Land In Upper Beas Basin (Jointly with Mohammed Baber Ali) in Sustainable Development of Mountain Environment in India and Canada, Ed. R.B.Singh, New Delhi: Oxford & IBH Pub., 1998, pp.126-141.

138. Environmental Stress of Forest Degradation on Erosional Intensity: A Micro-level Study of Solan District (Jointly with G.N.S.Gurudatt) in Sustainable Development of Mountain Environment in India and Canada, Ed. R.B.Singh, New Delhi: Oxford & IBH Pub., 1998, pp. 113-125.

139. Environmental Stress of Horticulture on Sustainable Livelihood and Food Security in Kullu Valley (Jointly with Kirti Bas and Paramita Datta) in Sustainable Development of Mountain Environment in India and Canada, Ed. R.B.Singh, New Delhi: Oxford & IBH Pub., 1998, pp. 98-112.

140. Environmental Implications of Land use Change and Forest Degradation In Upper Beas Basin (Jointly with D.K.Mishra) in Sustainable Development of Mountain Environment in India and Canada, Ed. R.B.Singh, New Delhi: Oxford & IBH Pub., 1998, pp. 76-85.

141. Biophysical and Anthropogenic Systems Interface in Kullu District (Jointly with D.Tingal), Sustainable Development of Mountain Environment in India and Canada, Ed. R.B.Singh, New Delhi: Oxford & IBH Pub., 1998, pp. 36-52.

142. Environmental Disasters in Marginal Areas: A Study of Road Construction Induced Impacts in a Himalayan Yamuna Catchment, (jointly with S. Ghai), in Development Issues in Marginal Regions, eds R.B.Singh and R. Majoral, Oxford & IBH Pub.,New Delhi, pp. 311-20, 1996.

143. Environmental Degradation and its Impact on Livelihood Strategies in the Urban Fringe of Delhi-Some Theoretical Reflections, (Jointly with P.C.J.Druijven), in Disasters, Environment, and Development, ed. R.B.Singh, Oxford & IBH pub, New Delhi, pp. 355-68, 1996.

144. Disasters Monitoring and Mitigation, ed. R. B. Singh, in Disasters, Environment and Development, Oxford & IBH Pub., New Delhi, pp.31-62, 1996.

145. Hazard Zone Mapping and Risk Assessment Analysis in the Upper Beas Basin: CIDA-SICI Experience jointly with B.W. Pandey in Disasters, Environment and Development, ed. R.B. Singh, New Delhi, Oxford & IBH Pub., pp. 121-35, 1996.

146. Disasters and Environment in India: An Introduction in Research in Geography, Disasters and Environment: Monitoring and Forecasting eds. R.B. Singh, New Delhi, APH Pub., Vol. II, pp. 3-7, 1996.

147. Perspectives for Managing Rural Water Resources: A Study of Ganga-Gomati Doab, jointly with B.W. Pandey in Research in Geography, Land Use Changes and Sustainable Development, eds. R.B. Singh, New Delhi, APH Pub., Vol. I, pp. 307-18, 1996.

148. Sustainable Development Initiatives in Indian Geosystems : Case of Social Forestry in in Research in Geography, Land Use Changes and Sustainable Development, eds. R.B. Singh, New Delhi, APH Pub., Vol. I, pp. 127-38, 1996.

149. Pattern and Prospects of Service Centres in Spatial Context, Case of the Siwan District (With Rana P.B. Singh et al). , Transformation of Rural Habitat, NGSI Res. Pub. 1978, pp.108-116.

150. The Importance of Natural Hazards Research to the Himalaya Region: Himachal Pradesh, a Case inPoint (Jointly with D. Rowbotham) in Global Environmental

Change Perspectives of Remote Sensing and Geographic Information System Ed. R.B. Singh, Oxford & IBH Pub., New Delhi, 1995 pp. 185-99.

151. The State of Environment and Resource Management in Drylands of North India, in Resource Mangement in Drylands, published in Indian Desert: Resources and Perspective Development ed. T.S. Chauhan, Jaipur, Printwell: pp. 254-62, 1995.

152. Identification of Critical Environmental Zones in Kullu District, jointly with D. Tingal in Sustainable Reconstruction of Highland and Headwater Regions, eds. R.B. Singh and M.J. Haigh, New Delhi, Oxford & IBH Pub., pp. 627-38, 1995.

153. Landuse Change and Land Information System in Kullu Valley, jointly with G.S. Chauhan in Sustainable Reconstruction of Highland and Headwater Regions, eds. R.B. Singh and M.J. Haigh, New Delhi, Oxford & IBH Pub., pp. 639-47, 1995.

154. Dynamics of land Use and Cropping Patterns: Historical Prognostication in Upper Beas Basin, jointly with D.K. Mishra in Sustaianble Reconstruction of Highland and Headwater Regions, eds. R.B. Singh and M.J. Haigh, New Delhi, Oxford & IBH Pub, pp. 649-61, 1995.

155. Common Resources and Sustainable Livelihood of Mountain Environment: A Micro-Level Experience of Upper Kullu Valley, jointly with B.W. Pandey in Sustainable Reconstruction of Highland and Headwater Regions, eds. R.B. Singh and M.J. Haigh, New Delhi, Oxford & IBH Pub., pp. 663-81, 1995.

156. Land Use Change, Diversification of Agriculture and Agroforestry in North-west India, Proc. of Indo- British Seminar, ed. A. Mukerji, 1994, New Delhi: Concept Pub. Also Pub. in Global Environmental Change, op.cit.,1995,pp.313-21. Also published in Land Use Planning in India, Ed. T. Haque, New Delhi, NCAEPR, 1999, pp. 122-130.

157. Land Use Pattern, Land Degradation Hazard and Sustainability in the Indian Desert Region: in Afroforestry

Systems for Degraded Lands, Eds. P. Singh et al. New Delhi: Oxford & IBH Pub., pp. 13-19, 1994.

158. Sikkim Himalaya: Geo Ecological System, Socio-Economic Change and People's Perception, in Sikkim -Society, Polity Economy and Environment, Ed. M.P. Lama, pp. 241-254, 1994.

159. Anthropogenic Impact on Landscape Synthesis in Kashmir Himalayas: A Study of Land Suitability and Capability Classification(jointly with A.A. Pirazizy) in Frontiers in Environmental Geography, Eds. Onkar Singh et.al., Concept Pub., New Delhi, pp. 93-104, 1993.

160. Micro-level Monitoring of Land Use and Agriculture in Himalayas, Landuse and Agricultural Planning, Ed. N. Mohammed, Concept, New Delhi, 1992, 227-41.

161. Forest Energetic and Environmental Anomaly in Temperate Himalaya, in Dynamics of Mountain Geosystems, Editor R.B. Singh, 1992, op.cit. 87-100.

162. Environmental Geography in India: Challenges and Opportunities, in The Roots of Indian Geography: Search and Research, NGJI Vol. 38 (1-2), 1992. Also pub. In Emerging Frontiers in Environmental Geography: An Introductory Overview, in Environmental Geography, ed. R B.Singh, New Delhi, Heritage Pub., pp. 1-23, 1990.

163. Geographical Monitoring of Himalaya Highland - Lowland Interactive Systems, NGJI, Vol. 37 pt. 3, 1991, 272- 81 also published in Environment-Yesterday, Today and Tomorrow, eds. H.P. Bhatt and S. Bhatt, Galgotia Pub., New Delhi, 1992, pp. 238-57.

164. Human Resource Base for Rural Development in a Mountainous Region, Human Resource Development Vol. 1, Ed. A. Kumar, Anmol Pub., New Delhi, 1991, pp. 207-22.

165. Drought - Prone Areas in India: Regional Planning Issues and Resources Management Strategy in Planning Development and Disparities in Rural India, Editor: Ashok Kumar, New Delhi: Commonwealth Pub. 1990.

166. Integrated Rural Development: Concepts and Experience in Regional Development, Cont. to Indian Geog. Vol.X, Ed. Moonis Raza, New Delhi: Heritage Pub. 1988, pp. 544-68.

167. Spatial Perspective of Population and Resource Management in Arid Environment : A Case Study of Western Rajasthan, India, Resource Management in Dry Lands, eds. H.G. Mensching *et al.* 1984, pp. 206-14.

168. Settlement Planning and Integrated Rural Development in Micro-level Unit : A Case Study of Daraundha Block, (with S.L. Kayastha) Proc. Or Nat. Seminar on Regional Planning and Rural Development, ed.L.R. Singh, Allahabad, 1984, pp.147-50.

169. Energy Resources and Technology Transfer in the Management of Rural Environment (with S.L. Kayastha), Environmental Management eds. L.R. Singh et al., Allahabad, 1983, pp.-89.

RESEARCH SUPERVISION

A. Ph.D. Thesis Supervised

1. Anthropogenic Impact on Environmental Changes in Temperate Zone of the Himachal Himalayas by Ali Akbar Pirazizy.
2. Integrated Block Level Planning in Agriculturally Developed Region, (A Study of Kapurthala Block) by Lakhvinder Singh.
3. Monitoring and Forecasting Hydrological Parameters in the Mahanadi River Basin by Romila Prasad.
4. Man-Forest Interactions in the Himalayan Yamuna Basin - A Geoecological Perspective by Sarita Ghai (1998).
5. Atmospheric Quality in Metropolitan Cities of India by Bhuwan Kumar (1998).
6. Environmental Implications of Farmland Diversification in Rural-Urban Fringe of Delhi by Tejbir Singh Rana (1999)
7. Hazard Risk Assessment and Sustainable Land Development in Upper Beas Basin by B. W. Pandey (2000).
8. Squatter Settlements of Delhi Metropolis- A Human Ecological Study by Ravi Shekhar (2001).
9. Environmental Degradation and Development in the Metropolitan Fringe of Delhi by Naval Prasad Singh.(2001)

10. Changing Land use and Land Degradation Problems in the Indian Desert Region by Gambhir Singh Chauhan (2002).
11. Impact of Land Use Change on Biophysical and Socio-Economic Systems of Upper Beas Basin by D.K Mishra.
12. Geographic Dimensions of Common Land Resources in the Santhal Parganas Region by Mohammed Baber Ali.
13. Population, Land Use and Environmental Sustainability in Haryana, by Suman Chauhan, Joint Supervisor at M.D.University, Rohtak (2005).
14. The Changing Pattern of Domestic Energy Consumption and its Impact on Quality of Life: A Case Study of National Capital Region (NCR) by V.K.Dixit.(Joint Supervision) (2006).
15. Traffic and Environmental Assessment A Case Study of Delhi by Ms.. Rakhi Mukherjee (2007).
16. Cyclone Disaster Vulnerability and Mitigation along the East Coast of India by Ms. Ushvinder Kaur (2007).
17. Management of Municipal Solid Waste in Urban Delhi: A Geographical Analysis by Subhash Anand (2007).
18. Impact of Land Use Change and Industrial Development on Atmospheric Quality in Agra Metropolitan City by Vishwa Raj Sharma (2008).
19. Influence of Environmental Degradation on Women in Rural Delhi by Monica Rana (2008)
20. Land and water Resource Management in Son-Karamnasa Interfluve in Bihar by Dilip Kumar.(2009).
21. Spatial Analysis of Water Quality in River Yamuna and its Interrelationship with Human Activities in NCT of Delhi by Vaneeta Chandana (2010).
22. Forest Resource Assessment: A Case Study of Dehradun Valley by Ms. Poonam Kumaria.
23. Environmental Implications of Agricultural Development in Patiala District, Punjab by Neha Arora (2011).
24. Hazards and Associated Risks in Upper Bhagirathi Basin by Subodh Kumar .

25. Monitoring Land Use Change and its Environmental Impacts in Jaipur City using Geoinformatics by R. K. Gupta.

B. Ongoing Ph. D. Thesis

26. Assessment of Integrated Watershed Development Programme in Haryana Shiwalik by Mehtab Singh.
27. Glacier Lake Outburst Flood and Associated Geohazards in Himachal Himalaya by Pankaj Kumar.
28. Influence of Climate Change on Nanda Devi Biosphere Reserve by Suraj Mal.
29. Climatic Variability and Farming Flexibility: A Case Study of Uttarakhand Himalaya by Swati Thakur.
30. Agriculture Vulnerability to Climate Change in Rajasthan 1950-2010 by Ajay Kumar Gurjar.

C. M.Phil Thesis Supervised

1. Environmental Effects of Highland Tourism in Kashmir Himalayas by Ali Akbar Pirazizy (1987).
2. Forestry and Rural Development in the Upper Yamuna Valley - A Study of Jaunsar Bawar by Sarita Ghai (1988).
3. Rural Development in Mountainous Region: A Case Study of Mandi Tehsil by A.R. Khan (1989).
4. Non-farming Activities and Rural Transformation: A Case Study of Alipur Block (Delhi) by Tejvir Singh (1989).
5. Assessment of Existing Environmental Data Base and Research Stations for Geographical Monitoring and Forecasting in India by Romila Prasad (1990).
6. Environmental Transformation of Ganganagar - Impact of Indira Canal by Seema Gupta (1989).
7. Flood Hazard in North Bihar by Arun K. Singh (1992).
8. Environmental Aspects of Rural Water Supply in Cholapur Block, Varanasi District By B.W. Pandey (1992).
9. Waterlogged Areas in Karnal District: A Geographical Analysis by Mehtab Singh (1994).
10. Ecological and Socio-Economic Impacts of River Basin Development: A Case of Sardar Sarovar Project by Umesh Kumar (1993).

11. Assessment of Erosional Hazards and Critical Zones in Chamoli District by Meeta Sharma (1994).
12. Spatial Pattern of Non-Conventional Energy Use in U.T. of Delhi by Ritu Malik (1994).
13. Ecological Implications of Land Use Changes and Forest Renewal in Solan District by G.N.S. Gurudatt (1994).
14. Transformation of Socio-Economic Environment in a Metropolitan Fringe Area - A Case Study of Kanjhawala Block by Sanjiv (1995).
15. Lanscape Fragility and Environmental Change in Kullu Valley by D. Tingal (1995).
16. Industrial Growth and Environment Degradation in Gurgaon District by Spandan Naneet (1996).
17. Environmental Implications of Rural-Urban Fringe of Delhi - A Case Study of Najafgarh Block by Shilpa Chikara (1996).
18. Degrading Delhi Ridge Forest: Changing Scenario and Conservation Efforts by Rakhi Mukherjee (1996).
19. Village Common Land and its Management in the Upper Beas Basin by Mohammed Baber Ali (1997).
20. Ecological Dimensions of Desertification in Jhunjhunun District by Anoop Kumar (1997).
21. Indigenous Knowledge in the Protection of the Mountain Ecosystem- A Case Study of Kullu Valley by Ambika Chugh (1997).
22. Women Empowerment and Sustainable Development of Mountain Environment in Kullu Valley by Monica Rana (1997).
23. Land Degradation with Special Reference to Spread of Brick Kiln Industry in Kanjhawala Block (Delhi) by Jag Mohan (1997).
24. Socio-Economic and Spatial Mobility of Street Children in Delhi by Hemant Kumar (1997).
25. Geographical Analysis of Water Pollution in Hooghly River by Debasri Mukherjee (1998).
26. Varanasi- A Study in Impact of Development on the Environment by Tripta Singh (1998).

27. Sustainable Livelihood and Food Security in Kullu Block by Kirti Bas (1998).
28. Impact of Rural Development Programmes on Socio-Economic Conditions in Rural Delhi by Subhash (1999)
29. Environmental Monitoring and Mitigation of Cyclone Disaster in Coastal Region of Andhra Pradesh by Ushvinder Kaur (2000).
30. Air Pollution and Open Cast Coal Mining in Talcher Region (Orissa) by Hema Gupta (1999).
31. Environmental and Socio-Economic Implications of Tala-Hydro Electric Project in Wanchu Valley, Bhutan by Smirti Kumari (2001).
32. Climatic variability, Agrohorticultural Production and extreme Events in Mountainous areas-A Case Study of Kullu District by S. Sen Roy.(2002).
33. Spatial Analysis of Air pollution-A Case Study of Agra City by Vishwa Raj Sharma (2003).
34. Flood impact and Management in the Bhagalpur district, Bihar by Nilesh Kr. Singh (2003).
35. Role of ENGOs towards Environmental Improvement in Delhi by Kiran Dabas (2003).
36. Urban Water Problems in Delhi Metropolitan Region by Sheel Kumar (2003).
37. Impact of Flood on Biodiversity in Kaziranga National Park, Assam by M. Bortamuly (2004).
38. Spatial Analysis of Wastewater Reuse in Delhi by Reena Singh (2004).
39. Impact of Super Thermal Power Station on Land Degradation: A Case Study of Kahalgaon by Arun Kumar Singh (2005)..
40. Retreat of Gangotri glacier and related geo-hazards in Upper Bhagirathi basin by Subodh Kumar (2005).
41. Integrated Watershed Management in Upper Yamuna-Tons basin by Nawal Prakash (2005).
42. Human Impact on landuse Change and Biodiversity in Sariska Tiger Reserve by Nidhi Gandhi (2005).

43. Ecotourism in Pokhara Valley by Baasanti Shahi (2006).
44. Land Use/Cover Change, Landslides and their Influence on Biodiversity in Nanda Devi Biosphere Reserve by Suraj Mal (2006).
45. Human Impact and Indegenous Knowledge on Biodiversity in Rajaji National Park by Abhineety Goel.(2006).
46. Sustainable Mountain Ecotourism Development in Uttarkashi District of Uttaranchal by Pankaj Kumar (2007).
47. Climate Variability, Extreme Events and Land Use Change- A Case Study of Doon Valley, by Swati Thakur (2007).
48. Integrated Coastal Zone Management along Visakhapatnam by Shalini Sharan (2007).
49. Disaster Vulnerability and Mitigation Plan for NOIDA and Greater NOIDA by Swarnima Singh (2007).
50. Environmental Vulnerability of the High Altitude Wetlands of Changthang Valley by Anupam Anand (2007).
51. Integrated Coastal Zone Management of Goa by Sujeet Kumar (2008).
52. Spatial Distribution of Urban Heat Islands in Delhi by Manoj Kumar (2008) .
53. Urban hydrology of Mithi River in Mumbai by Kanishk Kumar Singh.(2009).
54. Depletion and Degradation of Biodiversity in National Capital Territory of Delhi by Parul Jain (2009).
55. Herbal Home Remedies in North-West District of Delhi and Haridwar District of Uttarakhand by Nitu (2010).
56. Industrial Development and Environmental Degradation in Panipat City by Ritu (2010).
57. Spatial –Temporal Variability in Traditional Water Bodies of Delhi by Sarita (2010).
58. Vulnerability to Climate Change in Arid Region of Rajasthan by Ajay Kumar Gurjar (2011).
59. Urban Climate Change in Kolkota Metropolitan Region by Rituparna Nath (2011).

D. Ongoing M. Phil Thesis

60. Status and Planning for Low Carbon Resilient Mega City- A Case study of Delhi by Vidhi Saluja
61. Impact of Industrial Change on Hydrology in Agra City by Ruchita Tirkey.
62. Agricultural Vulnerability to Climatic Change in the Dry Regions of Haryana By Rajesh Kumar.
63. Integrated River Basin Management (IRBM) of the Kosi River, India By Manikant Manish.

Pursuing M. Phil Part-I

64. Ashok Kumar
65. Bhanu Prakash Singh
66. Ranjeet Kumar
67. Abhay Shankar Prasad
68. Santosh Kumar
69. S. W. Bhutia
70. Atithi Pant
71. Ravikant Meena

SECTION - I

ENVIRONMENT

Climate Change and Water Resources in Haryana-An Analytical Perspective

INDER JEET

Introduction

For the purpose of current and future impacts, the term "climate change" has become synonymous with modern global warming (Demeritt, 2001). Latter, in turn, refers to post- industrial revolution changes in global mean surface air temperature that are hypothesized to have caused by increased atmospheric concentration of carbon dioxide, an active greenhouse gas during the eighteenth, nineteenth and twentieth centuries. In 1765, the carbon dioxide atmospheric concentration was about 280 parts per million by volumes (ppmv); in 2000, it is near 364 ppmv (Loaiciga, 2003). The atmospheric concentrations of other gasses (methane, nitrous oxide, and chlorofluorocarbons) have also risen as a result of accelerated economic activity and energy use in the last century. The rapid rise in fossil fuel combustion as an energy source since the late eighteenth century has caused the observed increase in atmospheric concentration of carbon dioxide beyond doubt. Recent estimates indicate that temperature has increased between 0.3°C and 0.6°C during the last 150 years (IPCC, 2001).

In the prevalent circumstances, the science of climate change is now well established and recognised. It is the result of hard work of thousands of scientists drawn from across the globe since last two decades. These scientists assessed every aspect of climate change for the benefits of the humanity. The Fourth Assessment Report (AR4) of the Intergovernmental Panel on Climate Change (IPCC) was produced in the year 2007 and highlighted the basis of careful observations extending over long period of time that warming of climate system is unambiguous as is now evident from observations of increase in global average air and ocean temperatures, widespread melting of snow and ice and rising global average sea level. AR4 observes that most of the increase in global average temperatures since the mid-20th century is very likely due to the increase in the concentration of anthropogenic Green House Gasses (GHG). It is important to remember that changes in climate are not limited merely to an increase in temperature, but involve several impacts such as an increase in intensity and frequency of floods, droughts, heat waves and extreme precipitation events. Therefore, these pose serious implications for the availability of water in several parts of the world and could have negative impacts on the yields of several crops.

There is broad agreement that greenhouse warming will have major impacts on water resources of the study area. Possible impacts that may especially affect water planning and project evaluation include changes in precipitation, runoff patterns and landuse change. Most models project enhanced precipitation during the monsoon season, particularly over the northwestern parts of India. Warmer temperature will accelerate the hydrologic cycle altering precipitation, the magnitude and timing of runoff and intensity and frequency of floods and droughts. Being a part of Indo-Gangetic basin, Haryana State experiences larger climatic variations. Rainfall varies from 300 mm in the semi-arid areas in the southwest to more than 1100 mm in the north along the Shivalik hills. The State is highlighted, even in the global measurement scale as one of the highest water scarcity and stress region both today and as well into the future. Many north Indian rivers such as the Ganga and the Yamuna have shown a sharp decline in the summer discharge in the recent past, possibly due to shrinking of the Himalayan glaciers that feed them. During the past 40 years an estimated reduction of appreciable quantity in the Himalayan ice volume has occurred (Gosain and Rao,2003). These observations pose an important question: Is this a consequence of global warming, or is it only a part of the low frequency climate variability inherent to the system?

Research Problem

The evidences of last decade of researches show the climate change is real, is happing now and is likely to increase in the coming decades. While a number of national trends are clear, particularly in temperature and precipitation, but the regional implications of climate change are likely to differ widely. There is not yet complete understanding of the likely regional impacts of climate change. Moreover, there are key knowledge gaps regarding the impacts of climate change on water availability. Following the publication of the Third Assessment Report (TAR) of the Intergovernmental Panel on Climate Change (IPCC) in 2001, it is evident that the climate system of the planet Earth is about to change significantly. It is obvious that anticipated changes will not be evenly distributed; rather abrupt and shocking events are to take place in different parts of the world. The TAR also prognosticated that a general warming superimposed on erratic monsoon activity in the dry season over the semi-arid (western) parts of Indo-Gangetic Plain (IGP) would lead to more droughts, while increased monsoon activity over the eastern parts of IGP would cause higher floods in terms of both extent and frequency.

Substantial work has been done at global and national levels to evaluate climatic impacts, but so far less information is available on regional and local scales. Therefore, this investigation initiates to find out some of the consequences of climate change on water resources and water system in Haryana.

There is no doubt that agricultural demand for water in Haryana is in excess of its availability from all sources, including rainfall. In the State, according to 'Irrigation Statistics, 1993' to cover all the cultivable land by canal, the requirement is 4.13 million hectare meter (ham) as against 1.85 ham from all sources including groundwater (.80 ham). It is estimated by the groundwater agencies that only 507,000 hectare meters of groundwater is available in the State, out of that 65 per cent is saline and hence, unfit for irrigation. In this way, the contribution of groundwater to total water supply is maximum in the fresh quality zone, comprising mainly, north and eastern parts of the State including Ambala, Yamunanagar, Karnal, Panipat, and also parts of Faridabad, Gurgaon, Rewari and Mahendragarh districts. According to Groundwater Cell of Haryana Agriculture Department, the maximum fall in groundwater table was more than 10.24 meters in Mahendragarh district followed by 4.36 meters and 4.03 meters in Rewari and Panipat district respectively from June 1974 to June 2004. This study, however, has shown that average rate of depletion due to rapid use of groundwater through tubewells is nearly 0.87 cm/ year (Jeet, 2009). This has created an alarming problem among farmers because of increasing cost of tubewell installation, low groundwater yield, change in cropping pattern, well abandonment, saline water ingress due to high cone of depression in fresh water zone, threat of land subsidence and food security. Moreover, climate change, will affect the water use particularly and also by change in population, lifestyle, economy and technology, which drives irrigated agriculture, the largest water- use sector. Significant changes in water use or hydrological cycle (affecting water supply and floods) require adaption in the management of water resources.

In this context, impact of climate change on water resources in Haryana are determined by assessing rainfall and temperature trends on different time and spatial scales with the baseline year, i.e., 1961-1991 as adopted by Intergovernmental Panel on Climate Change. Haryana State, which highly depends upon irrigated agriculture, needs a thorough study of climate change impacts on water resources.

Methodology

Present paper is completely based on secondary sources in terms of data. Therefore, the database used for this work is downloaded from India Water Portal, that extended the data from Climate Research Unit (CRU) TS 2 out of the Tyndall Centre for Climate Change Research, School of Environmental Sciences, University of East Anglia in Norwich, UK. This published dataset consists of interpolated (on a 0.5 degree latitude and longitudes grids). The original dataset is available at http: // www.cru.uea/~timm/grid/CRU_TS_2_0.html. In addition, the secondary data relating to climatic parameters like, rainfall and temperature are obtained from Regional Meteorological Centre (RMC) of India Meteorology Department (IMD), from revenue records of Revenue Department of Haryana State, Groundwater Cell of Agriculture Department, Irrigation Department of State and Bhakara Beas Management Board (BBMB).

In the present study, time-series analysis is used to observe the specific trends in rainfall and temperature. Simply stated, time-series analysis is defined as analysis of data in which time is an independent variable. A time series is analysed for the purpose of formulating and calibrating a model that can be used both to describe the time- dependent characteristics of a climatic variable and to predict future values of the time-dependent variable. In the present study, linear regression (least squares method) and polynomial regression are used for establishing the trends.

Equation for linear regression:

$$Y = a + bx$$

Equation for polynomial regression

$$Y = a + b1\,x + b2\,x2 + b3\,x3$$

For this purpose, study area has been divided into two agro-ecological zones, namely, arid and semi-arid zones. Arid zone comprises Menderagarh, Rewari, Bhiwani, Hisar, Sirsa and Fatehabad districts. Semi–arid zone consist Kurukshetra, Karnal, Panipat, Yamunanagar, Ambala, Punchkula, Kaithal, Sonipat, Jind (northern and central tract) and Rohtak, Jhajjar, Gurgaon, Faridabad (southern tract).

Result

Variability in Rainfall and Temperature

Climatic variables such as precipitation and temperature are essential inputs to water resources. Different combinations and seasonal pattern have direct consequences on the water regime of a region. Several studies have evaluated past trends in rainfall extremes of various durations, mostly at national and global scales. The concern for global warming had arisen from the prediction of General Circulation Model (GCM), which indicated a gradual warming of the atmospheric temperature, resulting in dramatic changes, in rainfall increase, melting of polar ice and drowning of low coastal areas in near future. Global climate change studies have been focused on temperature, the recent upward trends, which can be ascribed with reasonable confidence to anthropogenic influences. Trends have also been identified in precipitation amounts with notable increase over middle and high latitudes of Northern Hemisphere.

In the State, where climate is arid and semi-arid, rainfall is the main determinant of groundwater regime. Rainfall is markedly restricted in season and dry conditions are usual occurrence in the State. The short rainy season is succeeded by long dry season. The distribution of rainfall is irregular in both time and space and water deficiency is a problem everywhere. The main source of rainfall is the monsoon rainfall (northeast and southwest), which encompasses the entire State between the third week of June and middle of July. The amount and distribution of monthly rainfall are not uniform in space. Similarly, a good amount of rainfall is also received from a few depressions in January to March in the northeastern parts of Haryana. The monsoon rainfall decreases from east to west and northeast to southwest and cyclonic rainfall decreases from north to south. Consequently, there is a significant variability in rainfall from month to month. The intensity of rainfall at any station can be very high to very low in the northeastern parts where the monsoon winds are strong as compared to the northwest on account of the fading frequency and intensity of cyclones.

It is evident from the Table 1 that coefficient of rainfall and dispersion from the normal is more in Narnaul and Hisar districts. These districts are situated in the southwestern and western parts of the State. As one moves towards the north and southern side of the State, the values of both standard deviation and coefficient of variability are smaller. But in Gurgaon district again these values are abnormally high. This abnormality might be due to the location of this district, which is situated at the outcrops of the Aravalli hills. Moreover, everywhere rainfall is restricted in its seasonal distribution. Therefore, in the wide areas where average rainfall is marginal in amount this variability is crucial. The coefficients of variability in monsoon and cyclonic rainfall are not identical in Haryana as the latter is highly variable. The cyclonic rainfall is least reliable and the same is true for monsoon rainfall in areas bordering the Rajasthan. Variability in excess of 20 per cent implies great risk in farming. The annual, seasonal and weekly coefficients of variation of rainfall are over 20 per cent in the State. The significance of variations in the amount of rainfall from year to year or season to season is difficult to assess unless these are related to defined water requirements of the area. This can be done by estimating the probability of specified amount of rain being obtained within a year. It is obvious from the foregoing analysis that the variability of rainfall is inversely related to amount of rainfall. The areas of maximum variability experience the least reliability of rainfall. On the contrary, areas of least variability witness greater reliability of rainfall. As the State receives 80 to 90 per cent of its annual rainfall during the southwest monsoon season, variability of rain during this season indicates the trend of its annual variability.

Table 1
Mean, Standard Deviation and Coefficient of Variability of Rainfall

Sr. No.	*Station*	*Mean*	*Std. Error*	*Std. Deviation*	*Coefficient of Variability (%)*
1	*2*	*3*	*4*	*5*	*6*
1.	Ambala	875.1810	50.0814	229.5017	34.2
2.	Karnal	628.5857	49.5224	226.9403	36.1
3.	Gurgaon	764.5095	54.1031	247.9317	35.9
4.	Narnaul	536.7714	24.1563	110.6979	42.6
5.	Rohtak	588.0381	24.4871	112.2138	39.0
6.	Hissar	468.8238	20.8049	95.3399	40.3

Source: Calculated by Author

Rainfall Trends in Arid Zone

Generally speaking, the longer the period covered the more significant the trend. Change in the climate and weather conditions such as rainfall and temperature show seasonal variations but to find out significant trends longer period is required. In the present study, 45 years span (1961-2005) is considered to analyse for rainfall and 42 years (1961-2002) for temperature. Figure 1 shows the mean linear trend and 2nd order polynomial regression for the arid zone of Haryana. Figure shows that rainfall have been decreasing in all seasons on annual basis. Annually, the rate of decrease is -0.44 mm/year. The rate of decrease is very slow and imperceptible. The result of fitting a linear model to describe the trend of rainfall also indicates that there is not a statistically significant relationship at 90 per cent or higher confidence level with R^2 =.0024. Equation for the least squares model for rainfall is:

$$Y = 466.0-0.446x$$

Similarly, second order polynomial regression also indicates insignificant change in the rainfall during last four decades. The equation of the fitted model is:

$$Y = -219427.0+222.224x-0.0561447x^2$$

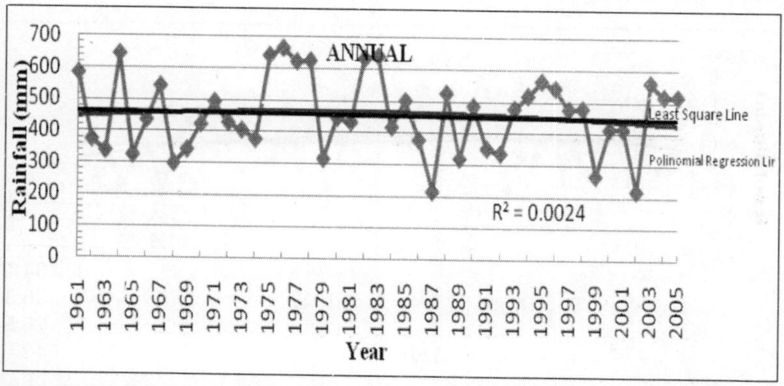

Figure 1: Arid Zone Regional Average Annual Precipitation Trend with East Squares and Polynomial Regression (1961-2005)

All Semi-Arid Series

Central and Northern Tract

Figure 2 shows all semi-arid annual series of rainfall trend in central and northern tract. Both least squares and polynomial regressions are establishing slightly downward trend in the rainfall. Total decrease in rainfall since 1961 to 2005 with the least squares analysis is nearly 58.05 mm. The rate of decrease is 1.29 mm/year. Equation for the least squares is:

$$Y = 625.6 - 1.298x$$

Similarly, polynomial regression of the 2nd order also indicates a downward trend in the rainfall in this tract of semi-arid zone of Haryana. The equation for the polynomial regression is:

$$Y = 565.56 + 6.367x - 0.166x^2$$

Southern Tract

About 80 per cent of the annual rainfall in this tract is received during the southwest monsoon. Normally, July and August are the rainiest months. There are chances to have some rainfall in the pre- monsoon month of July. The variation in annual rainfall from year to year is

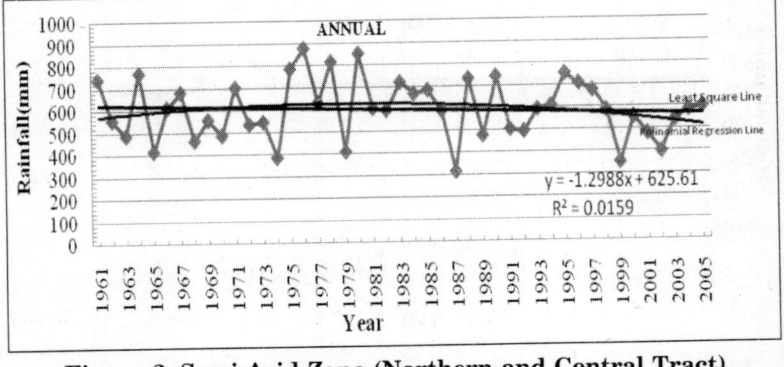

Figure 2: Semi-Arid Zone (Northern and Central Tract) Regional Averaged Rainfall Trend with Least Squares and Polynomialregression (1961-2005)

appreciable. In the 100 years period 1901-2001, the highest annual rainfall which was 21.5 per cent of the normal occurred in 1917. But in the next very year in 1918 the annual rainfall was lowest in the 100 years period amounting to only 25 per cent of the normal.

The analysis of least squares and polynomial regression establishes the downward trend of rainfall in this tract during the last four decades in terms of annual rainfall (Figure 3). Equation for least squares is:

$$Y = 625.6 - 1.29x$$

Equation for the polynomial regression is:

$$Y = 615.4 - 0.967x - 0.02x^2$$

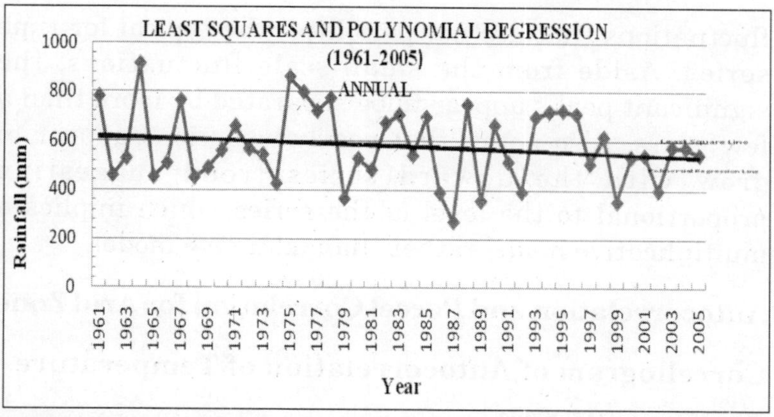

Figure 3: Semi-Arid Zone (Southern Tract) Regional Averaged Rainfall Trend with Least Squares and Polynomial Regression (1961-2005)

Temperature Trend

Spectral Density as a Function of Period

Figure 4 shows a strong signal which appears at about period 12, corresponding to a yearly cycle in the data.

All the series exhibits a number of peaks, but they do not appear to be equally spread. The output suggests that if the series has a periodic component, it also has

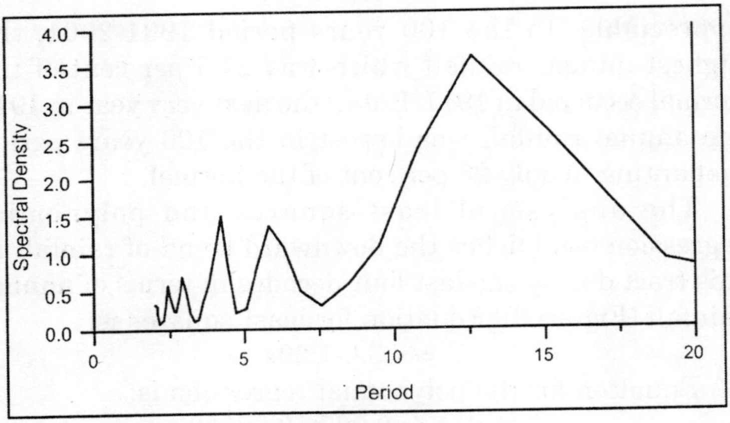

Figure 4

fluctuations that are not periodic – the typical for time series. Aside from the small-scale fluctuations, the significant peaks appear to be separated by more than a few years. Also notice that yearly variations appear to grow with the upward series trend suggesting proportional to the level of the series which implies a multiplicative model rather than addictive model.

Autocorrelation and Parcel Correlation for Arid Zone

Correllogram of Autocorrelation of Temperature

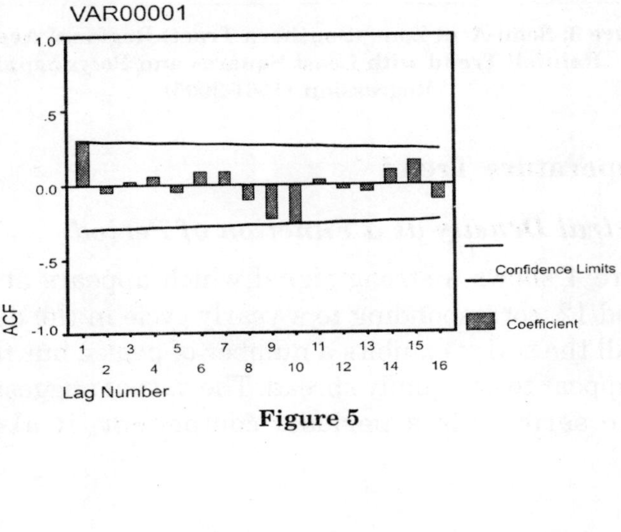

Figure 5

Correllogram of Parcel Correlation of Temperature

The first step in this analysis is to generate a run sequence plot of the response variable. A run sequence plot indicates stationarity (constant location and scale), the presence of outliers and seasonal patterns. Data should have constant mean variance and autocorrelation through time. In this analysis Box- Jenkins (1976) model is fitted to the differenced data or to the residuals after fitting a trend curve.

Both these correllogram's functions show that lag value is significant at first lag and decreases constantly as the lag numbers increase. This step of the ARIMA MODEL indicates of the fact that temperature does not have any seasonality and abnormality throughout the study period in any part of the study area. These correllogram (Figure 5 and 6) clearly manifest that there is neither positive nor negative relationship in temperature data from one year to another. The number of peaks are evident, but they appear at random intervals, showing no evidence of annual pattern. Therefore, any kind of yearly trends are completely absent in this time series temperature data.

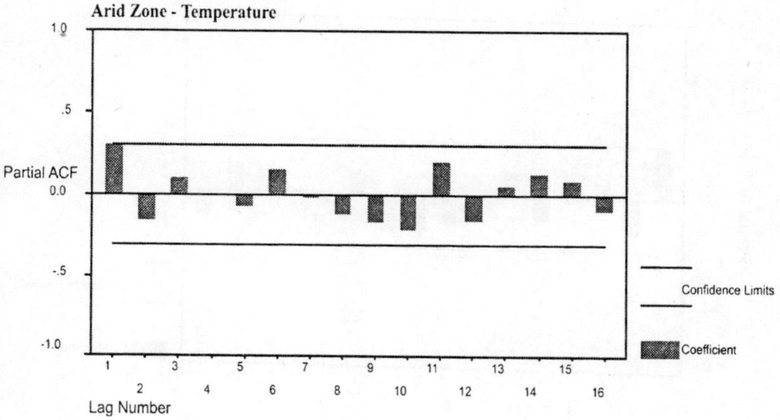

Figure 6

Similarly, in the Semi-Arid Zone temperature data exhibits more or less same picture as in the Arid zone. Temperature variations and trend in the series are not clear. In both the sub-divisions namely, central and northern parts and southern parts, (Figure 7 and 8) temperature peaks are available with non annual pattern.

Correllogram of Autocorrelation of Semi-Arid Zone

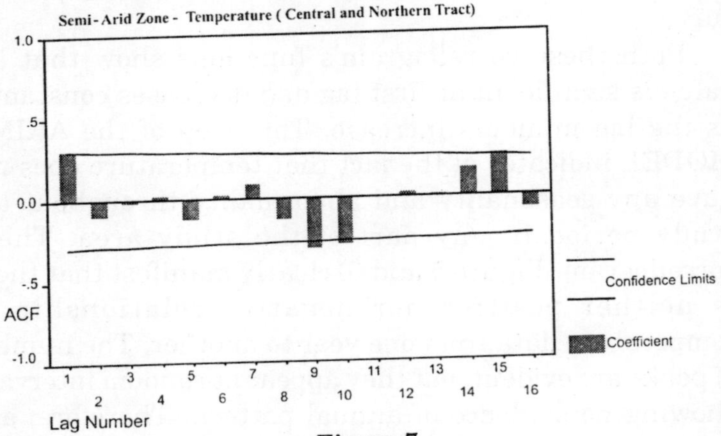

Figure 7

Correllogrm of Parcel Correlation

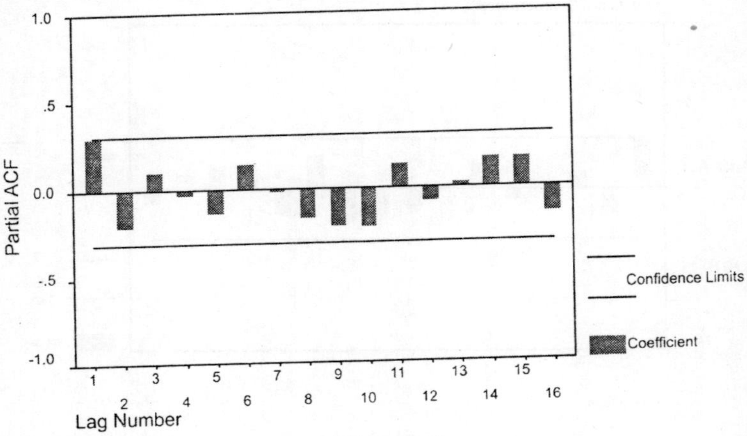

Figure 8

Conclusion

Time series analysis is an important tool in modelling and forecasting. ARIMA (1,0,1) model gives information that can help the decision makers, irrigation engineers and hydrologists to establish strategies, priorities and proper use of water resources. We can make the following conclusions from the run sequence:

(i) The data shows strong and positive autocorrelation.

(ii) There does not appear any significant trend or any pattern in the data.

(iii) Variability in rainfall is clear from one year to another and from one region to another.

(iv) Some trends are visible which manifest a little decrease in rainfall in southern tract of semi-arid zone.

(v) In the future arid and southern tract of semi-arid zone are more likely to face the acute shortage of water resources in the absence of proper groundwater recharge through rainfall.

(vi) Likewise, the northern and central tract of semi-arid zone will experience the more variability in rainfall. Higher the variability in rainfall, the farmers will depend more on groundwater for irrigation. Consequently, some parts of this zone will face groundwater depletion problems.

References

Demeritt, D. (2001), "The Construction of Global Warming and the Politics of Science", Annals of Association of American Geographers, 91 (2): 307-337.

Devid, R. A., and Hayley, J. F. (2004), "Spatial and Temporal Variation in Precipitation in Upper Indus Basin, Global Teleconnections and Hydrological Implications", *Hydrology and Earth System Science*, 8 (1): 47-61.

Gosain, A. K. and Sandhya Rao, (2003), *Climate Change and India, Vulnerability Assessment and Adaptation* (eds. Shukla, P. R. *et al.*), Universities Press, Hyderabad, pp. 159–192.

Government of India, Climate of Haryana and Union Territories of Delhi and Chandigarh.

Groundwater Cell, (2004), Agriculture Department, Government of Haryana.

Irrigation Statistics (1993), Government of Haryana.

Jeet, I. (2009), Rainwater Harvesting, Mittal Publications, New Delhi.

Loaiciga, H. A. (2003), "Climate Change and Groundwater", Annals of the Association of American Geographers, 93 (1): 30-41.

Williamson, A. V. (1925), "Irrigation in Indo- Gangetic Plain," *Geographical Journal*, 65(2): 141-153.

Community Perception and Environmental Behaviour

RAKHI PARIJAT

The study of man's perception is the key to understand societal behaviour and also gives an explanation to the working of the 'real world'. It is the process by which an individual or a community gains knowledge of the world, by receiving stimuli from the environment through his senses and creating an image of that environment. Individuals and groups relate to their environment through their perception and accordingly take decisions within the framework of their perceived set of elements. The usual image that he/she creates is filtered down based on his/her reasoning and past experiences. Geographers have shown more concern with the spatial pattern of behaviour rather than the behavioural process (Davis, 1972). Spatial behaviour is understood to be a product of personal views, experiences and preferences.

As man behaves differently in different places and situations, and even in similar situations, it is very difficult to generalise a set pattern. Many times, one fails to understand whether the human specie is scientifically motivated, and 'logic' that has derived out of his own innovations do really have a place in his life. Therefore, behavioural theory has been developed to explain deviations from the theoretical norms in the real world and to take account of human motivation and behaviour (Knowles and Waring, 1978). The level of perception is generally affected by personal views as well as the influence of culture in shaping their view. Yet within one's given environment, perception may differ due to the sub cultural factors, such as class, community and language, to which man is subjected to. With this background, the image produced is 'filtered' through one's needs and desires.

Preferences and Behaviour

Every individual forms a mental map produced by the individual's experience and the information, one has received, that may vary from person to person. Despite this, some common elements are identified to understand some common action over a space. As this perception translates into behaviour, it is important to assess both behaviour as well as the consequences of that behaviour. This gives a fair understanding of many decisions which were thought to be irrational. Thus, many planning decisions or policies often fail as they do not suit the community over which it is applied and thus can totally go wrong. At the same time irrespective of faulty characteristics some others may be quite attractive.

Perception may vary with age, sex, class, education within a set of environmental condition creating problem to arrive at a common definition. However certain commonalities have to be explored in order to find a widespread practice. Frequently one finds a wide gap between planning and public ideologies. But it has to be seen that one is not trapped in either of the extreme philosophy. There has to be a balance of opinions from both sides to reach the right decisions and in doing so many planning costs can be optimised if preferences and choices are known.

The perception based relationship of man with his environment can be reconstructed again and again due to the constant process of interactions within his social environment. This includes individual's or group's behaviour within an economic, political and social milieu. This results in developing a psychological identity which recognises facts and fallacies that may contribute immensely towards his social behaviour and actions. Thus application of psychological or behavioural, quantitative and fieldwork oriented techniques are needed to understand the system of human organisation and behaviour in space. The geographic space perception based on cognizance, evaluation and preferential grading of spatially differentiated objects of space have much significance in such studies.

Any space-specific environmental problem concerns the people of that place. Physical measurement is important to quantify the stimuli, but it is meaningless unless it is related to the human reaction. It may be often argued that the purpose of planning is to regulate and encourage changes in such a way as to maximise the benefits to the society. In order to do this, it is clearly necessary to be able to predict the effects of all alternative new schemes on people. It should also be able to set acceptable environmental standards for a better living.

The present study on traffic, environment and community assessment done in the metropolitan city of Delhi also presents many facts that reflect on the understanding of the society about this issue. The plans and policies enacted by the government from time to time can be more successful if an insight into the public opinion is given enough importance. Similarly the government should give emphasis on ethics and education to bring about significant change in the society. *Attitudes have to be shaped through society building process without questioning the prevailing thinking. A culture of positive behaviour has to be developed.*

Community Perception about Traffic

Traffic based perception studies are designed to obtain predictive information on the relationship between physical factors such as air or noise pollution, traffic flow, road conditions and some measures of subjective reaction like annoyance, needs, demands, values and preferences. Information about the attitudes, feelings, opinions or preferences of people is usually obtained by questioning a sample of the population in which interest lies. This is obtained by interviewing them at home or streets through structured questionnaire or informal discussions. Designing questions which yield such information in a valid and reliable way requires a great deal of care. This is designed to gain insight into how people think about a particular subject, the kinds of words they use, and the similarity or otherwise between their and the researcher's point of view.

The present perception study primarily focuses on all aspects related to the mobility of the people. The environmental issues arising out of this mobility in form of traffic and transportation to some extent have been assessed. All the behavioural outcomes were considered to appreciate the general pattern of community's understanding and derive logical explanations. The study took into account all possible view points of the people without imposing any restrictions in their free expressions. At the same time, all the responses, positive or negative have been evaluated to understand environmental consequences of such behavioural pattern. Observations in the field and physical measurements of various problems have given insight to many issues which can be of prime concern for both the citizens and the government.

The study was designed to obtain predictive information on the relationship between physical factors and subjective reactions. Since behavioural studies in

Geography deal with how one thinks and feels about the environment, and how one behaves with it and alter it. The study aimed:

- To identify the responses of the people to environmental problems.
- To determine whether attitudes concerning the threat from environmental discomfort have any influence on travel behaviour.
- To understand the effectiveness of incentive and disincentive schemes aimed at reducing environmental pressure.
- To determine the factors that can influence public opinions.
- To determine whether behavioural intention expressed in terms of people's willingness to take positive steps to improve their environment is conducive to implement environment-friendly activities.

It has been observed that responses vary from one socio-economic group to another. Since attitudinal responses are related to the understanding of their environmental systems, they have been grouped and classified. This enabled the researcher to understand the opinion of the people on one hand and also the behavioural aspect arising out of it on the other hand. These opinions can be mobilized to generate positive reactions. In a democratic set up such opinions should definitely be heard and given due weightage.

The study was conducted at selected traffic intersections and localities in Delhi, through questionnaire survey and physical observations. It brought about significant facts that are worth considering. The analysis helped in understanding the perception regarding traffic problems, attitude towards environment and their genuine concerns. Expert opinions have also been sought wherever deviations were felt between perceptions and objectivity.

The major findings were derived out of the interviews of people and their responses, expert opinions and observations made during the period of study. Each category of response has been grouped as following:

- Almost all the respondents, irrespective of their residential location from the road, felt bothered by road traffic. 'Bothered' here refers to "annoyance, irritation, troublesome, offensive" or "simply worry". Dust/smoke/ fumes have been directly related to problems arising out of motorised vehicles. However it is quite surprising to know that a large percentage do not hold traffic related inconveniences a major threat as compared to other more relevant issues such as shortage of water, electricity and inflation.
- The degree of botheration faced because traffic was directly proportional to visible smoke, fumes or odour. Excessive and visible smoke which is a result of improper engine maintenance is not quite well understood by the people. Also odour is not given top priority as far as traffic related problems are concerned. Thus, the parts of emission (e.g., lead and CO) that is not picked up by eye or nose, did not seem to bother the respondents.
- Irrespective of high volume of users of personal modes of transport, people still demand for public transportation. Most people feel dissatisfied with the present mode of transportation which they use due to personal reasons. Most are troubled by the dilemma of travelling every day due to the hassles on the road. But there is not much concern about the rising traffic volume on the streets and its externalities even among educated people.

- The roads of Delhi are increasingly showing that they are visually un- aesthetic, but most people have not given any constructive opinion. It clearly indicates that they do not associate the problems of visual intrusion, as it is correctly related to traffic or otherwise to be of any serious nature.
- Air quality is of major concern to most residents and understanding is primarily about pollution, which is seen as undesirable, unhealthy and dangerous. Dirty windows/curtains and their own faces are indicators. While people do understand the impact of pollution on health such as runny noses and sore eyes. But traffic could be a major factor responsible for ill health is either not very well understood or simply ignored. Impacts of pollution on the population is not a thought provoking issue at all.
- Although everybody is not clear about the factors which are increasingly affecting health, but some do complain of irritation to eyes, stress, headaches and discharges from their mucus membranes commonly attributed to problems of pollution in Delhi. People have complained about breathing problems also. Children in Delhi are particularly prone to asthma. This shows that although most people are subject to some or the other health problems which is associated to be arising out of road traffic. But they do not find a voice to generate strong public opinion.
- The general perception is that it is the duty of the Centre or the State Government to take the lead rather than the individual. The notion of the community's role in improving air quality is less clear. Willingness to use personal mode of transport is quite high and is not thought to be

irrational. Air quality is seen as a national (even global) issue rather than a local issue and residents feel helpless and unable to do anything individually about the situation. Not much concern has been shown towards their own behaviour which can drastically change the environmental situation.

- Congestion has emerged as a major problem due to the delays caused. Apart from this, parking is definitely an issue worth giving a thought. Most people have never thought about the externalities associated with congestion such as fuel loss or more emission.
- Over speeding and lane changing has not received enough importance which shows people do not associate it as a major problem. On the contrary, it is a very important issue which should be addressed from the point of road safety.
- Residential streets are exposed to thorough traffic to avoid congestion on main roads but it does not emerge as a major threat probably because people have got used to it. People have mentioned about parking within colonies as a major problem. Some have mentioned about parks etc being encroached upon by vehicles and lack of playing spaces. Few people staying near the main road have remarked about problem of noise from trucks and heavy vehicles disturbing their sleep at night. While others have complained about safety of their vehicles to be a major issue. But the question of residential wellbeing from the health point of view and safety of residents not given enough consideration.
- When asked about violations of traffic rule most know that they have violated rules from time to

time. Infact observations points out to a completely different situation which shows violations taking place every minute. There is a clear indication of how violations in the roads are increasing by the day and accident rates are shooting up. Everybody seems to be in a hurry and do not care for others, neither their own safety. People constantly change lanes, speed up their vehicles at wrong places, honk unnecessarily, jump traffic signals, do triple riding on two-wheelers, do not wear proper helmets, do not give proper indications and so on. If respondents are to be believed, the commuters who use public transport or walk probably do not consider themselves making any violations. Observations reveal that violations have become part and parcel of their daily lives and do not consider their own mistakes to be of any consequence. Also pedestrians who otherwise look innocent are often the biggest violators at times.

- When people were asked to describe the vehicles which are of maximum nuisance value, slow moving vehicles have been ranked highest due to the space they take and also lead to too much chaos on the road in congested areas. The other vehicle which has been causing problem is trucks and tempos due to smoke and noise. Buses are also culprits but nevertheless useful. Again nobody gives a thought to their own vehicles to be a cause of any problem.
- Most people have humbly spoken of their irrational behaviour due to stress and anger caused by another commuter's irrational behaviour. They often react violently and have done more often in different points of time. Some

people have talked about being violent within their vehicles that is, they have not shown anger against the defaulter but stressed themselves. Many have even fought with their family members in order to vent their anger. Some have even tried to react in the same manner as the defaulter. This shows how stress is slowly building in our lives. This undefined problem not ranked higher by people but is worth considering is the increased incidences of Road Rage and its psychological impact.

- Most people think violations are common on Delhi roads as people here simply do not bother. Besides violations become even easier as traffic police are not that vigilant and also corrupt. Many of their drives are quite superficial, temporary and deal with smaller problems. They remove encroachments, it comes back; they fine speeding vehicles especially buses and trucks, it still persist; polluting vehicles are still visible, but they never seem to notice; the trucks still move on the roads freely irrespective of bans and so on. Issue of illegal licenses are mostly overlooked by people whether private or public transport users. It is however quite evident that most people forego bad practices.
- Another interesting fact is that presently very few use car pool option, cycling or walking. Even when they were told about the positive impacts of such behaviour on the health as well as the environment, most people have shown inability or unwillingness to follow it regularly. Similarly people are ready to walk a reasonable distance only if they have enough time at disposal. None has talked about favourable traffic policies and

Plate 1: Pedestrians Walk all over the Main Road

Plate 2: Lane Driving not Followed

Plate 3: Typical Traffic Scene

Plate 4: Motorists use Pavements to Beat the Traffic

Plate 5: Subways or Foot Overbridges are Not Preferred by Pedestrians (Instead they Take Risk by Crossing Roads from Busier Sections; Subways are Either used as Eateries or are Deserted even if they are Well Maintained)

better pedestrian options to help them follow such mobility pattern.

- The present survey shows that a high degree of agreement exist in their willingness to use public transportations if cheaper as well as efficient options are available. Their willingness is still favourable among a large section of people in service sector even if cost is little higher but provides hassle free travel. A very small few have shown their willingness to use bicycle if proper tracks are built. It is not popular among both high income group as well as middle income group besides elderly and women. The young adult also refuses to use such an unglamourised mode as it affects their status symbol. Not much support has been found towards environment friendly practices like many European countries who have taken to healthy biking and low congestion.
- There is a positive reaction with respect to the improvements which has taken place in late 1990s and especially post 2000. Most respondents strongly believe that improvements have been possible mainly due to CNG and Metro Rail wherever it operates. While construction of

flyovers (underpasses) have been listed by most respondents to have provided partial improvement, as they have allowed free movement at many congested points. When they have been quizzed about what really have improved, the general feeling is now there is less of smoke on the road. Most have failed to see the environmental costs of many policies and have rarely questioned their usefulness.

- Public perception varies with individual requirements. The private mode is appreciated by most due to the series of attractive characteristics, such as speed, flexibility, comfort, and the ease of carrying along one's burdens. The choice of mode also depends on individual's paying capacity and time at disposal. The negative effect is not perceived because of environmental problems but more due to high cost, safety issues or parking difficulties.
- Policies like congestion pricing or high parking fees may result in more aggression. People are also suspicious of plans that otherwise may be good for the environment. Some changes in travel pattern introduced for the benefit of the residents themselves may be frequently opposed. This may also instigate the political parties to back step due to fear of protests. In reality, these disincentives can trigger changes in behaviour of the people to indirectly work towards bettering their surroundings by acting responsibly. But they may not support drastic solutions which will adversely affect their mobility and accessibility.
- Another problem that frequently arises is the differing perceptions based on their roles as a resident or a commuter. Within their own

residential localities, residents are angered by the trespassers. There is a difference in opinions when they change roles from residents to commuters. A resident of one locality may be quite responsible while totally disregarding traffic rules when travelling through another locality as a commuter.

- On the whole, the respondents have reacted variedly with respect to the nature of problem in general about traffic and transport. Maximum people feel lack of efficient public transportation is the major problem. This is followed by congestion and bad road condition as an issue which is a cause of concern as it both increases driver's difficulty as well as unsafe condition. The violations of traffic rules or more number of vehicles on road does not bother them so much.

Many suggestions have also poured in for bettering the problems related to traffic and transportation. The major suggestions have been:

- ➢ Traffic rules should be made stringent, traffic police should be more honest, heavy fines should be imposed to check over speeding and other violations.
- ➢ Public transportation should be increased—increase the number of DTC buses, limit the number of passengers in buses, more public buses than private, high capacity buses should be introduced wherever possible.
- ➢ Pick and drop vehicles and chartered services should be increased to avoid use of private vehicles, car loans policy assure phasing of old vehicles.
- ➢ Metro have been appreciated by most people, although there have been complains of rush during peak hours but it is still considered most

safe. There is a need for wider network of metro, and metro stations be made accessible.

- Good governance is required.
- Creating awareness had been the most common suggestion, pedestrian awareness to reduce accidents and strong driving rules.

Very few have actually talked about some very pertinent issues that can drastically bring positive environmental changes such as:

- Clearing traffic on major roads, removing encroachments, improved roads conditions, good pedestrian options and rules or lane systems to be followed.
- One day off to private vehicles, zero tolerance zones.
- Reducion in number of vehicles.
- Heavy fines on violations on both drivers and pedestrians.
- Congestion fee and heavy parking fee in congested locations.
- No suggestion has been made about self responsibility or their own behaviour.

Common Perceptional Issues and Guiding Principles

- *Responses of community vary according to their socio-economic status. Increased income and education does not guarantee declining use of personal modes of transport or positive environmental behaviour.*

There is a wide difference in the opinion of people with respect to the traffic related problems faced by commuters of different class and community. The degree of the problem perceived by different sections of population does result in significant differences in the pattern of travel

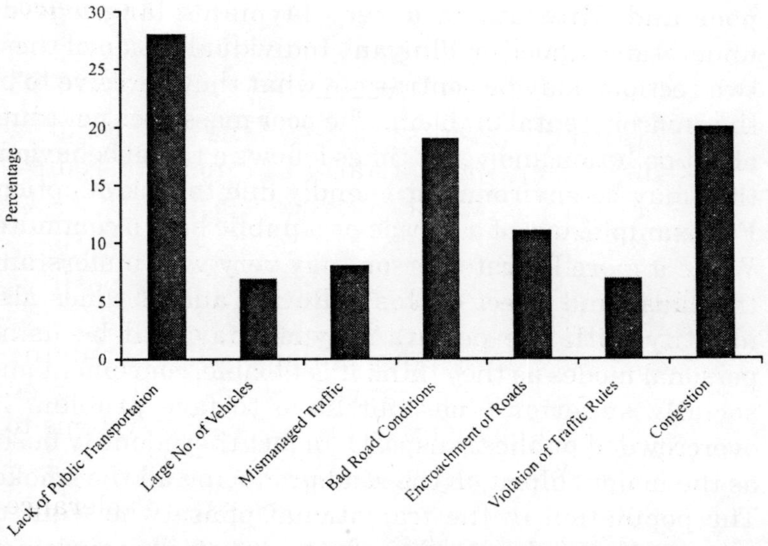

Figure 1: Perception of General Traffic Problem in Delhi

behaviour of the people. This further complicates the issue of understanding the environmental impact caused from such behaviour which depends on the factors that are normally overlooked or not classified. The factor of mobility is the most important priority in the minds of people rather than noise or air pollution. In the city environment where occupations are diversified and so are the workplaces the major focus of attention goes to easier and faster travel to their respective destinations. Most of this commuting population is more concerned about following a travel pattern that suits them even if it amounts to risking their health. The major intention is to earn a better living that uplifts their status and consumerism comes in the way of any positive action they can potentially take for the betterment of their surrounding.

'Environment' or pollution may be a much understood phenomenon among the educated classes but it is never in their minds when actions are undertaken. Even the poor and illiterate in a very layman's language do understand 'dhool' or 'dhuyan'. Individual action of these two sections may be contrary to what they perceive to be the environmental problem. The poor mass does not think about pollution and yet at times follows a travel behaviour that may be environment friendly due to lack of option. For example, use of a bicycle or a public bus to commute. While a more literate person may very well understand the cause and effect of this pollution and at times also identify with the pollutant type, may still be using personal modes as they think it is flexible, convenient and socially superior (does not have to face problem in overcrowded public transport). In fact they identify buses as the major culprit on the road producing all the smoke. The population in the transitional phase who want to upgrade themselves still prefer to use an ill maintained two-wheeler or an old car. He is either incapable of maintaining his vehicle due to the expenditure involved or simply not bothered as he feels his action is not harming him or his family. 'Others' for him is non existent.

It is also sometimes very difficult to categorise the population in terms of education and status. The rigid old definition of low, middle and high class population is often transient. With increased income and opportunity there has evolved many transitional classes which have greater degree of variation in perception as well as behaviour. The very poorest or the stinking rich do have homogenous behavioural patterns to some extent, while the various sub categories of so called middle class population project different attitudinal pattern.

The above facts states that there is an increased use of private vehicles as educational levels increase and people get better salaries. This is typical of developing economies where people start to acquire more once they achieve a better status. As the buying capacity increases and choices become wider, they do not restrict themselves and become buyers of major luxuries in life. Cars increasingly become larger. Indirectly it has its own side effects, such as increase in personal modes on the road in this particular case.

However, educated people can be thought off as reasonable individuals. By giving them the right opportunity, this section of the population can also be motivated to use public transport. They should be constantly reminded of their responsibilities for bettering their environment. They should be provided good and efficient options of travel even if it is slightly expensive.

➢ *Male and female reactions vary*

Many responses vary significantly with males and female. Females are better users of public transport due to known reasons like lack of personal transport and less preferences for private mode of travel. Their preferences often vary as per their travel needs. The road rage is more prevalent among men and rarely found in women drivers. However certain women drivers are not very familiar with driving rules and also often lack fast responses required while handling situations in the midst of traffic. This population should be provided with safe and hassle free travel and given special driving assistance training.

➢ *Public opinion is subject to visible gains*

Most people make perceptions based on what they see. The investigative capacity is generally lacking. Many times the administrators take advantage of such situation by giving them option that looks attractive and lucrative. For example people find that pollution has improved largely as they are not able to see the dark smoke. But they are not aware of the invisible harm which is still going on as CO or smaller particulate matter is still higher and life threatening. Similarly the attractiveness of the metro has appealed more rather then its quality in siphoning of vehicles from the road. This can be positively used by making all public transport facilities attractive, efficient and clean. This would indirectly bring down pollution as people would leave their cars and go for hassle free transportations. The present situation where most roads in Delhi have been dug up and has resulted in tremendous load of particulate matter, there is hardly any concern shown by people or the media.

➢ *Convenience over rules healthy practices which may lead to inconvenience in mobility*

Many people are aware of the unhealthy state of Delhi's air which is poisoning the city. Yet they still depend upon practices which go against their health due to convenience attached to particular practices. They use more private vehicles, avoid car pools when it does not suit them and thus increase pollution. They avoid walking and even in congested market places take their vehicles right next to the places of utility without bothering about others and also how they cause trouble to the pedestrians. They spoil their own residential localities by parking improperly, driving carelessly within residential streets causing severance and encroaching on other's spaces. They have somehow accepted the lawlessness and think it is fine. The psyche of the people has drastically changed towards individualism and in non committal thinking. The environmental friendly travel pattern is not favourable with the public if it involves sacrificing on their comfort as well as status. Pedestrianism and use of bicycle is not much favoured as they are getting status and comfort conscious. People like to showcase their flashy vehicles.

Policies thus should not only favour the rich or acquiring luxuriant cars but should ensure that better and cheaper electric engines and other environment friendly travelling options are utilised to the maximum.

Bigger companies and role models can favourably encourage such alternatives.

➢ *Policy changes and enhancement of transport system affect travel behaviour*

The strict policies should be suitably complemented by improvements in public transportation. For passenger's positive response there should be fewer delays, slightly faster speeds, comfort, security and a reasonable fare structure. The alternatives will be accepted if they are superior and cheaper. While providing services to move them away from personal modes, different socio-economic groups should be targeted as per their needs and not universally applied.

Policymakers should bear in mind the socio-economic background when developing specific policies and also understand how those policies might be received. Educating the public about the impacts of vehicular traffic through mass awareness programme and campaigns can also target specific groups. Right from formal education system to media campaigns to circulation of messages through leaflets and SMSs can all address the problem in variety of ways. Since all problems have multifaceted causes, thus the solution should also be multidisciplinary rather than focusing on individual agenda.

The metro is a live example of one such policy decision which has inspired many people to drop their cars or two-wheelers at home. It has provided mobility in a descent manner and many people do not mind paying parking fees everyday when they park their vehicles at the station. It has also provided safer travel to women too many of whom have moved from three-wheelers to this mode. But it cannot be taken as transportation for the masses as the authorities often project. However, improvement in the bus system is much needed to address the problems of weaker sections that look for cheaper travel.

Conclusion and Recommendations

A perception study brings out the man-environment relationship in which individual and collective understanding of the environment is seen as a major force in shaping the environmental consciousness. It contributes to the solution of practical problems of integrating human dimensions into environmental management and policy making. Understanding people-environment interaction is best achieved by looking at both the psychological processes through which individual "actors" come to know of the environment, and the way in which these processes influence the nature of resultant behaviour.

The study which tried to review perspectives of community on environmental issues, affected by transportation and traffic gave an understanding of the context of public perceptions towards transportation. From driving to work, to doing the shopping, daily lives consist of a myriad of spatial behaviours — movements across and within a spatial environment. Each day one makes hundreds of complex spatial choices and spatial decisions. In the vast majority of cases one relies upon a previously acquired spatial understanding of the world in which one lives i.e. upon mind's spatial representation of the environment, so-called 'cognitive map'. Individuals are not of course sovereign decision-makers: many have personalities and attitudes that impose severe restrictions on the amount of information that they are capable of handling, and many find that the information available to them is severely constrained by social class, technology, culture, and the political system.

It has been well highlighted that there exists a gap between perception and reality in some cases. Rather, 'reality' is seen as something that is constructed in the minds of individuals while they are engaged in an indivisible relationship with their environment. The notion of the communities' role in improving air quality is less clear as peoples' over-dependence on cars or other personal modes is clearly visible. There is a distorted perception among many people in the high middle and high income group that use of public buses are beneath their dignity and used only by the 'poor'. Even the young population which are hooked to two-wheelers are averse to the idea of using bicycle as alternate means of transport. Everyone's on a short leash of time limit nowadays, and thus, mobility is seen as an important issue that should not only serve the need to travel quickly but also suits ones comfort level and image.

While using private mode seems to be the choice of most people, this behaviour is increasing pollution and congestion. Every day one wastes precious time in traffic intersections. If the average person spends about an hour of each working day in traffic, then by the end of the year that person would have wasted a total of 10 days waiting in traffic. If this is multiplied to all mobile population in Delhi, it will have a cumulative impact. This indirectly leads to loss in income and, since people don't actually see this money being squandered, they don't think it's a problem, but the reality is that money and time that are being wasted. There is a need to change peoples' perceptions on how they use their cars and how it affects them and their children's health.

The design of an efficient public transportation system that gives an alternative to one's car involves the use of buses, bicycles and one's own legs to bring about a decrease in automobile emissions, a reduction in time spent stuck in traffic and increased national savings in petrol, automobile and health expenses. When it comes to implementing public transport solutions, residents do not think about day to day information about air quality. In addition, travellers are moderately interested in being involved in the transportation planning process but feel they are not particularly well informed about or encouraged to become involved. However they strongly feel it is the responsibility of the authorities alone to bring about positive changes and their own efforts do not have any relevance.

Another disturbing fact is that people are not conscious of traffic rules, even as callousness and disregard of the safety of others and one's own self is on the rise. Everybody seems to be smitten by the 'rush bug' and how motorists, cyclists and pedestrians behave or rather misbehave on road is nobody's concern. Whether parking wrongly, overtaking from the wrong side, unnecessary honking, using pressure horns, using highway lights in the city or driving/walking on the wrong side of the road, seems to be all ok with them. Violating traffic rules seems to be a habit now rather than an exception. Then there are another set of people who easily get stressed on using public roads and suffer psychological distress.

It is realised that sustainable transport planning both in terms of mobility and decreased environmental problems can only come when people are given the best of alternatives. Mass awareness programme at grass root level is pertinent. Public is most likely to support expanding programs that provide alternative modes of travel — public transportation, bicycle and pedestrian system only when strict measures are coupled with strong incentives. In addition, there is strong support for additional traffic information. Finally, travellers agree that existing roads should be expanded along with well integrated metro network. Many developed countries in Europe have actually been able to develop a positive culture of road behaviour.

Efforts to *mitigate congestion* and *improve traffic flow* should continue to be a focus for improvement in the future. Other areas of concern include better traffic signal timing to improve traffic flow and developing a system that provides alternate routes. At the same time, it is important to provide a hassle free travel. In developing society perception can be shaped by positive intervention. An enlightened education system can definitely influence the young minds in generating positive behaviour. Apart from this educating people about environment is the topmost need in order to develop positive environmental responses.

References

Davis (1972), The Conceptual Revolution in Geography, University of London Press, London.

Efrat J. (1981), Planning A Favourable Environment For Bicycle Use In Towns in G. Godwin (ed) Traffic, Transportation and Urban Planning, Vol I, The Pitman Press, London.

Hillman M., Henderson I. and Whelley A. (1973), Personal Mobility and Transport Policy, PEP, London.

Knowles and Wareing (1978), Economic and Social Geography, Rupa & Co, Calcutta.

Marks H. (1981), Traffic Protection for Residential Neighbourhoods in G. Godwin (ed) Traffic, Transportation and Urban Planning, Vol. I, The Pitman Press, London.

Small Keneth A., Jose A., Gomez I. (1997), Road Pricing for Congestion Management: The Transition from Theory to Policy in Oum, Dodgson et al., (eds.) Transport Economics Selected Readings, Harwood Academic Publishers, The Netherlands.

Thompson J. M. (1974), Modern Transport Economics, Penguin, Harmondsworth.

Watkins L. H. (1981), Environmental Impact of Road and Traffic, Applied Science Publishers, London.

Ionic Liquids: Green Solvents for the Future

PRADEEP PRATAP SINGH AND AMBIKA

Introduction

Environmental concerns in research and industry are increasing with the increasing pressure to reduce the amount of pollutants produced, including organic solvents whose recovery is mandated by ever more strict laws. Hence the challenge for a sustainable environment calls for the use of clean procedures which can avoid the use of harmful solvents. Since there is no remedy for the chemicals, it is imperative that chemicals are produced in a sustainable way and that they are safe to use. One of the most important principles of the green chemistry is the elimination of hazardous solvents in chemical synthesis which, avoids using toxic solvent and the generation of waste. Ionic liquids have gained wide popularity in recent years[1,2] for their increasing use in the two important fields of chemistry-synthetic and biochemical and their concept and history has been well documented.[3,4,5,6] They are 'designer solvents' as they are made up of at least two components which can be varied (the anion and cation), the solvents can be designed with a particular end use in mind, or to possess a particular set of properties.[7,8] They have essentially no vapour pressure and thus serve as potential replacements for volatile organic compounds in the chemical industry.

Reactions in ionic liquids have different thermodynamic and kinetic behaviour, good thermal stability, ability to dissolve a wide range of organic, inorganic, organometallic compounds and gases such as H_2, CO, O_2, CO_2. They have high potential for enantioselective reactions, chiral ionic liquids have been used to control the stereoselectivity. Most of the ionic liquids can be stored without decomposition for a long period of time. Therefore, an attempt has been made to present a detailed and comprehensive review of the versatility of ionic liquids as environmentally friendly green solvents, as well as reagents for various chemical and biochemical transformations.

Classification of Ionic Liquids

There is significant interest in the development of task specific ionic liquids, where the anion and/or cation is designed to impart specific properties to the ionic liquid, such as chirality, or improved extraction or separation capabilities. The design and choice of ionic liquids commonly focuses on physical properties such as water-miscibility, conductivity, viscosity and solubility properties, although how the chemical structure of the ionic liquid affects these various characteristics is still poorly understood. However, there is another chemical property that imparts a variety of physical characteristics to the ionic liquids that has been little investigated; the relative acidity or basicity of the componentions.[9]

Neutral Ionic Liquids

Typical ionic liquid anions are those that can be described as neutral in the acid/base sense or very weakly basic; these exhibit only weak electrostatic interactions with the cation and thus impart advantageously low melting points and viscosities. The common anions which lead to neutral and stoichiometric ionic liquids are given in 'Table 1'. Ionic liquids formed from these anions typically exhibit good thermal and electrochemical stability and thus are often utilized as inert solvents in a wide range of applications.[10,11]

Acidic Ionic Liquids

The simplest examples of slightly acidic ionic liquids are those based on the protic ammonium, pyrrolidinium and imidazolium ions, of which many are known. The well known $AlCl_3$ based ionic liquids are Lewis acidic when they contain an excess of $AlCl_3$ (Table 1).[12] There is another class of polynuclear anions, which are air and water sensitive such as $Al_2Cl_7^-$, $Al_3Cl_{10}^-$, $Au_2Cl_7^-$, $Fe_2Cl_7^-$ and $Sb_2F_{11}^-$.

Table 1
Different Types of Ionic Liquids and their Anions and Cations

Ionic liquids	*Anions*	*Cations*
Neutral	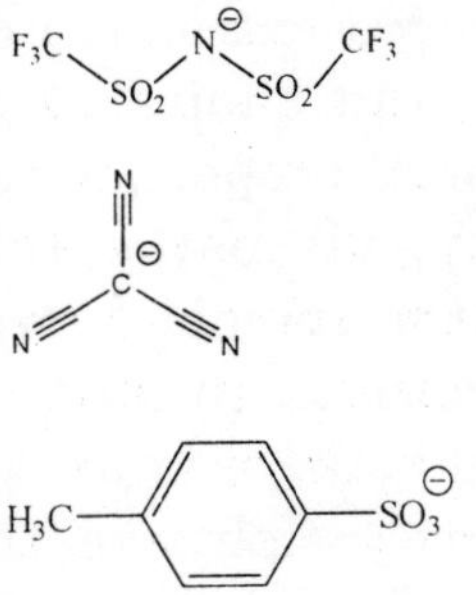 BF_4^-, PF_6^-, NO_3^-, SbF_6^-, $ZnCl_3^-$, $CuCl_2^-$, $SnCl_3^-$, $N(CF_3SO_2)_2^-$, $N(C_2F_5SO_2)_2^-$, $N(FSO_2)_2^-$, $C(CF_3SO_2)_3^-$, $CF_3CO_2^-$, $CF_3SO_3^-$, $S{=}C{=}N^-$ and $MeSO_3^-$	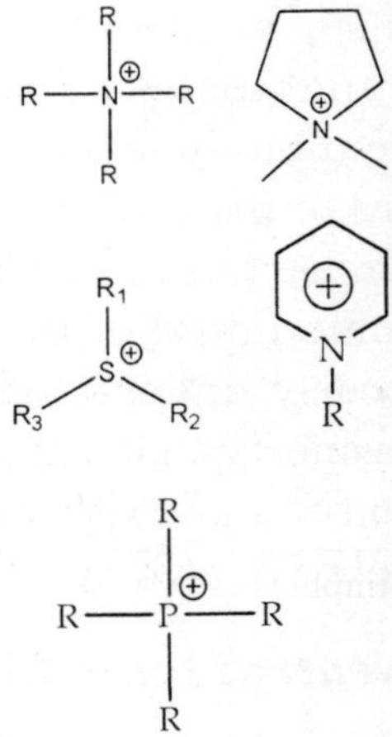
Acidic	$Al_2Cl_7^-$, $Al_3Cl_{10}^-$, $Au_2Cl_7^-$, $Fe_2Cl_7^-$ and $Sb_2F_{11}^-$	

		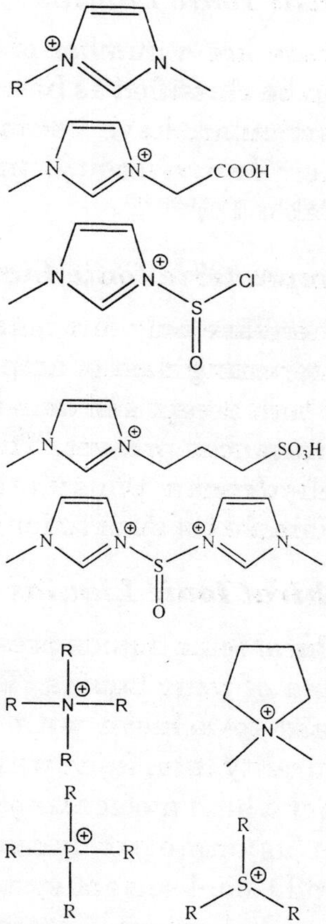
Basic	$HCOO^-$, OH^- 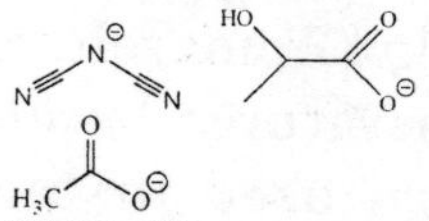 $HCOO^-$, OH^-	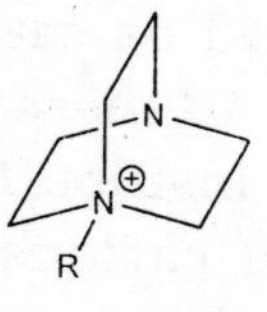
Amphoteric	HSO_4^- $H_2PO_4^-$ 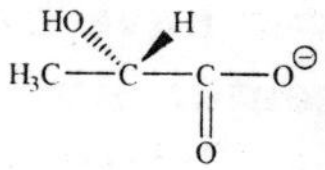	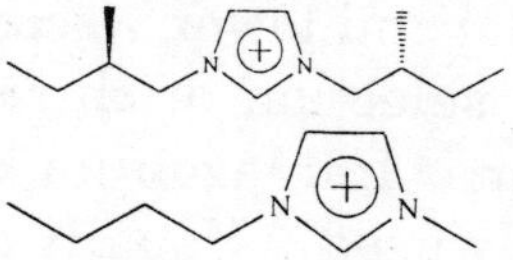
Chiral	Br^-	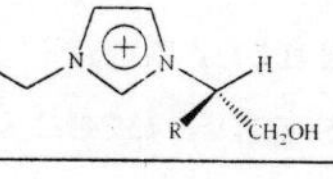

Basic Ionic Liquids

There are a number of ionic liquid forming anions that can be classified as basic (Table 1). The dicyanamides, in particular, have become readily available because of their low viscosity, including as phosphonium salts (Table 1).[13-16]

Amphoteric Ionic Liquids

There are only few ionic liquid anions that fall into the interesting class of amphoteric anions, with the potential to both accept and donate protons depending on the other substances present. The hydrogen sulphate (HSO_4) and dihydrogen phosphate (H_2PO_4) anions are simple examples of such anions.[17]

Chiral Ionic Liquids

Chiral ionic liquids present the most complex application area of ionic liquids (**Table 1**).[18] Chirality in solvent is hard to achieve with simple molecular solvents, but chirality is an issue with complex liquids. It is speculated that a high molecular order might have a beneficial effect on asymmetric induction in ILs compared to the traditional solvent systems. Amino acids have been used as chiral anions in ionic liquids.[19,20] Dialkyl imidazoliums have been used as cations, mainly in 1-ethyl-3-methyl imidazolium. All these dialkyl imidazolium amino acids have glass transition temperatures lower than 6°C. If phosphonium cations were used instead of dialkyl imidazoliums better thermal stability (typically over 250°C) and lower viscosity could be achieved. There is a wide selection of chiral ionic liquids which use chiral cations. Imidazolium derivatives have been the most widely used.[21] Usually a chiral center is introduced close to the quaternary nitrogen. Chiral imidazolium derivatives have been used as a reaction medium in Diels-

Alder reactions, Michael-addition but no significant improvement in enantioselectivity was found.[21]

Applications of Ionic Liquids

As Solvent

The use of ionic liquids as solvents for a broad range of chemical processes is potentially one of the most significant developments in the last decade. There are a number of reactions reported in ionic liquids including esterification, carbon-carbon bond formation, carbon-nitrogen bond formation, epoxidation, hydrogenation etc.[22-26] The final role that can be played by an ionic liquid in a catalytic process is simply as the solvent for the reaction. It is in this role that ionic liquids have been most widely used. Most studies in this area have, so far, been of the screening of reactions type, with a single ionic liquid being used as the solvent for some catalytic process without optimization or systematic investigation.

Oxidation

Selective oxidation of alkyl arenes and alcohols to the corresponding carbonyl compounds with $NaBrO_3$ promoted by [bmim]Br in excellent yields has been reported.[27] The oxidation of alcohols into aldehydes and ketones can also be carried out by using a catalytic system based on $RuCl_3$ dissolved in ionic liquids with molecular oxygen. This method involves the use of a new fluorinated ionic liquid, 1-*n*-butyl-3-methylimidazolium pentadecafluorooctanoate, which demonstrated better performance than the other ionic liquids.[28] The primary and secondary alcohols can also be converted into carbonyl compounds using $PhI(OAc)_2$-derived hypervalent iodine reagent as an oxidant in ionic liquids with good to excellent yields (**Scheme 1**).[29]

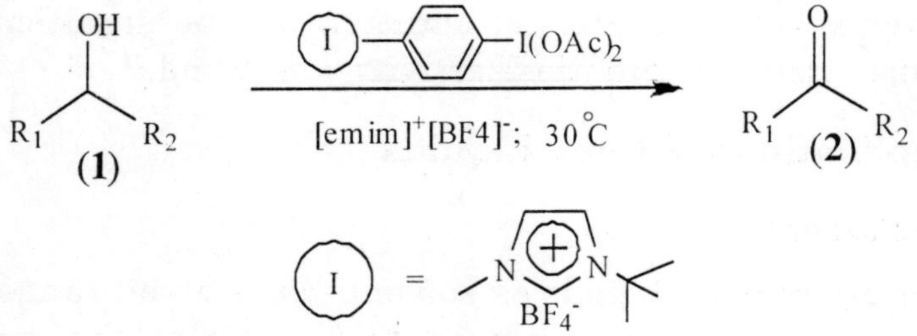

Scheme 1

Alkene epoxidation can also be carried out using lipid-soluble[18] manganese(III)porphyrins and water-soluble[19] iron(III) porphyrin immobilised in ionic liquids as an efficient catalytic system for the epoxidation of alkenes using hydrogen peroxide using iodobenzene diacetate and H_2O_2 as the oxidising agents (**Scheme 2**).

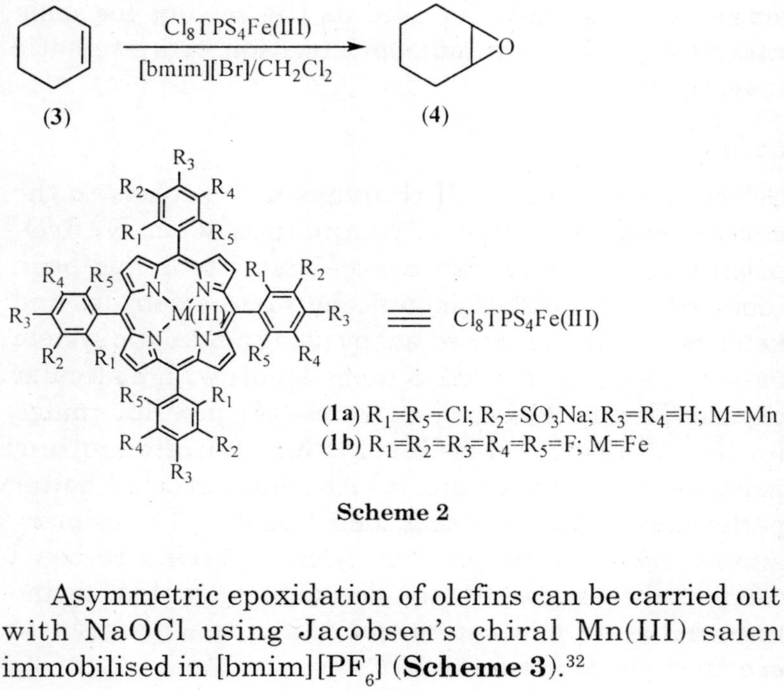

Scheme 2

Asymmetric epoxidation of olefins can be carried out with NaOCl using Jacobsen's chiral Mn(III) salen immobilised in [bmim][PF_6] (**Scheme 3**).[32]

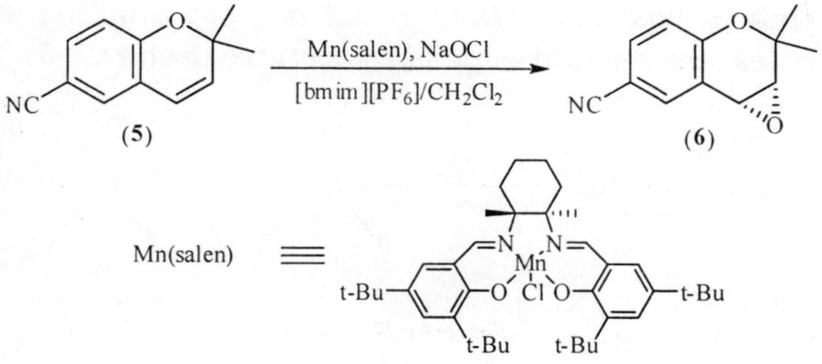

Scheme 3

Complexes of the type (dimethyl-bpy)MoO_2Cl_2 and Schiff/Lewis-base complexes of methyltrioxorhenium (MTO), being efficient homogeneous catalysts for the epoxidation of olefins, have been reported in high yields in room temperature ionic liquids (RTILs) (**Scheme 4**). The epoxidation of cyclooctene was observed to be strongly dependent on the water content of the system and the catalyst solubility in the RTIL.[33]

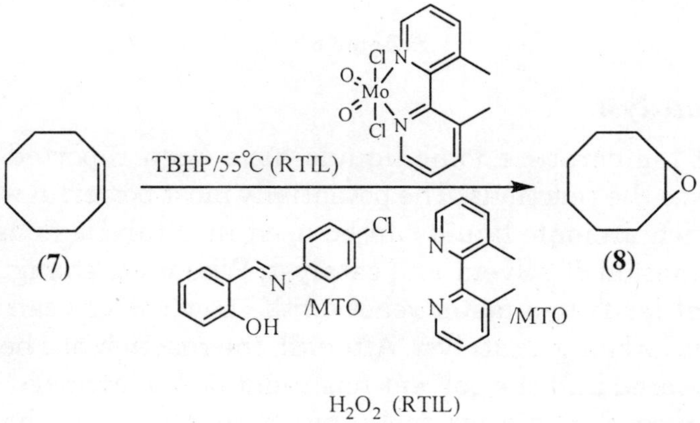

Scheme 4

Condensation

Bismuth (III) triflate has been used in ionic liquid as a recyclable catalytic system for the synthesis of *cis*-

aziridine carboxylates through the one-pot coupling of aldehydes, amines and ethyl diazoacetates (**Scheme 5**).[34]

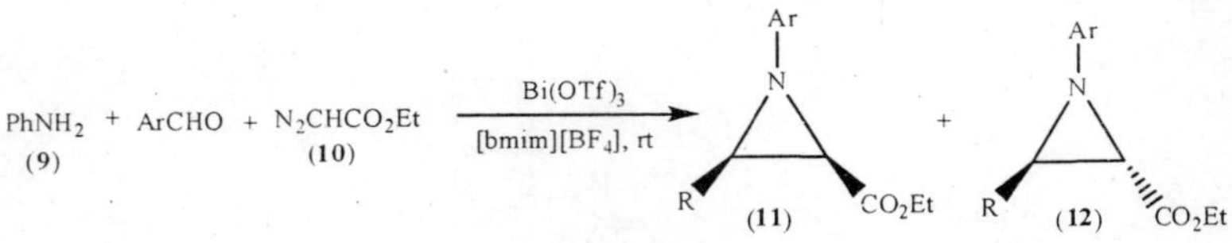

Scheme 5

The conversion of substituted 1-(2-hydroxyphenyl)-3-phenyl-1,3-propane diones (**13**) to the corresponding 2-phenyl-4H-chromen-4-ones (**14**) under microwave irradiations using ionic liquid $[EtNH_3][NO_3]$ has been reported in high yields (**Scheme 6**).[35]

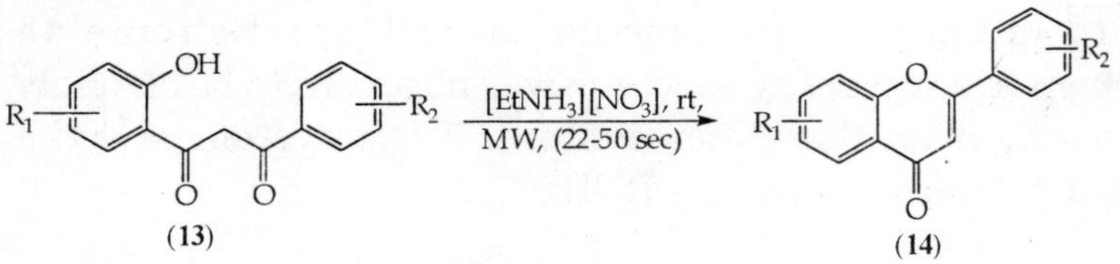

Scheme 6

As Catalyst

Room temperature ionic liquids have been reported to catalyze the reactions. The potentially most powerful way in which an ionic liquid can be used in catalysis is as a combination of solvent and catalyst. Whenever changing solvent leads to a faster reaction the new solvent can be viewed as being a catalyst. After all, the reaction has been accelerated and the solvent has remained unchanged by the process. Different task specific ionic liquids have proved to be effective catalysts for a variety of reactions.

The synthesis of 2-aryl benzimidazoles, benzoxazoles and benzthiazoles has been reported in excellent yields using the ionic liquids, as reaction media and promoters (**Scheme 7**).[36]

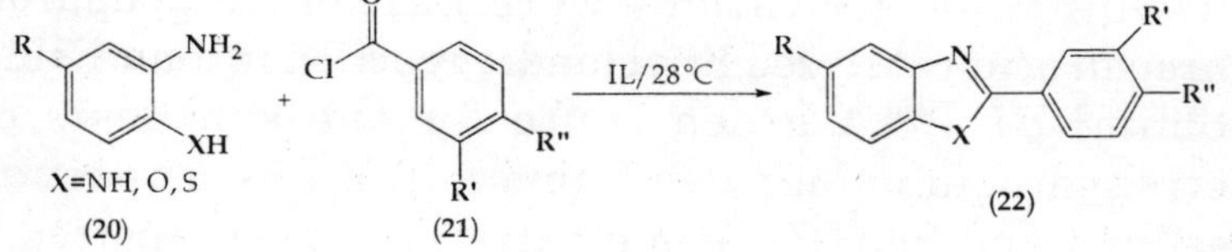

Scheme 7

The synthesis of 2,4,5-triaryl imidazoles can be carried out in room temperature ionic liquid, without addition of any catalyst in excellent yields (**Scheme 8**).[37]

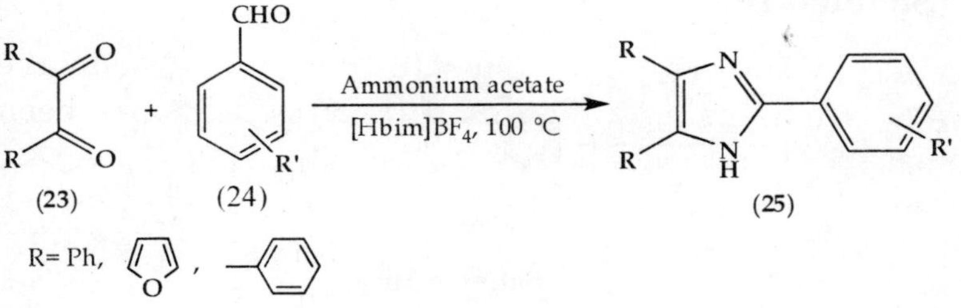

Scheme 8

The one-pot, three-component coupling reaction of homophthalic acid (**28**), aldehydes, and amines (**29**) to afford the corresponding cis-isoquinoline acids (**30**) in excellent yields and high cis-selectivity (**Scheme 9**). Compared to the conventional methods, this avoids the preparation and isolation of unstable imines prior to the reaction. Moreover, the use of ionic liquids as reaction media avoids the use of moisture-sensitive reagents or high-temperature reaction conditions to promote the reaction.[38]

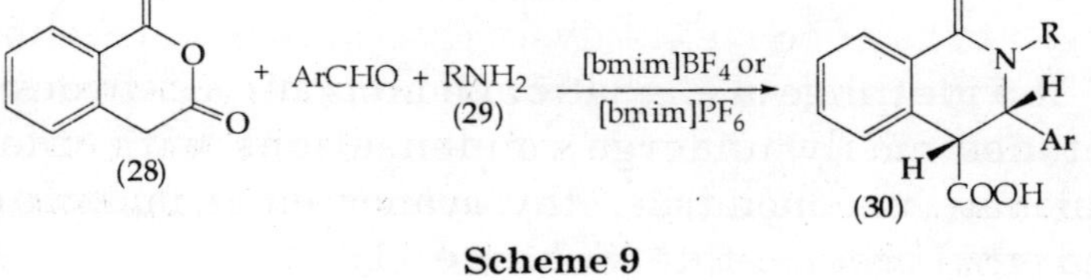

Scheme 9

Quinoline derivatives were efficiently prepared through acid-catalyzed Friedlander reaction in ionic liquid [bmim][BF_4].[39] A green route for the synthesis of tetrahydroquinolines using [bmim][PF_6] as a reaction medium and $Sc(OTf)_3$ as a catalyst has been reported.[40] Both 3,4-dihydro-2*H*-pyran and 2,3-dihydrofuran gave products in good to excellent yields, with *endo*:*exo* ratios often greater than or equal to 11.5:1. This route gives products in higher yields and shorter reaction times.[41] The ionic liquid/catalyst phase is recoverable and reusable (**Scheme 10**).

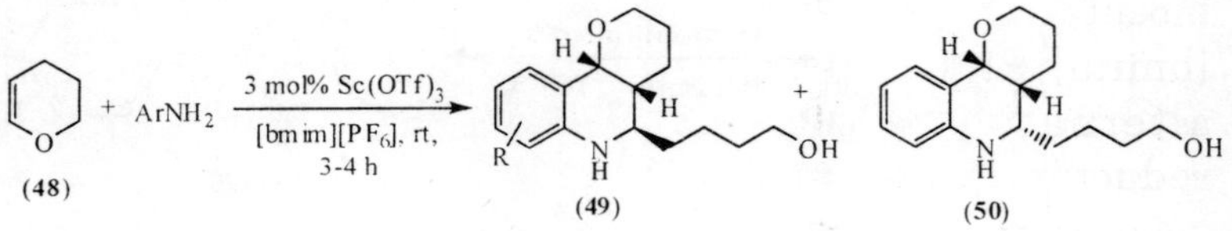

Scheme 10

The basic ionic liquid 1-butyl-3-methylimidazolium hydroxide, [bmim]OH, efficiently catalyzed Knoevenagel condensation without requirement of any organic solvent at room temperature.

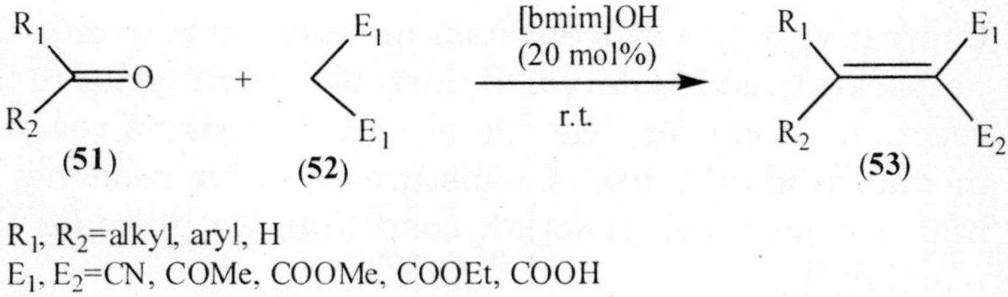

Scheme 11

A wide range of aliphatic and aromatic aldehydes and ketones easily undergo condensations with diethyl malonate, malononitrile, ethyl cyanoacetate, malonic acid and ethyl acetoacetate (**Scheme 11**).[42]

Miscellaneous

Enzymatic Reactions

Oxidation of anthracene, catalyzed by oxidative enzymes, laccase C and assisted by a number of mediators, has been increased to several-fold by addition of 4-methyl-*N*-butylpyridinium tetrafluoroborate ionic liquid.[43] The high yields of the oxidation products of 2-methoxyphenol (guaiacol) with Fe(III)protoporphyrin(IX)chloride (hemin), microperoxidase-11 (MP-11), and cytochrome-*c* (cyt-*c*) systems showed that ILs are suitable media for bioelectrocatalysis.[44] Baker's yeast is an important biocatalyst in synthetic organic chemistry. An ionic liquid [bmim][PF_6]-H_2O (10:1) system has been used as an alternative solvent or the Baker's yeast-mediated reduction of ketones.[45] The yields of the products were good and the reduction of aromatic ketones such as 4-bromoacetophenone was not observed. The enantiomeric excess obtained in an ionic liquid was comparable to those obtained in other media.[45]

Material Chemistry

Ionic liquids have been used as electrolytes in electromechanical actuator systems based on conducting polymers[46] and in P-conjugated polymer electrochemical devices.[47] Trialkylimidazolium-based ionic liquids have been developed as novel, thermally stable treatments for layered silicates and graphite for the purpose of preparing high-quality polymer nanocomposites by Aida and co-workers, who discovered that carbon nanotubes and room temperature ionic liquids can be blended to form gels that may be used to make novel electronic devices, coating materials, and antistatic materials.[48]

Biodegradability and Green Aspect of Ionic Liquids

The green credentials are not sufficient for design of sustainable chemical products, and it is necessary to refer to the concept of eco-toxicological risk profiles, which comparatively assess substances according to the five risk indicators: release, spatiotemporal range, bioaccumulation, and biological activity. As the popularity of ILs with industrial chemists increases, the probability that ILs will find their way into water sourses through effluent discharges or accidental spills will also increase, and they may affect the aquatic environment for a long time because of their poor biodegradability. Concerning the release of and from ionic liquids, applications and processes have to be designed to take into account the toxic potential of the substances employed.

- The biodegradability of ILs has focused on making the cation more biodegradable. Because of the need for hydrophobic ILs, it seems to be of primary importance to develop anions that introduce sufficient hydrophobicity but are at the same time biodegradable.
- Biodegradation rates increases with longer alkyl chain length, which may be related to enhanced selection of a microbial community. For rapid biodegradability, pyridinium cations and longer alkyl substitutions should be considered during the synthesis and design.[49]
- The T-SAR directed design for low toxicity has been employed. Certain micro-organisms are enriched by ILs used as a carbon source.[50-54]
- Even if high bioaccumulation and toxicity might be observed for the parent compound, its ready degradation, starting in wastewater treatment facilities, would decrease the joint bioaccumulation of the parent compound and the transformation products, and critical body residues would not be reached.

- The toxicity of the degradation products of ILs should be less toxic than the original products.
- ILs with fluorine containing anions were suggested to be relatively toxic because the anions were hydrolyzed to fluoride in the aqueous solution and the fluoride had a toxic effect.

Conclusion

Intensive research over the last few years in the area of ILs has resulted in the development of a number of processes which, for some of them, are more efficient than those carried out in organic solvents. It is important to note that the reaction efficiency can be greatly dependent on the nature of the IL and, furthermore, the IL can be a non- innocent solvent. For some studies carried out using IL/organic solvent mixtures, we suspect that the IL could be a phase-transfer catalyst. Most ILs are expensive and, often, their toxicity as well as their stability remain to be determined. With regards to economic and ecological concerns, of particular interest are the methods allowing the recovery and the re-use of both the catalyst and ILs. Although significant progress has been made, further work is necessary to improve the recyclability of the catalyst/IL associations. The use of ILs in organic synthesis is still in its infancy and new ILs are emerging. Consequently, it is likely that new and more efficient procedures will appear in the near future.

Notes

1. D. Zhao, M. Wu, Y. Kou, E. Min. *Catal. Today*, 2002, 74, 157.
2. R. Sheldon. *Chem. Commun.* 2001, 2399.
3. F. H. Hurley, T. P. Wier. *J. Electrochem. Soc.,* 1951, 98, 207.
4. J. Fuller, R. T. Carlin, H. C. De Long, D. J. Haworth. *Chem. Commun.,* 1994, 299.
5. P. Wasserscheid, A. Boesmann, A. Jess, L. Datsevitch, C. Schmitz, A. Lauter. *PCT Int. Appl.,* 2003, 18.

6. C. F. Poole, B. R. Kesten, S. S. J. Ho, M.E. Coddens, K. G. Furton. *J. Chromatography,* 1986, 352, 437.
7. M. Freemantle. *Chem. Eng. News,* 1998, 76, 32.
8. R. Hagiwara, Y. Ito. *J. Fluorine Chem.,* 2000, 105, 221.
9. D. R. MacFarlane, J. M. Pringle, K. M. Johansson, S. A. Forsyth, M. Forsyth. *Chem. Commun.,* 2006, 1905.
10. Z. Duan, Y. Gu, Y. Deng. *J. Mol. Catal. A: Chemical,* 2006, 246, 70.
11. R. Kumar, A. Sharma, N. Sharma, V. Kumar, A.K. Sinha. *Eur. J. Org. Chem.,* 2008, 5577.
12. K. Qiao, H. Hagiwara, C. Yokoyama. *J. Mol. Cat., A: Chemical,* 2006, 246, 65.
13. C. Ye, J.C. Xiao, B. Twamley, A.D. LaLonde, M.G. Norton, J.M. Shreeve. *Eur. J. Org. Chem.,* 2007, 5095.
14. J. M. Xu, Q. Wu, Q. Y. Zhang, F. Zhang, X. F. Lin. *Eur. J. Org. Chem.,* 2007, 1798.
15. A. Wykes, S.L. MacNeil. *Synlett,* 2007, 107..DABCO-
16. A. R. Hajipoura, F. Rafiee. *J. Iran. Chem. Soc.,* 2009, 6, 647.
17. D. R. MacFarlane, J. M. Pringle, K. M. Johansson, S. A. Forsyth, M. Forsyth. *Chem. Commun.,* 2006, 1905.
18. J. Ding, D. W. Armstrong. *Chirality,* 2005, 17, 281.
19. W. Bao, Z. Wang, Yuxia Li. *J. Org. Chem.*, 2003, 68, 591.
20. X. Chen, X. Li, A. Hua, F. Wang. *Tetrahedron: Asymmetry,* 2008, 19, 1.
21. K. Bica, P. Gaertner. *Eur. J. Org. Chem.,* 2008, 3235.
22. Ambika, P. P. Singh, S. M. S. Chauhan. *Syn.Comm.*, 2008, 926.
23. S. M. S. Chauhan, A. Kumar, K. A. Srinivas. *Chem. Commun.*, 2003, 9, 2348.
24. J. Howarth, P. James, R. Ryan. *Syn. Comm.*, 2001, 31, 2935.
25. C. Angrish, A. Kumar, S.M.S. Chauhan. *Ind. J. Chem.,* 2005, 44B, 515.
26. N. Jain, A. Kumar, S.M.S. Chauhan. *Tet. Lett.*, 2005, 46, 2599.
27. A. Shaabani, E. Farhangi, A. Rahmati. *Monatsh Chem.,* 2008, 139, 905.
28. R. F. De Souza, J. Duponta, J. E. de L. Dullius. *J. Braz. Chem. Soc.*, 2006, 17, 48.
29. W. Qian, E. Jin, W. Bao, Y. Zhang. *Angew. Chem. Int. Ed.*, 2005, 44, 952.

30. Z. Li, C.G. Xia. *Tet. Lett.*, 2003, 44, 2069.

31. K. A. Srinivas, A. Kumar, S.M.S. Chauhan. *Chem. Commun.*, 2002, 2456

32. C. E. Song, E. J. Roh. *Chem. Commun.*, 2000, 837.

33. D. Betz, W. A. Herrmann, F. E. Kuhn. *J. Organometallic Chem.*, 2009, 694, 3320.

34. J.S. Yadav, B. V. S. Reddy, P. N. Reddy, M. S. Rao. *Synthesis*, 2003, 1387.

35. S. R. Sarda, M. Y. Pathan, V. V. Paike, P. R. Pachmase, W. N. Jadhav, R. P. Pawar. *Arkivoc*, 2006, 14, 43.

36. R. N. Nadaf, S. A. Siddiqui, T. Daniel, R. J. Lahoti, K. V. Srinivasan. *J. Mol. Cat. A: Chem.*, 2004, 214, 155.

37. S. A. Siddiqui, U. C. Narkhede, S. S. Palimkar, T. Daniel, R. J. Lahoti, K. V. Srinivasan. *Tetrahedron*, 2005, 61, 3539.

38. J. S. Yadav, B. V. S. Reddy, K. S. Raj, A.R. Prasad. *Tetrahedron*, 2003, 59, 1805.

39. X. Y. Zhang, X. S. Fan, J. J. Wang, Y. Z. Li. *Chin. Chem. Lett.*, 2004, 15, 1170.

40. J. S. Yadav, B. V. S. Reddy, K. U. Gayathri, A. R. Prasad. *Synthesis,* 2002, 17, 2537.

41. J. Zhang, C. J. Li. *J. Org. Chem.*, 2002, 67, 3969.

42. B. C. Ranu, R. Jana. *Eur. J. Org. Chem.,* 2006, 3767.

43. G. Hinckley1, V. V. Mozhaev, C. Budde, Y. L. Khmelnitsky. *Biotech. Lett.,* 2002, 24, 2083.

44. J. A. Laszlo, D. L. Compton. *J. Mol. Cat. B: Enzymatic,* 2002, 18109.

45. J. Howardh, P. James, J. Dai. *Tet. Lett.,* 2001, 42, 7517.

46. J. Ding, D. Zhou, G. Spinks, G. G. Wallace, S. A. Forsyth, M. Forsyth, D. R. MacFarlane. *Chem. Mater.,* 2003, 15, 2392.

47. W. Lu, A. G. Fadeev, B. Qi, B. R. Mattes, G. Wallace, J. Ding, G. Spinks, J. Mazurkiewicz, D. R. MacFarlane, S. Forsyth, M. Forsyth. *Science,* 2002, 297, 983.

48. T. Fukushima, A. Kosaka, Y. Ishimura, T. Yamamoto, T. Takigawa, N. Ishii, T. Aida. *Science,* 2003, 300, 2072.

49. M. Kathryn, D. JaNeille, K. Dixon, C. F. Kulpa Jr. *Biodegradation*, 2007, 18, 481.

50. J. Ranke, K. Molter, F. Stock, U. Bottin-Weber, J. Poczobutt, J. Hoffmann, B. Ondruschka, J. Filser, B. Jastorff. *Ecotoxicol. Environ. Saf.,* 2004, 58, 396.

51. K. Docherty, C. F. Kulpa. *Green Chem.*, 2005, 7, 185.
52. P. Stepnowski, A. C. Składanowski, A. Ludwiczak, E. Laczynska. *Hum. Exp. Toxicol.*, 2004, 23, 513.
53. R. P. Swatloski, J. D. Hobrey, S. B. Memon, G. A. Caldwell, K. A. Caldwell, R. D. Rogers. *Chem. Commun.*, 2004, 668.
54. R. J. Bernot, E. E. Kennedy, G. A. Lamberti. *Environ. Toxicol. Chem.*, 2005, 24, 1759.

Global Warming and Solution through Yoga for Environment Sustainability

Suresh Kumar and Paramjit

Introduction

The word *Yoga* derived from the Sanskrit word for **yujirayoga** or *yoke* meant join together. Essentially, it means union. It is the science of uniting the individual soul with the cosmic spirit through physical disciplines (postures) and mental disciplines (meditation). Patanjali explained this word in short that *Yoga is the cessation of mind.* Yoga is the austerity of mind. The Yoga is the end product of history of ancient India spirituality and here one question arises that how the yoga technique is helpful to solve the issue of global warming? Before one move further to elaborate this query, it is better to understand the meaning of global warming in present context.

Industrialization led to colonialism that came to an end with the result of nuclear bombing on Hiroshima and Nagasaki. The outcome of cold war period (1948-1990) led to Arms race and the industrialization in the developed countries. As a result, there is increasing stress on the Earth's ecosystem (60% of the Earth's ecosystem has been degraded in the past 50 years) and consumption of natural resources is expected to rise to 170% of the Earth's bio-capacity by 2040. The environment alarm bell has now sounded about this nexus of business houses and promotion of consumerism (through media and print advertisements) and the positive news is that after decades of denial or inaction, many governments have now made low-carbon and inclusive growth a national priority.

Brundtland Commission defines the term sustainable environment and mentions, "Humanity has the ability to make development sustainable-to ensure that it meets the needs of the present without compromising the ability of future generations to meet their own needs (WBCSD. 2009:4)." This is the time to understand the real meaning of Climate Change because of progressively degrading environments and iniquitous social structures of the global community seriously felt the threat to mankind's future. More than 4.840 billion people, comprising more than two-thirds of the world's population, live in poor nations and bear the severest brunt of these environmental and societal imbalances. The change in weather, increasing drought situation, rising of the water level in the Oceans, less or no snow fall on the mountains, sudden floods due to heavy rains, hot weather (up to 50 degree) in summer and less cold in winter and melting of the glaciers around the Africa and Asia can be termed as *Climate Change*. It further threatens flora, fauna, forests and wildlife may lead to over industrial, vehicular air & chemical pollution, deforestation and disappearance of graded land areas. The impact of climate change on rain fed agricultural sector is quite complex, especially the adverse effect on soil, biodiversity and ecological regimes in the developing countries of Africa, Asia and Latin America.

It is the time to drive forward with a shared understanding of what is meant by sustainable, responsible consumption from the yoga perspective. The global community faces the rising fuel prices, serious issues of hunger and poverty, scarcity of water, arable land and other natural resources questions human consumption and how one may deal with it has never been more crucial. Jai Gopal Sharma stated that 'progressively degrading environment and iniquitous social structures seriously threaten the future of mankind. Exacerbating these challenges today is the new danger posed by global warming and climate change. The developing countries bear the most severe brunt of these environmental and societal imbalances. It warns the post 1990 unstoppable competition for survival in the liberal market economy either to stop and think for the future of the coming generations or face the natural destruction of the Mammalian Class and natural environment. The yoga should attract all the democratic governments in the world towards more sustainable choices and sharing this information with common man. This information should act as a trigger for innovation and ultimately set the trend towards sustainable environment through yoga among global community.

Yoga concerns around the world should address global challenges related to shortage of resources, water scarcity, climate change and loss of biodiversity and planning, promotion, co-ordination and overseeing the implementation of environment and forestry programmes by connecting it with yogic sutras.

Today, there is an urgent need to analyze the drivers of consumption patterns & impacts on the body, heart, mind and soul. The definition of sustainable products and lifestyles to formulate actionable responses towards this challenge should be explained through yoga. Keeping climate change in mind, the global population should adopt the sustainable production and consumption policy learning from yoga and should practice in the society as a whole.

Yoga and Issue of Climate Change

Yoga senses and controls the ambitions and physical matter of life and provides the platform for self-discipline. The climate change is the result of uncontrollable consumerism that promotes indiscipline and no regulation in life. It's now a well known fact that none of us can prosper in an environment that is polluted with noxious gases. Climate change is indeed making temperature rise and the negative impact is felt in agriculture, livelihood, forests, water bodies and the health of people. Since the problem is global in scale and impact, every society can play the saviour in their region and can tackle the problem on multiple fronts. The Yoga has eight elements such as:

Yamas

In the *Yoga Sutras,* Patanjali defined five yamas or ways to moral conduct with universe. They are nonviolence, truth & honesty, non-stealing, meditation, and non-possessiveness.

(i) Nonviolence: The yamas persuades everybody to get rid of painful life and when a person is free from all pains, he or she attains the position of yamas. Today, the world community faces War on Terrorism that has resulted in bombing in Somalia, D R Congo, Afghanistan, Pakistan, India and Sri Lanka. The use of thousand pounds of bombs is exacerbating the violence in the society that leads to climate devastation. Mahabharata mentions,

तस्मात् प्रमाणतः कार्यो धर्मः सूक्ष्मो विजानता।
अहिंसा सर्वभूतेभ्यो धर्मेभ्यो ज्यायसी मता।।

This means that the humanity should respect the feelings of all living creatures and behave with love and grace. Non-violence is the biggest weapon to solve all the disputes in the world. The distribution of military technology includes chemical weapons, biological weapons, nuclear warhead missile technology, nuclear waste from power plants for nuclear bomb technology, nuclear trigger, etc. It is the time to respect non-violence and invite all these terrorist group leaders on the world stage to put their opinion that will help to establish a non-violent world and environment sustainability. Moreover,

"Law and love then, are not alternatives. Mercy and forgiveness are binding and obligatory, not optional extras" (Winn: 28-30). There is need to understand the relation between the violent terrorism and the weapon industries of the developed world. As a student of political economy, this terrorism market strengthens oneself to show the reality to the global society by understanding the thumb rule playing behind it. This thumb-rule is:

(War + Terror)^2 = War^2 + Terror^2 + 2War Terror that suits the contemporary environment of capitalist monopolization. The political economy of Post 9/11 here refers to

- War technology in partnership with Terror vigorously working in the global market acting as a whole square.
- It is equal to the War^2 means boosting the weapon industries. One may check the annual turn over (Balance Sheet) of these industries, which strengthen this argument transparently.
- Terror^2 a label or brand name appropriately is using for the market advertisement. All kinds of media, different web sites and information technology uses in promoting their market.
- The 2 WarTerror promotes the war industries in lieu of wide advertisement in the international media gets approval for the US Armaments markets on the globe.

It signifies that more the (War + Terror)^2 technology work as war against terrorism formula, enhancing the armament market as a result equalizes (=) War^2 + Terror^2 + 2War Terror.

(ii) Truth & Honesty: The global humanity listens, speaking and reading about truth since their tender age but failed to adopt in real life. The sole reason behind it that everybody is running behind to collect unlimitedly worldly material things whole heartedly twenty four hours, which leads to think, speak and implement a true lie in life. Simply speaking tells no lies for instance, cheating on your income taxes falls into this category. There is no single example in the world history that glimpses the person attains Moksha on the basis of true lie.

Speaking honestly does not consider a noble act that raises the question that how, where and when to use the truth. The best way to initiate the truth sutra is the home. It is the time to speak truly with parents, husband, wife, children, family and lead gradually towards society. Any human being by practicing truth in life will able to know the real importance of any worldly physical material thing and self-knowledge. By attaining any material thing and explaining its real importance in life, is a real difficult task for any human being that is accepted by all.

Comparing to it, a liar person never become a brave person in life. One fearful person cannot concentrate in life. One cannot become a wise person without full concentration. How cleverly a liar person behave, his acts are caught by a honest person in this world only. It is rightly pointed out that

सत्यमेव जयते

A person can achieve anything any life by speaking true and practice honesty in life. Generally, the people lie before the government that spoils the environment of that concerned offices. It further leads to attract towards illegal means to earn more and more money. Similarly, the nexus of liar-ness between the big industrial houses and the government today reveals the secret of destroying the Climate due to last fifty years blind race for endless run. The repercussions of climate change are visible more or less in the whole world. There are increasing reports of drought, rising sea levels, eroding coastlines, melting glaciers, dry rivers and complete breakdown of ecosystems and human sustenance in the developing world. The consequences of global warming have led to rapid climate change. These impacts are inextricably linked to the broader concern on environmental degradation, natural resource consumption and sustainability. It's now a well known fact that the impact of climate change knows no borders and businesses, governments and individuals across countries. The issue of environment sustainability needs to implement a true and honest policy by involving big business houses and governments, which will seek to deliver solutions to climate change problems through adaptation and mitigation strategies, recommending policy changes, showcasing innovative research and industrial best practices. The Yoga programmes suggest to the well-aware learned community to adopt a comprehensive policy such as:

- Sustainable use of natural resources, including water and environment conservation, and reduction of direct impact on environment or reduction of carbon footprint through yogic sutras.
- Responsible control on production of industries that rely heavily on natural resources.
- Yogic application to ensure reduction in carbon footprint through walk in daily life that will help in minimum use of patrol or diesel vehicles.

(iii) Non-stealing (Asteya): It means that never touch any body material things, don't provoke others to take the things and never have any desire to keep anybody's things is non-stealing. This may pose problem in implementing in real life in today's world. The owner's consent is must before taking anything is the real solution (that includes the material things between husband and wife). With consent, one may take the material things even in his or her absence. Following are the benefits of non-stealing that directly deals with the environment sustainability such as:

- A person practices non-stealing is full of peace, happiness and lives patiently. This practice is against the consumer culture that directly hit the blind race of industries products. One may live happily, peacefully and prosper life by practicing non-stealing in life.
- The consumer culture leads to stealing. Numbers of youth are caught red handed stealing shirts, trousers, shoes, mobile phones and other material things in the big malls through out the world. Non-stealing gives importance to honesty and curbing corruption in life.
- The practice of non-stealing explains the efficient and proper use of the natural things in a best way. It will stop the misuse of natural material and help in sustaining the environment.
- A person of non-stealing gets respect in the hearts of everybody that will help to control the consumer cum market culture.

(iv) Brahmacharya: Brahmacharya relates to purity, chastity and non-lust. The purity of a brahmachari (a person who adopts brahmacharya) includes heart, words and body. A brahmachari possesses:

1. A strong body, attractive and healthy.
2. The body power goes through out the life.
3. The person work efficiently in minimum time and shows his/her wisdom.
4. The body resistance against any disease increases.
5. The body power strengthens soul power.
6. The body attracts towards non-violence.

All these points are prerequisite to control this attraction towards this physical materialistic world among human beings through out the world that will helpful for the environmental sustainability. The yoga should develop this awareness programme among youth and explain their body and mental power to fight against the challenges of the climate change and spread this message in the school children, colleges and universities and society. The yoga initiatives or steps to mitigate the harmful effects of climate change should be advertised in the press and media. Mahrishi Ved Vyas defines brahmacharya in short by saying:

ब्रह्मचर्य गुप्तेन्द्रियस्य उपस्थस्य संयमः

It means to control on sexual desires and patience to उपस्थ इन्द्रिय. Those who cannot have self-control are known as *Vahbichar* (a person full of lust). The misuse of science and technology through internet pornography, advertisement of power capsules and unlimited power through it has worsened the natural environment. The misuse of Sc. & Tech. in the industries is seen everywhere such as the Japanese are changing tooth brush every week because of over production of it. It leads to degradation of environment. The present young generation never uses the Neem, Eucalyptus and plant branches for tooth brush. As a result, one may observe their sitting with the dentist in their tender ages. Similarly, all the metropolitan and big cities organizes weekend night dance and drink (Liquor) party and waste money, human resource, thousands watts of electricity and getting nothing. The practice of brahmacharya is the only solution to keep away this physical material worldly attraction. It is truly stated that सर्वेन्द्रियाणां जरयन्ति तेजः It means all these attraction weakens oneself will power. All these lusts cannot give physical and mental peace in anybody's life. The life full of limitless greed & more greed leads only to grab more and more material things in life and one can see the consequences of it through climate change.

(v) Non-possessiveness (Aparigraha): It means that all unimportant physical material as well as ideas should be kept away from the heart, words and body is known as Aparigraha or Non-greed. The non-possessiveness explains that a human being should free from greed, hoarding and collecting. Does human being really need more shoes, another car or to hog the conversation every time you see your friends? Make human being life as simple as possible.

There is a big debate on need and greed. A person need is other person's greed. A human being essential need relates to Bread, Cloth and Shelter. The role of education and other qualification strengthens the human being's professional life but the essential needs are the same. Similarly, the agricultural land is used for the production of wheat, rice, cotton, fruits, flowers and other items but the over production on that land leads to infertility of land. The use of more and more land for the rapid urbanization is a danger to the forests land and leads to climate change. This change promotes high weather temperature, no or less rain, winter and other seasons and shortages of every natural environment. This is the result of possessiveness. The possessiveness towards violent solution to all the problems in the state and society in the developing world leads to chaos and devastation of humanity on the one hand and ruining the natural climate here on the other hand. Overall, the mental satisfaction over greed and possessiveness is the only yogic solution that will save and promote the environment sustainability.

Niyamas

In the *Yoga Sutras,* Patanjali defined five niyamas or observances relating to inner discipline and responsibility, how one treat oneself. They are purity, contentment, self-discipline, study of the sacred text, and living with the awareness of God. Today, the niyamas is the best self rule to the present before the industrial sector technology innovations that have driven the over-growth standing as challenge to the climate change in the world. Peter Senge's book on *Necessary Revolution* highlights that a collective awakening to new possibilities that changes every thing over time—is the *Renaissance of Industrial Revolution*— what is starting to happen around the world today (CII. 2009:9). There is an urgent need to see a new industrial revolution that will be Clean, Lean and Mean. Clean because of climate consequences, water stress, and economic impacts. Lean, because it will be characterized by a low carbon economy and eco-efficiency, uncertainty around access to energy and resources, high resources prices and government regulations on resource efficiency. Mean, because transformational change and destructive innovations will create winners and losers. Bjorn Stigson of World Business Council for Sustainable Development (WBCSD) beautifully articulates the challenge for companies. He says that business cannot succeed in societies that fail and has a strong interest in helping build thriving societies which are good places for doing business (WBCSD. 2009: 7).

The rule of self-discipline for the business community is making investment decisions that will impact the climate issue for next few decades into the future. Companies therefore need clarity on the policy framework within which they do business will lead to contentment. An important element for the post-2012 international framework will be the establishment of a quantifiable, long-term (50-year) emissions pathway for the management of global Green House Gas (GHG) emissions (CII. 2009:14), which needs purity, contentment and self-discipline. The magnitude of change required demands of yoga participation by all stakeholders of society – from government to business to civil society. To succeed, they must work together. The next few years are critical for establishing policies to deal with energy security, competitiveness, mitigation and adaptation to climate impacts, and they will determine our faith in self-discipline, study of the sacred text, and living with the awareness of God. It will help in proper utilization of energy infrastructure and emission patterns for many decades.

Moreover, instead of looking at climate changes as only an environmental problem, where a maximum company can do is to reach zero pollution is not sufficient. It is time to look at how a company also can promote yoga discipline and become part of the solution by proving self-discipline. It is important to the developing world business community to understand and adjust the requirements of future consumers who increasingly demands for ethical choices. Overlooking this trend would be shortsighted and a risk for the industries in the developing world. Kirsi Sormunen, Vice President & Head of Environmental Activities, Nokia highlights, "Companies can also influence consumer behaviour towards more sustainable choices, both through product development and consumer information. It is our aim that this trend analysis on sustainable consumption will act as a trigger for innovation and ultimately for more sustainable consumption (WBCSD.2009:5)." It means that the Aparigraha or non-greed and self-discipline in life will lead to sustainable choices. This development offers great potential as a driver for innovative product among the educated global civil society not merely in the sense of green products, but it offers consumer related solutions that link product quality to the shared responsibility of producers and consumers and promote need among consumers and not greed. The self-discipline to environmental footprint of the product in terms of production and disposal in the developing world will become more important with the proper use of the product with respect to its environmental impact that step will be decisive to promote purity and contentment in the lives of individuals, society and states. The industrial groups of whole world should focus on the use of Yoga in real life and work more closely with consumers, communicating top performance, added value of sustainable products and enabling behavioral changes accordingly. WBCSD shares that the leading businesses of the future will be those whose core businesses directly address global challenges and should include propagating *Yoga Sutras*. The investment of the *Yoga Sutras* and the industrial growth need to define their role in tomorrow's society. The leadership of *Yoga Sutras* should work with the industrial sector in driving forward a shared understanding of what is meant by sustainable, responsible consumption from a business perspective and addressing the global challenges related to shortage of resources, water scarcity, climate change and loss of biodiversity.

Asanas

It means to fix the body like statue, having happiness position and doing meditation is known as Asanas in the different Yoga postures. Patanjali said, "The posture of yoga is steady and easy. The ancient yogis used asana to prepare the body for meditation. To sit for a lengthy time in contemplation required a supple and cooperative body. If you are free of physical distractions, control the body then you can also control the mind. Posture is mastered by freeing the body and mind from tension and restlessness and meditating on the infinite" (Patanjali. 2010). The asanas strengthens the physical and mental strength that helps the human body to live peacefully, happily, living with minimum material needs in this world. The asanas strengthens all the five yamas that persuades the human beings to live their lives having moral behavior with the creator yama. Probably the most famous pose is the lotus position is pronounced as Ah-sah-nahs.

Pranayama

The *Prana* means air and *Ayama* means to stop. It is a method of controlling prana or life force through the regulation of breathing. The pranayama keeps all the evils away from human body and even finished evil-greed such as Avidya, Asmita, Raag and Dwesha and Abhinivesha. 'The **Avidya** means ignorance, which is forgetfulness. It helps in forgetting of the self-attachment to the material things around the world. The ignorance is the misconception of human being's true reality having myth-faith in the temporary eternal, the impure is the pure and pleasure to be painful. This false representation of reality is the root evil that produces the four others' (Yoga site: emphasis mine).

The **Asmita** 'means the ego, individuality and I-am-ness. Asmita is the identification of oneself only with his/her ego. Human beings create self-image of themselves that believe on us; means I is Us. This self-image can contain both external (I am poor) and internal (I am a bad person) false projections. The **Raaga** and **Dwesha** relates to (the attachments and aversions) the physical objects remain in the senses. One should not come under the control of these two as indeed both are two stumbling blocks on one's path of Self-realization. When the expectation is not fulfilled, the attraction turns into repulsion and hatred (**dwesha**). In either case one suffers from insecurity, **stress**, anxiety and fear of various sorts. Finally, the **Abhinivesha** means survival instinct. Abhinivesha or love for earthly life is the main cause of all fears that is clinging to life and **body**. The attachment to physical objects, name and fame causes fear' (Yoga site: emphasis mine).

All these evils need to control in this free market global world. Dennis Pamlin suggests that indirectly, a company becomes carbon positive when it does something that allows others to directly reduce their emissions. A supplier could, for example, sell pellets that allow a utility to move from coal power to bio-energy. A company could also sell energy efficient appliances and renewable energy equipment like solar panels. Looking at the positive contributions requires a perspective where the whole value chain is included and then both the emissions and the contribution must be assessed (Dennis Pamlin. 2008). Some companies have also realized the potential in forestry for their carbon positive status that is the slogan of **Pranayama**. Well thought forestry and carbon credits generate livelihood opportunities for communities managing those forests, and create carbon sinks for local emission that is the basis of *Yoga Sutras*.

The industrial sector technology innovations have driven the over-growth that stand as challenge before the **Pranayama**. The developed world needs to redefine the value and corporate value in line with its *Yoga Sutras* vision of a sustainable and prosperous society on the one hand and decide an optimal balance between self-regulating market mechanisms and legislative initiatives on the other hand. It is the time to unite sustainable production with sustainable consumption that needs respect to **Pranayama** and active participation of implementing it in the activities of their industrial houses. This participation linkage will understand current and future consumption patterns, harnessing innovation to develop more sustainable products, services and behavioral change initiative through **Pranayama**. The whole world community should adopt the following points that will take care of producers, consumers and society at large such as:

- Human social systems and well being does not necessarily rely on high levels of consumption and should practice *Yoga Sutras*.
- More and more Consumers are concerned about environmental, social and economic issues, and increasingly willing to act on those concerns. Consumer willingness needs awareness programme through **Pranayama** and motivate them to act as sustainable consumer behavior. The young generation should understand brahmacharya and keep away these elements such as availability, affordability, convenience, product performance, conflicting priorities, skepticism and force of habit.
- Yoga idea of sustainable lifestyles should be based on informed purchasing decisions and changes in behavior of consumers that should get the support of active participation of business, governments and civil society in implementing *Yoga Sutras*.
- Yoga and five niyamas should formulate a consensus on the legal, fiscal and cultural environment for sustainable business to flourish and implement in the world market.

Swami Dyanand writes that there is a need to tolerate the pains to move ahead in life. It helps everybody not to be over-dependant on all scientific gazettes in their livelihood and practice pranayama. Those who don't believe in pranayama, cannot become a successful man in life and the yoga sutras is the best meditation. Pranayama keeps heart, mind in a peaceful position and all the impurities of life are vanished by enhancing the knowledge through it. Every human being despite their religious background may practice Pranayama because it is rightly mention that

यथा ऊर्णनाभिः स जते गृह्यते च यथा प थिव्याम् ओाध्यः संभवन्ति।
यथा सतः पुरूषात्केशलोमानि तथा अक्षरात् संभवति इह विश्वम्।।

It explains the web surrounding the life of spider, herbal medicines grown on earth, hairs grows on the body of a living person and this nature grows with the blessings of God. It means that God is one but is known with different names in different parts of the world. On the name of Devine power, there is no need to destroy the environment.

Pratyahara

It means that withdrawing the senses in order to still the mind as in meditation. *Pratyahara* is withdrawal of the senses. It occurs during meditation, breathing exercises or the practice of yoga postures. The best example of pratyahara is that honey bee house in which it is controlled by Queen Bee. When she is working, all the bees are on work, and when she sits on the house, all bees come around her and work. The same method should be used to concentrate the heart and mind, is known as pratyahara. It is further explained that

संध्या- पूर्वा संध्या जपन् तिष्ठन् नैशम् एनः व्यपोहति।
पश्चिमां तु समासीनो मलं हन्ति दिवार्क तम्।।

The human being should sit in the morning and remove all the mental evils came during night and do the same by sitting in the evening. It controls the lust, greed and ambitions towards physical material things among human beings.

Dharana

It is derived from the word *dhri* meaning to hold firm means to concentrate or holding the mind to one thought. The *dharana* involves teaching the mind to focus on one point or image. "Concentration is binding thought in one place," says Patanjali. The goal is to still the mind gently pushing away superfluous thoughts by fixing the mind on some object such as a candle flame, a flower, or a mantra. In dharana, concentration is effortless. You know the mind is concentrating when there is no sense of time passing. One says that

यथा यथा मनस्तस्य दुष्कृतं कर्म गर्हति।
तथा तथा शरीरं तत्तेनाधर्मेण मुच्यते।।

It means that a human being can remove all evils by criticizing it. By doing it, the evil be removed from the body and mind. This is the practical yogic method to get rid of all lust and greed and concentrate on dharana.

The needs towards sustainable environment may be achieved through dharana in public life. There are still differences on post Copenhagen summit among Afro-Asian-South American countries. This region envisages long-term cooperation in terms of enhanced action on reducing greenhouse gas emissions (Mitigation), and increasing the capacity to meet the consequences of climate change that had already taken place and is likely to continue to take place (Adaptation). These objectives must be supported by sufficient financial resources (Finance) and technology transfers (Technology) from developed to developing countries (Public. 2009: 2). The developing countries focus on climate change action cannot just be on current emissions. There is the equally important issue of Adaptation to Climate Change that has already taken place and will continue to take place in the foreseeable future even in the most favourable Mitigation scenarios. This region is already subject to high degree of climate variability resulting in droughts, floods and other extreme weather events which compels all states to fix over 2 % of its GDP on Adaptation and may go up accordingly. It is suggested that the developing countries should devise a global package that:

1. commits developed countries to significant reductions in their Green House Gas emissions;
2. achieves the widest possible dissemination at affordable costs of existing climate-friendly technologies and practices; and
3. puts in place yoga sutras directing the collaborative Research & Development effort among developed and major developing countries, to bring about cost-effective technological innovations and transformational technologies, which can put the world on the road to a carbon-free economy.

Dhyana

It means meditation or contemplation. Uninterrupted meditation without an object is called *dhyana*. Concentration (dharana) leads to the state of meditation. The goal of meditation is not unconsciousness or nothingness. It is heightened awareness and cneness with the universe. There is a difference between concentration and meditation? If there is awareness of distraction, one is only concentrating and not meditating. The calm achieved in meditation spills over into all aspects of one's life during a hectic day at work, shopping for groceries, coordinating the Halloween party at one's child's school.

Samadhi

It means that state of absolute bliss, super-consciousness that is known as the S*tate of God-union*. The ultimate goal of the eightfold path to yoga is *samadhi* or absolute bliss. This is pure contemplation, super consciousness, in which a humanbeing and the universe are one. Those who have achieved samadhi are enlightened.

During the Samadhi all *yoga sutras* work together: The first five steps such as yama, niyama, asana, pranayama, and pratyahara are the preliminaries of yoga and build the foundation for spiritual life. They are concerned with the body and the brain. The last three, which would not be possible without the previous steps are concerned with reconditioning the mind. They help the yogi to attain enlightenment or the full realization of oneness.

Suggestions and Conclusion

The major challenge before the world community is to sustain biodiversity by recognizing the biodiversity-related risk and the value of ecosystem services. The developing world industrial track record in managing biodiversity should access through retention of land and other natural resources, capital, markets and skilled staff. The responsible management of biodiversity concerns can result in benefiting directly to meet consumer demand for environmentally sustainable products. This applies not only to companies that depend on primary production, such as forestry, fisheries, agriculture, oil and gas development, and mining, but to others too in the developing world. Most businesses use natural resources, such as paper, food products and clean water, in their operations. These resources are a function of a stable environment, maintained by complex interactions between diverse ranges of species. 'Agenda 21 in the Convention on Biological Diversity (CBD) framed in the Earth Summit, aims to bring about sustainable development in the 21st century that is, development which respects the environment while meeting presents and future social and economic aims. India was among 160 countries to sign the CBD. Both the CBD and Agenda 21 call for the private sector's active engagement' (CII.2009: 17). All the governments should work

यथोद्धरति निर्दाता कक्षं धन्यं च रक्षति।
तथा रक्षेन्नृपो राष्ट्रं हन्याच्च परिपन्थिनः।।

It means that the rice cover provides security to rice grain at the time of separation from plant. Similarly, the ruler should punish the corrupted, cruel, black marketers, dacoits, thieves, rapists and provide security to the society. The global community should promote **firstly**, the leading businesses seeking to incorporate biodiversity concerns through yogic sutras into their investment decisions, notably into the design and implementation of their business practices and supply chain management systems. It is impossible for anyone to replicate the actual ecosystem but businesses need to maintain characteristic biodiversity so that the right conditions can be created through the complex interactions in sustainable way.

Secondly, all the governments should promote the scope of the biodiversity business case extending beyond the marketing departments of companies promoting the environment and yoga as a means of improving the images of corporate brands. The areas of opportunities in Asia relates to yoga is rich in biodiversity.

The industries of Asian region can engage with biodiversity issues and make a contribution to the objectives of the CBD ranging from raising employee awareness of biodiversity issues and managing environment impacts responsibly enhancing the conservation value of habitats on landholdings, helping to fund conservation initiatives and adhering to principles of sustainable use of biodiversity and promoting yoga.

Third, there is a need to form uniform biodiversity policy under Yoga for leading companies in the world that will strengthen the approach towards biodiversity issues (Figure 1). All the major industrial houses of the world need high-quality corporate policy on board level and implement biodiversity action plans connecting with the environmental management system will work under the guidance of Yoga. 'India has developed a number of policies and laws to promote the conservation and sustainable use of biodiversity and protect the environment. The challenge now is to assess and plug the gaps in these policies and laws, and in their implementation, so that economic development processes are reoriented to protect the natural resource base and associated livelihoods. A real solution to the biodiversity challenge must include contributions by all stakeholders, from government agencies to businesses.

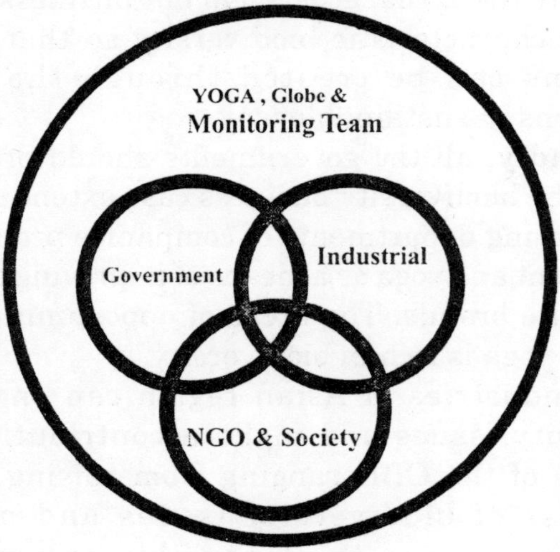

Figure 1

Fourth, all the states need to work under Yoga sutras with the mutual coordination with the industrial houses. To fight against natural challenges, the governments should initiate an effective corporate team having strong footings on environmental management systems on the one hand and should focus on the proper governance practices like धन्यं च रक्षति। on the other hand. All shareholders and financial analysts will increasingly assign value to companies that prepare for and capitalize on business opportunities posed by these challenges whether from Green House Gas regulations, direct physical impacts or changes in corporate reputation.

To sum up, while climate crisis covers most of the headlines today, the simple fact that the wealth of the 200 richest people in the world exceeds the combined annual income of the world's 2.5 billion poorest people is the result of limitless greed. Almost half of the world's population lives on less than $ 2 per day while the average American earns $ 130 per day. The belief that economic growth alone will solve the problems of poverty is a myth created by greed and lust of the capitalist world. The climate change in the world poses a well-rehearsed complexity of challenges and opportunities, from the tangible physical constraints of changing and extreme weather patterns and sea levels, to the reputation and legal context of which boards and brands are held accountable and liable for climate disasters, or rewarded for their foresight and insight in the innovation of new market solutions. The developed world should come forward and work with one of the most populous Afro-Asian region and share the budget, R& D and transfer of technology. It is rightly said,

यच्छक्यं ग्रसितुं ग्रस्यं ग्रस्तं परिणामे च यत्।
हितं च परिणामे यत् तदाद्यं भूतिमिच्छता।।

It means one should chew the eatable things, digestible and provide good health once it is digested. It applies on the developed world that does the things, which should be good for the society and nature and ultimately provides yogic life.

Note

This chapter is the revised version of the paper that was presented in the second International Bali-India Yoga Festival on **Global**

Warming and Solution through Yoga in Afro-Asia Region from 3-10 March 2010 in Bali, Indonesia.

References

CII. 2009. 4th Sustainability Summit Asia. CII-ITC. New Delhi.

Candace Ron deaux, "A Pipe line to Promise, or a Pipeline to Peril", *the St. Petersburg Times*, May 15, 2005.

Dennis Pamlin (2008), Winners in a Low Carbon Economy. *Sustainability Tomorrow*. October-December.

EurasiaNet, April 13, 2005.

Jeremy Scahill (2007), *Blackwater*, Serpent's tail. UK.

Joanne Kimberlin and Bill Sizemore (2006), *Blackwater: On American soil.* the Virginian-Pilot. July 27, 2006.

Patanjali. *http://www.yogamovement.com/resources/patanjali.html:* accessed on 21 February 2010.

Public Diplomacy Division. 27 February 2009. The Road to Copenhagen. Ministry of External Affairs. Rebuilding America's Defenses: *Strategy, Forces and resources for a New Century*, the Project for the New American Century, September 2000.

Tariq Ali (2003), *The Clash of Fundamentalisms.* Virso. London.

United States Code, Title 18, Section 2331 (18 USC 2331).

WBCSD Global Scenario 2000-2050. 2009. Exploring Sustainable Development. World Business Council for Sustainable Development.

WhiteHouse press release, "Caspian Pipeline Consortium Fact Sheet", November 13, 2001.

Winn Craig, (2003), *Tea With Terrorists*, Cricksong Books, Canada.

Yoga Site. *http://en.mimi.hu/yoga/dwesha.html:* accessed on 21 February 2010.

Joint Forest Management: Perspectives of Forest Dependent People

Poonam Kumria

Introduction

The concept of people's participation in management of forests is not new to India. Historically, India has great traditions of protecting and managing forests as common resources. Every village, hamlet and community ensured that the utilization of natural resources including forests should not exceed the ecological carrying capacity consequently renewing the net primary productivity of biodiversity. The economic and political colonization of the country adversely affected the traditions of conservation of sustainable utilization of resources. The forests and the people, which grew under the mutually beneficial relationships, suffered together, as the growing population put ever-increasing demands on the resources. The forest wealth gradually depleted and the people & forests were caught in the vicious circle of ecological, economic and social deprivation. Post independence, Central as well as State's forestry practices also continued to neglect the need for forest conservation practices and people's involvement in the process. Diversion of vast stretches of forestland was made for non-forest use; the biological diversity was considerably affected as also the good tree cover and the forest area.

A sizeable proportion of Indian population uses natural resources for subsistence living in the form of timber, fibers, grasses, oils, fruits, tans and dyes, gums and resins, medicinal plants and other non- wood forest products (Pandey, 1998). Due to heavy dependence and exploitation, the importance of forests for the ecological and economic stability of the country was realized by the foresters, as well as the government, which necessitated the reexamination of the policy, laying emphasis on the conservation and sustainable utilization of our forest resources than just commercial exploitation. The National Forest Policy of 1988 recommended significant changes from the existing forest management policies and practices. One of the main recommendations was the involvement of people in forest conservation and management as the major means of putting off the impending ecological crises and providing the benefits of these efforts directly to the people. It was realized that conservation and proper management of forests was not possible without active participation of the local people. Efforts were accordingly initiated in the seventies to involve local people in this task. This approach of involving the local people in the protection and management of forests is commonly referred to as Joint Forest Management (JFM).

Experience with JFM over the last decade has shown that it can be a successful approach in reducing conflicts between the state, represented by the Forest Department, and the local communities. JFM has also demonstrated the ability of Forest Department to forge alliances with communities to successfully regenerate and manage forest lands. There are numerous success stories of JFM; some of the better-known and well-documented ones are Arabari in West Bengal, Sukhomajri in Haryana, and Harda in Madhya Pradesh (Saigal *et al.*, 1996; Sarin, 1996; The World Bank, 2006). Experience from these areas has shown that JFM has vast potential to restore ecological balance by regenerating large tracts of India's degraded forests in a cost-effective manner and to empower local communities and improve their lives by giving them greater access and control over forest resources. It is perhaps due to these reasons that many of the enthusiastic supporters of JFM have tried to study, document, and highlight mostly the positive aspects of JFM. There is, however, growing experience and documentation to show that there are a number of problematic issues and conflicts that have emerged in the course of implementation.

The Study Area and Methodical Approach

The study area Dehradun valley, is a longitudinal valley lying between 29° 55' and 30° 30' N latitude and 77° 35' and 78° 24' E longitudes. Its length is 100 kms and width varies from 20 to 25 kms covering an area of 2250 sq. kms. The valley is a distinct eco-biome in the district of Dehradun, situated in the Himalayan foothills of the state of Uttrakhand. The valley is bounded on the north-east by the lesser Himalayan ranges, and on the eastern half of its south-west by the Shivalik ranges. The two most important rivers of North India, the Ganga and the Yamuna, demarcate its south-eastern and north-western boundaries, respectively. (Working Plan, 2000). Administratively, the Doon valley is a part of Vikasnagar, Dehradun, and Rishikesh Tahsils, which is further divided into four development blocks- Vikasnagar, Sahaspur, Raipur, and Doiwala (Figure 1).

The 'fragility' of the valley is further accentuated by the presence of a major boundary fault passing through the northern parts of the valley and by the unusually heavy rainfall of about 2,000 mm per year. The average width of the Valley is about 20 km, and the length is nearly 70 km. The valley eco-biome comprises two distinct sub-catchments, one formed by the drainage basin discharging

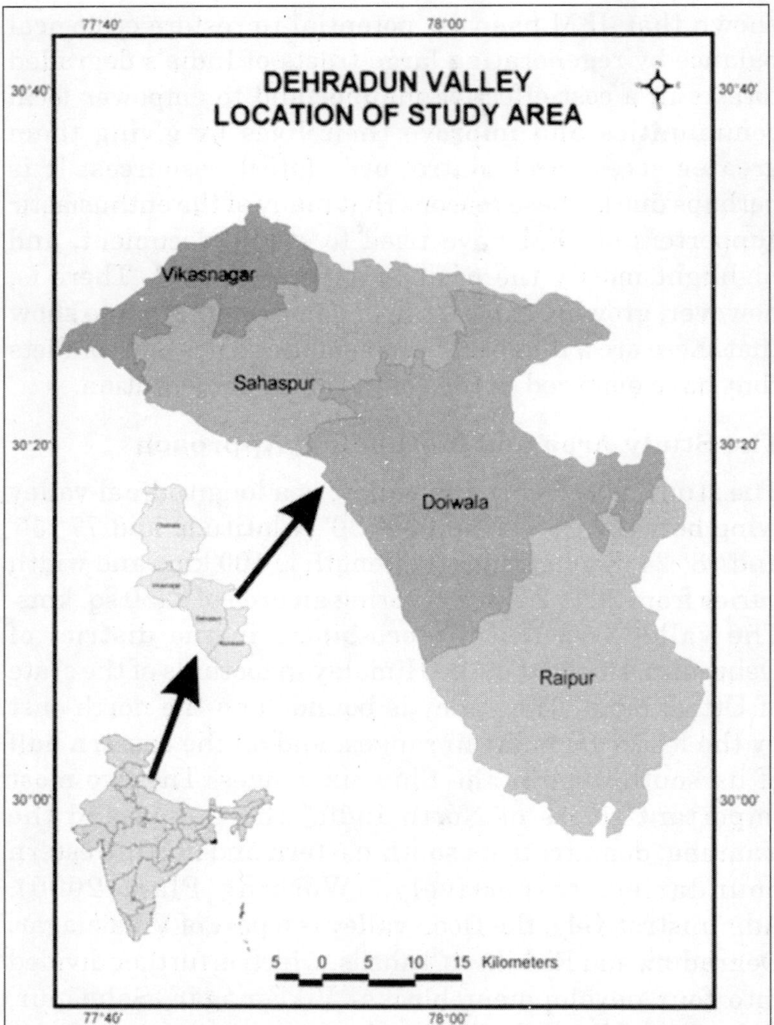

Figure 1: Dehradun Valley: Location of the Study Area

into the Ganga a little south of Rishikesh, and the other formed by the drainage-basin discharging into the Yamuna near Rampur Mahdi (just outside Dehradun district). Thus, the valley forms a sub-catchment for the Ganga-Yamuna rivers system which carries the vital water resources for the northern part of the Indian subcontinent. The Lesser Himalayan ranges, which form the northern boundary of the Dehradun valley, are part of the Great Himalayan range. The Shivalik ranges, which form the southern boundary of the valley, are alluvial formations that are younger than the Himalaya. They were formed by the debris which was swept down from the mountains. The Shivalik ranges present a stiff face to the plains, while a long and gentle slope meets the foot of the Himalaya to form a shallow longitudinal valley (Walton, 1911). These valleys or longitudinal depressions formed between the Shivaliks and the Himalaya are generally called 'Duns'. They are not continuous but are cut through by streams that drain the adjacent mountains. In some places, the Duns disappear with the merging of the Shivaliks and the Lesser Himalaya. The lower parts of these Duns are generally covered by a deposit of boulders, so that the floor of the valley is considerably higher than the level of the plains beyond the Shivaliks. Owing to this elevation of the Duns and the short distance over which the drainage from them meets the water courses in the plains, the landscape is marked by deep gorges and gullies, which cut through the unconsolidated strata that form the floors of these valleys. For the same reason, tapping of underground water through wells has not been as feasible in the valley as in the plains (*www.britanica.com*).

The Dehradun valley is well endowed with forest resources. Since the ancient times, it has attracted many migrating tribes and other people who later on settled down in this valley due to their dependence on forest resource. The vegetation cover of the valley is the result of physical forces that have produced it; although, increasingly this natural landscape has been modified by the human interference. The event of forest depletion and degradation in the valley has started long back. Large area in the valley is now, under secondary vegetation cover. The forest in Dehradun valley can be classified into following groups and sub- groups (Champion and Seth, 1968; and Saxena, 1975)- North Indian Tropical Moist Deciduous forest- Moist Siwalik Sal Forest, Moist Bhabhar Dun Sal Forest, West Gangetic Moist Mixed Deciduous Forest, Low Alluvial Savannah Woodland; Tropical Fresh water Swamp Forest- Sub-Montane Hill Valley Swamp Forest; Northern Tropical Dry Deciduous Forest- Dry Siwalik Sal Forest, Northern Dry Mixed Deciduous Forest, Dry Deciduous Scrub, Khair Sissoo Forest; Sub-Tropical Forest- Sub-Tropical Siwalik Chir Pine Forest, Sub-Tropical Himalayan Chirpine Forest, Himalayan Sub-Tropical Scrub; Himalayan Moist Temperate Forest-Ban Oak Forest, Himalayan Temperate Secondary Scrub.

The entire study is based on primary data collected through extensive and intensive fieldwork. Informal discussions were held with local villagers as well as with the forestry staff from the study area, to obtain general information regarding different issues related with forestry and joint forest management. After, discussions and field observations, a questionnaire has been prepared analysing the policy of joint forest management in the valley. On the basis of purposive random sampling method, five villages have been selected for the purpose of study. The criteria for selection of villages are proximity to forest, part of JFM project, and also part of National Afforestation scheme. The household in these villages have been chosen on the basis of random sampling. Door to Door survey has been conducted in the Tauli, Kherikalan, Bhattonwala, Kalimati and Kandhuli. The total sample size is 131 households. Tahuli and Nawada are the largest villages when it comes to area but on the basis of population Kandhauli, Nawada, Bhaatonwala and Tauli are big villages (Figure 2).

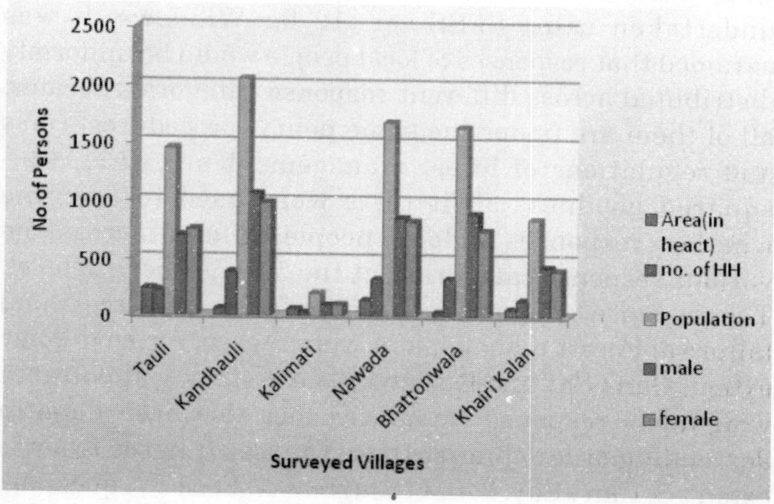

Figure 2: Demographic Characteristics of the Surveyed Villages

Source: Census of India, 2001.

The villages have subsistence economy. Most of the people are dependent on agriculture and livestock farming. On the one hand, population growth, less economic diversification, underdevelopment and poverty people are very much dependent on forest resource. On the other hand, proximity, accessibility, and approachability to forest areas increase their dependence on forest products. Consequently, forest resources in the valley are under immense pressure. With adoption of forest management techniques and policy of joint forest management, the deforestation has reduced, but giving way to high rate forest degradation. The study aims to look into different issues related with forestry, forest policy and planning and perspectives of local people about joint forest management policy.

The non-parameteric chi square test has been performed to analyse the responses forest depdent people on the success and failure of JFM. Data analysis was undertaken using SPSS ver. 12 for Windows. It was assumed that responses of local people would be uniformly distributed across different response categories because all of them are under the same policy, procedures, rules and regulations of forest management and JFM. Chi-squared goodness-of-fit tests were used to examine whether responses of local people differed across the various response categories, at the 5% significance level. The response categories chosen for factors of success and failure of Forest management were: Agree, Agree to some extent; Can't Say and Disagree. The assigning importance weights to response categories has also been done in descending order of importance (Agree: 4; Agree to some extent: 3; Can't Say: 2 and Disagree: 1). For each question, the number of respondents assigning each particular importance level were multiplied by the corresponding weights and these were summed to obtain the total score.

The responses of people about possible factors of success and failure of joint forest management for each variable are placed into four categories. The following hypotheses have been tested:

H0: The responses of forest dependent people are uniformly distributed between response categories (AG, ASE, CS and DA) for each response variable.

H1: The responses of forest dependent people are not uniformly distributed between response categories for each response variable.

Evolving Policy & Legal Framework

Historically forest had always been state's property or village institutional property. Earlier it was managed by the village people in the common interest, though the interest of kings overrides their needs. Due to less population pressure and abundance of forest resource, the right of local people has never realized. The importance of this resource, although can be seen in the historical accounts as it were part of life style. The Indian social structure is divided into four ashrams (life circles); the third is the - 'Vanprastha ashram'. It symbolises the life in forest a person has to live. It is written in the 'Aranyakas'- 'Book of the forest'. It is a later development of the Brhamanas or expositions of the Vedas, which were composed in India in about 700 BC. The First Forest Policy 1894, failed to lay down the guidelines of the proper utilisation of the Indian forests; inspite of this, it is still formed basics of the new forest policy after the independence of India (Jha, 1994).

In ancient India, it was generally accepted that forests and the communities living in the forest were not controlled by the rulers, because the forest was not seen as a source of revenue or commercialization. The effects of industrialization side by side with British rule in India in the 18th century brought about dramatic changes: the need to meet the growing demand for timber and a growing dissatisfaction with the legal restrictions imposed by previous legislation, led to the institution of the Indian Forest Act in 1878, according to which the nation state was recognized as sole proprietor of classified forest lands. State forest lands were loosely defined as lands which did not fall under 'continuous' cultivation or 'permanent' settlement. Traditional forest practices such as 'slash and burn' cultivation, rotational agriculture, grazing and gathering of forest resources were rejected as a basis for private property rights. Instead, according to a Forest Department resolution in 1890, previously defined rights of access and use were to be redefined by government as 'privileges' for specific tribes, castes, villages and organizations. A new Indian Forest Act in 1927 incorporated few substantive changes over the 1878 Act, and remains the legislative basis for state forest management today. The National Forest Policy 1952 laid guidelines that one - third of the geographical area of the country should be covered by the forest, in which forest coverage in the mountains and hills should be sixty per cent and the coverage in the plains should be twenty per cent (Table 1).

The country had great tradition of environmental conservation which taught us to respect nature and its resources. Our constitution also lay down provision for environmental protection. The Part IV of the constitution encompasses the directive principles of state policy which lays down the duties for state to protect and improve the environment and to safe guard the forests and wildlife of the country. Apart from this, the Constitution, under article 51 (g) states that- "it shall be the duty of every citizen of India, 'to protect and improve natural environment including forest, lakes, rivers, and wildlife and to have compassion for living creatures'." Objectives of forest management are normally determined on the basis of constitutional provision, various legislations, current forest policy and national conservation strategy and development. The National Forest Policy 1952 laid guidelines that one-third of the geographical area of the country should be covered by the forest, in which forest coverage in the mountains and hills should be sixty per cent and the coverage in the plains should be twenty per cent. The policy failed to provide guidelines for adopting time- bound programmes for attaining objectives. The next policy in 1988, laid down the guidelines for time-bound implementation programmes and need based afforestation was emphasized (Pandey, 1996). The afforestation was fast to augment the supply of fuelwood, so that the gap between the demand and supply of fuelwood can be narrowed down. The main objective of this policy was also to achieve environmental stability. The direct economic benefits were subordinate to environmental conservation. The policies and programmes were implemented by the forest department in each state, each district and in each

Table 1
Indian National Forest Law and Policy, 1878–2009

Year	*Law*	*Relevant Measures*
1878	Indian Forest Act	State is sole proprietor of classified forest lands.
1890	Forest Dept Resolution	Previous rights of access and use redefined as 'privileges' for specific tribes, castes, villages and organizations
1927	Indian Forest Act	Few substantive changes over the 1878 Act. It remains the legislative basis for state forest management today. The Indian Government adopted the 1927 Act after it gained independence in 1947.
1952	National Forest Policy	Set out guidelines which were, for the most part, directed towards the supply of cheap timber and non-timber forest products for state-sponsored industrialization and modernization.
1976	Indian Forest Act added to the concurrent list of the Constitution of India	Central government and states given shared control over forest matters.
1980	Forest Conservation Act	The central government reasserted some of its control over forest-based resources. The 1980 Act restricts the state government's power to de-reserve a forest, and it restricts the use of forest land for non forestry purposes without the prior approval of the central government.
1988	The National Forest Policy	Envisaged people's involvement in the development and protection of forests for the first time. , never translated into law.
1990	JFMC/VPC	Participation& involvement of locals
1992	National Afforestation and Eco development Board	Regeneration of degraded forest areas and lands adjoining forest areas,
1992	Eco Development Forces	Eco Task Force: raising nursery, plantation, protection
1998	'National Afforestation Programme	village level institutions
2000	Grants-in-Aid for Greening India	Planting trees and creating mass awareness
2009	Revised Operational Guidelines of the National Afforestation Programme	Building capacity of the institutional actors and institutions

Source: Compiled from MoEF, 1952, 1988, 2003 and various annual reports.

development block (Jha 1994; Ministry of Forest and Environment, 1990). Besides, Stockholm Conference, Rio Summit, Johannesburg Summit, other international meetings, conventions and workshops held in the last decade influenced forest management practices.

The state's role in forest management has always been one way process. With changing role of the state and people, the right and use of the forest is also changing. The commitment of the Indian state, dominated by the coalition of capitalists, rich farmers and political - bureaucratic alliance, to the modernisation pattern of development has been apparent. The state earlier did not recognise the industrialisation and urbanisation model of growth can lead to depletion of forest resources (Pathak, 1994). Chipko movement came to national prominence in the context of a rising environmental consciousness. Caught in the vortex of market economy and industrialisation, and faced with environmental hazards, hill society of Uttarakhand is in a state of continuing economic deterioration. Timber exploitation started in a bigger scale in the whole region. This led to construction of roads and other infrastructure facilities that helped in exploitation of inaccessible forest area. The forest department was assigned a role of revenue generating organisation. While largely oriented towards industrial needs, state forestry was unable to involve local people, which gave impetus to Chipko movement. The main turning point was floods in 1970. Villagers who bore the burnt of damage were beginning to perceive the links between deforestation, landslides and floods. It was observed that some of the villages most affected by landslides lay directly below forests where felling operations had taken place. In all about dozen Chipko agitations has been organised in different parts of Garhwal and Kumaun regions since its inception in 1973. While the last agitation occurred several years ago, the movement leaders still organise camps, marches at regular intervals to make people aware of problems of deforestation and environmental deterioration in the region. Chipko did not begin by making a representation to the centre but realised on the mobilisation of local people and especially, women in the villages close to forest areas. It was formed of layers at different levels-the peasants, the leaders of the movement, mobilises, the state government and the central government. Each layer has its historical context in which it shapes and is shaped by the others. This captures the dynamic relationship between the people and the government, forcing government to stop commercial felling in the Uttarakhand region. This movement is a symbol of peasant's environmental consciousness, unity of people with nature and the struggle from the peasants to preserve itself from erosion by urban - industrial development (Pathak, 1994; Banuri and Marglin, 1993; Guha, 1989).

Once the land comes under the management of the Forest Department, then only the Forest Department know what to do with the forest land. It becomes a forest offence for the local people to enter that area without the permission of the forest officer. 'The Forest Department, by evolution and training did not have the values and skills to work with the people' (Pathak, 1994). The Forest Department does not have connection at the local level. They are fully dependent on the information and help of block development officer who in turns is dependent on the village headman for the information of village problems. The visits to villages by these officers are rare making headman powerful and having monopoly over the local people in most of the cases. Forest ranges administratively fall under the development block, still there is no systematic co-ordination between the development officer and forest rangers in a development block. This has created a vacuum in the administration, which has lead to increase in the smuggling activities in the valley and corruption among forest officials and headman. There is no relation of development planning and presence of the resources in the valley. Regional development should be co-related with the forestry programmes in the valley. The state has always condemned grazing by the local people. Hill animals have been considered as unproductive. Government has banned the grazing activity in the forests but due to the dependence of the people it has not been materialised till now. The resistance from the gujars living in Rajaji National Park is the example of this. 'In day - today functioning, forest dwellers accept the domination of forest officials, who are feared and respected. At the same time, forest officials are abused behind their backs' (Pathak, 1994). The encroachment of the forest land is a comoon feature, which reflect the faulty government policies. With the view to have co operation from the local people the social forestry programme was started. Later on, it was viewed as a programme for the welfare of the local people but earlier social forestry referred to the market for non - industrial wood as opposed to production forestry, which supplied industries.

In recent years, there has been a shift towards participatory approaches in forest management and biodiversity conservation. The National Forest Policy (1988) declared that local communities were to be involved in natural resources conservation. Subsequently, in 1990 the Indian Ministry of Environment and Forests issued a circular for Joint Forest Management (JFM) and resource sharing. The JFM approach seeks to develop partnerships between state forest departments (as owners) and local community organizations (as comanagers) for sustainable forest management. In pursuance of the directions established in National Forest Policy, 1988, the Governement of India issued a set of broad directions to the state governments, on 1 June, 1990, about the involvement of village communities and village organizations in forest management. The circular established the broad principles for involvement of local people and voluntary organizations, and details were left to be worked out by the respective state governments. 25 Indian states have passed enabling resolutions for involvement of local people in forest management by June 2001, and this program is commonly known as Joint Forest Management (JFM).

The main objective of JFM as stated in the resolutions issued by different state governments are the protection and rehabilitation of degraded forests with the involvement of local people. The National Forest Policy, 1988 provided that requirements of forest products of the tribal and other villagers living in and near the forests are to be treated as the first charge on forest produce. It implies that meeting the requirements of the local people consistent with the productive capacity of forest should be the main objective of JFM. Besides meeting the forest products requirements of the local people, additional income generation needs to be recognized as very important to sustain the interest in JFM of the local people who are generally very poor. Additional income generation can be achieved by increasing employment opportunities and by establishing cottage industries for value addition to the forest products obtained by the beneficiaries under JFM. The National Environmental Policy, 2006 also focuses to conserve the biodiversity, inter-generational and inter generational equity, sustainable utilistion of biodiversity, integration of environmental, social and economic concerns and principles of good environmental governance (Singh, 2009).

Local Perspectives on JFM

The most important objective of the JFM has been the meaningful partnership between the Forest Department and the people in protecting and managing the forest resources. Traditionally, the Forest Department managed the forests by regulating the rights and concessions of the local communities. The Forest Department maintained a distance from the people and asserted its authority. The restrictions on the use of forests by local communities in the interest of forest conservation resulted in conflicts between the Forest Department and the forest dependent people. It became evident that the forests cannot be protected by Forest Department alone and active involvement of the local people in this task became necessary. This realization necessitated the adoption of JFM approach. Although, joint forest management initiated cooperation among officials and the villagers; there were many problems, which were later reduced with the adoption of changes in Joint Forest Management Policy of participatory eco development. The government realized that forest cannot be protected and conserved unless people are given responsibilities of these forest areas. Being the custodian of forests, the Forest Department initiates the process establishing contacts with the people. This work is very much facilitated by NGOs having contacts and good reputation with the local people. A series of meetings are held by the Forest Department staff to explain the details of the programme as contained in the relevant Government Orders. Doubts, if any, in the minds of the people are removed in these meetings. Having been convinced of the need and benefits of the programme, villagers used to held discussions at appropriate levels. The process ends with the constitution of Forest Protection Committee (FPC) or Eco Development Committee (EDC) to undertake the work of JFM in partnership of Forest Department. This involves trust building and enhancing capacity of local people, which in turn reduce the dependence of local people on the forest resources.

JFM in the state of is based on the premise that people are to blame for the destruction of the forest. The objective, therefore, is gradually to diminish people's dependence on forests. The Resolution to introduce JFM has been made unilaterally by the Forest Department. There is no scope for consultation with the people, no effort at consensus building, and no mechanism for dissemination of information. The resources required for villagers' needs are to be extracted from the protected forest according to the micro/annual-working plan. There is no mention of the existing rights of the villagers to use the produce from these forests. This is a gross violation of customary rights. The profits are from commercial productions, which are not based on local needs, thus indicating the focus on commercialization of forest. Clause 5, of the agreement provides "that if the beneficiary fails to carry out any of the directions issued by the Forest Officer for forest management or any of its binding obligations, the Forest Officer shall carry out at his discretion any or all the works regarding forest management departmentally". The Forest Department, thus, acquires the unilateral right to take over management of village forests brought under VFJM if they are not perceived to be well managed by it. The Forest Department itself did not fulfill any of its responsibilities towards Van Panchayats under the 1976 VP Rules.

The chi square analysis shows mix responses. In all the cases X^2 values allow the null hypothesis to be rejected which leads to the conclusion that distribution of the views between different groups and gender is not uniformly oriented across different response categories. Local people sited many reasons of the success of JFM in the valley. The analysis of the data shows chi square value is high in many cases like- frequent meetings with members of JFPC or EDCs, community participation, subsidies for fodder,

NGOs involvement, and plantation programmes. This indicates that people across different section and in various villages differ most in their responses. Many factors shows slight low X^2 value although these are higher than the critical chi square value (Table 2). The factors like-granting rights for forest use, presence of roads, employment opportunities and harvesting of NWFPs shows less differentiation in responses across different categories. Based on scores frequent meetings with JFPCs/ EDCs, community participation, awareness and workshop, and plantation programmes considered by the local people as most important factors for achieving successful implementation of the JFM, followed by subsidies for fodder, NGOs involvement, employment opportunities, and harvesting of NWFPs. The factors considered to be least important for success of JFM are presence of roads and granting right for forest use. People perceived that forest land belong to them so no question of granting rights to the local.

Within the JFM, under plantation programme the state had two objectives- accesses to vacant lands and needs to raise forests. Vacant lands were lands along the roads, canals, railway lines, revenue wastelands, village common land and village agricultural land. On the government owned lands, Forest Department raised plantation of quick growing species for the fuelwood and light timber market. On the private land, it provided trees to farmers for plantation on the bunds and farm boundaries. Extension of the forestry, inspite of addressing fuel, fodder needs of the local people was profit making venture. It was realised later on that the Forest department had neither the man power nor the capital to look after small scattered plantations; so the charge at the local level was given to panchayats. The Forest Department raise plantation of quick growing species on common land and revenue wasteland with the help of

Table 2
Factors for Success of JFM

Factors for Success of JFM	*Agree*	*Agree to some extent*	*Can't Say*	*Disagree*	*Goodness of fit*			*Scores*
					X^2	*df*	*p*	
Right for Forest Use	43	35	24	29	6.13	3	0.105	354
Harvesting of NWFPs	44	40	21	26	11.076	3	0.011	364
Employment Opportunities	45	38	26	22	10.344	3	0.016	368
Plantation Programmes	49	42	25	15	22.13	3	0.001	387
Subsidies for Fodder	36	58	21	16	32.573	3	0.001	376
Community Participation	50	55	12	14	48.084	3	0.001	403
Awareness & Workshops	48	50	17	16	32.328	3	0.001	392
NGOs Involvement	39	55	12	25	31.29	3	0.001	370
Frequent Meeting with JFPC/ EDCs Members	59	48	11	13	54.496	3	0.001	415
Presence of Roads	39	42	29	21	8.45	3	0.038	361

Source: Primary Survey.

panchayat. It manages the plantation for three to five years and then transfers the charge to village panchayat. The villages get grass, branches and twigs free for collaborating with the project (Pathak, 1994). The power structure and the social heterogeneity make impossible for the villagers too have the benefits of social forestry as they are ignorant of the rules and their rights.

Life is hard and the women have to cope with diverse situations, often alone. To bring about sustainable mountain development, the immediate need is to work towards making women equal partners in development. Instead of being a silent presence in the rural areas, they should be involved in the process of policy making. They also get preference in forestry operations undertaken by the forest department in the villages. They are also earning additional income for their houses. The involvement of women in planning and implementation of micro-level plan of the village is the need of the hour. They are people who are most dependent on forest as well as most information and knowledge on NWFPs. Participatory research therefore needs to take into account existing workloads of women, time budgets, priorities, facilities like day care for children etc. To use women efficiently, the work timings need to be flexible and as far as possible, the work place should also be easily accessible. In this way, the productivity of women can be increased.

The analysis of chi square shows that people do not differ much in various villages when itcomes to give reasons for failure of JFM in villages. The null hypotheses to be rejected all the cases as it more than critical value of X^2, i.e. 0.95 at the 5 per cent of significance level; except in the case of variables-inter village disputes, and illiteracy. In these two cases, the response of the local people is uniformly distributed across all categories (Table 3).

Table 3
Factors of Failure of JFM

Factors of Failure of JFM	*Agree*	*Agree to some extent*	*Can't Say*	*Disagree*	*Goodness of fit* X^2	*df*	*p*	*Scores*
High Dependence	43	34	32	22	6.802	3	0.079	360
Communication Gap	39	35	28	29	2.466	3	0.482	346
Illiteracy	34	35	33	29	0.634	3	0.889	336
Non-Involvement of Women	45	37	24	25	9.305	3	0.251	364
Caste Structure & Power Relations	49	45	18	19	25.061	3	0.001	386
Role of Panchayat and JFPC/ EDCs	40	37	26	28	4.237	3	0.237	351
Scarcity of Grazing Areas	40	43	25	23	9.551	3	0.023	362
Inter Village Disputes	34	35	30	32	0.45	3	0.93	333
Unemployment	28	33	34	36	1.061	3	0.786	325
No Benefits Sharing	49	38	24	20	16.026	3	0.001	378

Source: Primary Survey.

The analysis of different factors of failure of JFM show different result, then of factors of success, as chi square value is low in comparison to factors of success. The responses are almost uniformly distributed across different categories various villages. On two variables like – caste structure and benefit sharing, people's perception of factors of failure is not uniform. Factors like - caste structure and power relations, no benefits sharing, non involvement of women, scarcity of grazing areas, high dependence, and communication gap ranked as most important factors leading to failure. While least important factors are - unemployment, inter village disputes, and illiteracy. The true spirit of JFM gets translated only when forests are also managed to meet the people's needs. The JFM should become a permanent policy of the government with addressing the relevant community concerns being addressed. The JFM process needs to acknowledge historic forest rights of communities over the forest. All the stakeholders should be beneficiaries as well as active components for forest planning and management. With so much of mistrust with each other people and forest officials to some extent are coordinating with each other. But still there is a tussle between them, as short term goal of each factor work against each.

In Uttarakhand, there is a long tradition of voluntary agencies advocating the community importance in decision-making particularly in forest use and management. The existence of Van Panchyat provides strong support for arguing local community's involvement. Given the importance and variety of administrative control of forest in Uttaranchal (Reserved Forest, Civil/ Soyam Forest/Panchyat Van), it has been argued that there is need for integrated forest policy. The objectives of such a policy should be maintenance of ecological balance, to fulfill the basic demands of fuel, fodder and timber, of local people, to make raw material available for small scale industries, to develop social organizations for forest protection and to train the women in technology and forest management. Unless the local community is not involved in forest management, the forest couldn't be conserved only by forest department. Even after the implementation of programmes like JFM, the involvement of community and NGOs has been considered a must for the forest conservation and management. Since Doon Valley forests are under pressure from increasing urbanization, population in-migration and infrastructure expansion, the stake of NGOs present in large number in the area become very crucial. Community management systems have survived despite insensitive bureaucratic interference and controls in the past. Forest users continue to resist intrusions by the Revenue and Forest Departments through non- cooperation and withdrawal. In all the case study villages, women forest users' priorities were to increase the direct use values of their forests. An improved quality of life through easy access to fuelwood, fodder and water, and ecological security for them preceded considerations of income from sale of forest products. Through JFM introduced under a World Bank funded forestry project the Forest Department is being empowered to reassert control over *Van Panchayat* forests and civil/*soyam* lands, the only surviving village commons. Instead of validating the rich diversity of indigenous knowledge of local women and men, and the land use and management systems they have developed for supporting livelihoods and ecological security, JFM reinforces the Forest Department's claim to being the monopoly holder of technical forestry knowledge and forestry being the best land use even for the remaining commons. This is despite its historical lack of experience in bio-diverse forest management for enhancing livelihoods and ecological security.

Conclusion

The analysis explains that why JFM is not successful in order to address the problem of forest degradation in the valley. Due to afforestation programmes, the area under forest has increased as well as there is increase in dense forest area in the valley; but communication gap between forest department and people still exist. The forest dependent people perceived frequent meetings with JFPCs /EDCs member and community participation as most important factors for the success of JFM. They also blamed caste structure and power relations, and no benefit sharing as major factors leading to the failure of JFM progamme in the valley.

The forest department carried out JFM in Doon Valley with the objectives of reviving the forest area in and around the villages by involving villagers, creating local employment through regeneration of resources and to reduce the villagers' dependence on the forests. The project was conceived on the principles of resource sharing-fuel wood, fodder, medicinal plants etc but not the timber because that can prompt villagers to hasten the drying of good mature trees. The project has been successful because people in the villages could be made aware of JFM benefits. However it could bring still better results if it was taken up as a continuous activity and not as a scheme. It is very difficult to assess whether JFM is a failure or a success. For those who benefitted with this programme it is a success and those who have not got anything, it is a failure. But JFM is a milestone in the Indian forestry history, which has opened many ways to coordinate and work with local communities.

References

Banuri, T. and Marglin, F.A. (eds.) (1993), Who Will Save the Forests? Knowledge, Power and Environmental Destruction, Zed books Ltd., London.

Champion, H. G. and Seth, S. K. (1968), Revised Classification of Forest Types in India, Manager of Publications, Govt. of India, New Delhi.

Census of India (2005), District Census Handbook, Dehradun District, Town & Village Directory: 2001, Dehradun.

Guha, R. (1989), The Unquite Woods: Ecological Change and Peasant Resistance in the Himalaya, University of California Press, Berkeley.

Jha, L. K. (1994), India's Forest Policies: Analysis and Appraisal, Ashish Publishing House, New Delhi.

MOEF (Ministry of Environment and Forest) (1952), National Forest Policy, MOEF, New Delhi.

MOEF (Ministry of Environment and Forest) (1988), National Forest Policy, MOEF, New Delhi.

MOEF (Ministry of Environment and Forest) (2003), Terms of Reference for National Forest Commission, Delhi, MOEF, New Delhi.

MOEF (Ministry of Environment and Forest): Various Year Annual Reports, Delhi.

Pandey, D. P. (1998), Beyond Vanishing Woods: Participatory Survival Options for Wildlife, Forest and People, Himanshu Publications, Udaipur.

Pathak, A. (1994), Contested Domains: The State, Peasants and Forest in Contemporary India, Sage Publications, New Delhi.

Saigal, S., C. Agarwal, and J. Y. Campbell (1996), Sustaining Joint Forest Management: the Role of Non Timber Forest Products. In Proceedings of the Conference on Voices from the Commons, 5–8 June, International Association for the Joint Forest Management in India 281 Study of Common Property, University of California, Berkeley.

Sarin, M. (1996), Who is Gaining? Who is Losing?: Gender and Equity Concerns in Joint Forest Management. Society for Promotion of Wastelands Development, New Delhi.

Saxena, N. C. (1995), Administrative Issues in Forestry. In Forests, People and Profit: New Equations for Sustainability. Natraj Publishers, Dehradun.

Singh, R. K. (1996), Independence Once Again: A Training Story. In V. S. P. Kurup (ed.), New Voices in Indian Forestry. Society for Promotion of Wastelands Development, New Delhi.

Singh, R. B. (eds.) (2009), Biogeography and Biodiversity, Rawat Publications, Jaipur.

The World Bank (2006), India: Unlocking Opportunities for Forest – Dependent People, Oxford University Press, New Delhi.

Walton, H. G. (1911), Gazetteer of Dehradun, Natraj Publishers, Dehradun.

Working Plan Circle (2000), Management Plan for the Dehradun Forest Division 2000-2010.

Web References

www.britanica.com

www.fao.org/forestry/fra/fra2010/en/

www.fao.org/waicent/faoinfo/forestry/nwfp/nonwood.htm.

www.earthwatch.unep.net

www.envfor.nic.in/icfre

www.euttaranchal.com/uttaranchal/dehradun

Impact of Global Warming on Food Production

SURENDER MOR

Introduction

Global warming is one of the areas of greatest uncertainty for agriculture as well. The temperature of earth has increased by 0.6 °C in the 20th century and it may increase by 1.5-3.5 °C in the next 100 years, if continue with this rate. An increased global average temperature is likely to affect rainfall frequency, which is very crucial for agriculture growth. Human activities such as industrial process, fossil fuel combustion, deforestation etc. are responsible for accumulation of these gases. The atmospheric concentration of GHG's has increased due to the developmental process of nations. The concentration of $CO_{2,}$ CH^{-}_{4} and N_2O has increased by 34%, 153% and 17% respectively as compared to pre-industrial level[1]. Food and Agriculture Organization reported that nearly one billion people go to bed hungry and another 40 million people have been pushed into hunger in 2008 due to high food prices. This happened despite the collective action of the international community in bringing down the percentage of hungry people from 37 to 17. Higher temperature in the tropics can be expected to cut yields of the primary food crops, wheat, maize and rice, by 20 to 40 per cent and rising temperature also will dry up soil moisture, cutting yields even further. Climate-related food shortages will not be limited to the tropics, for instance, summer-long heat wave (in 2003) in France and Italy that reduced the production of wheat and fodder by one-third.

The effects of climatic change are not uniformly distributed across the world because many regions/parts of the Sahel, the Mediterranean, Southern Africa, have already become wetter, and parts of Southern Asia are becoming drier. Water scarcity will increase in many areas, particularly in the already-dry parts of Africa and in areas where glacial melt is an important source of irrigation water. Crop yields are expected to increase in temperate areas and decline in the tropics with moderate warming. The combined effects of higher temperature, greater variability of temperature and precipitation leads more frequent and intense droughts, floods, and reduced availability of water for irrigation can be devastating for agriculture in many tropical regions. Presently, about three billion people live in the tropics and subtropics and their number is expected to nearly double by the end of the century. The stresses on global food production from warming alone are going to be huge, and that doesn't take into account water supplies stressed by the higher temperature. In this background, the present study examines the impact of climatic change on the food production in the world vis-à-vis on India.

Future Food Projections

Researches have been conducted to project the future consequence of the global warming on agriculture world wide, reflecting both pessimism and optimism. Some found a trend of pessimism for most of the developing countries in the lower and mid latitudes whereas for high latitude countries these reflect a trend of optimism. Environment Protection Agency (1989) forecasted the impact on US agriculture just by doubling the atmospheric concentration of CO_2 above pre-industrial levels or benchmark levels[2]. This analysis, primarily based in crop model, which forecasts net losses of about $6 billion to $34 billion at 1982 prices without carbon fertilization[3] and $10 billion with carbon fertilization effect. Environment Protection Agency (1994) forecasted, for the year 2060, that global mean temperature will increases by 4 °C to 5.2 °C and the yield of soybeans, wheat and rice will decline by 21 per cent, 17 per cent and 6 per cent respectively. The study further forecasted that without carbon fertilization the global wheat yield will decline in the range of 16-33 per cent. Similarly at the global levels the production of maize will be severely a reduced by 20-31 per cent. The productivity of rice also forecasted to be decline by 15 to 25 per cent whereas the yield of soybeans 7-16 per cent. As a result of these global declines in food production the prices will rise by 100 per cent, which will put 941 million people at the risk of hunger globally.

Reily, Hohmann and Kane (1994) estimated the losses and calculated the economic effects due to global warming as compared to pre-industrial levels. They estimate that without carbon fertilization the damage would be in the range of $116 billion (1989 prices) to $248 billion and with carbon fertilization the range of loss would be between $7 billion to $38 billion. Due to decline in food production in tropics some exporting countries would gain in export value due to increased food prices in the world. Fischer *et al.* (2002) developed an ecological zone model that identifies suitability of land for agricultural production and stimulates the changes in the availability of suitable agricultural land that can be expected from climatic change. The average climatic impact estimated for 117 developing countries indicate that in 39 countries with a population of 2.5 billion (in 2080) would gain by 5 per cent or more in agricultural production, 29 countries with population of 1.1 billion would experience no change, 49 countries with a population of 4.2 billion would experience losses of 5 per cent or more, causing aggregate output net loss of about 89 millions metric tones of cereals for developing countries.

Mendelson, Dinar and Sanghi (2001) applied their crop model on India for 2080 on seasonal data for 20 years period (1966-86).Their findings are very pessimistic for India, forecasting a reduction in agricultural output without carbon fertilization by 60.9 per cent for Northeast region, 57.9 per cent for Northwest region, 31.3 per cent for Southeast region and 36.8 per cent for Southwest region. The corresponding decline in net revenue is 91.3 per cent, 86.9 per cent, 46.9 per cent and 55.2 per cent for all the four regions respectively. Tol (2002) shows that at 2.5°C warming levels Africa will gain 0.47 per cent of GDP, Central Europe and the former USSR will gain 2.65 per cent of GDP and 3.1 per cent of GDP gains for Central Asia. Jorgenson et al. (2004) forecasted that at 3.1°C warming, (by 2100), the average impact on agriculture would range from a decline of 2.6 per cent to an increase of 20 per cent. Parry, Rosenzwerge and Livermore (2005) estimated that at 3 °C of warming level agriculture production will increase by 3 to 13 per cent in developed countries whereas it will decline by 10 to 13 per cent in developing countries. IPCC (2007) forecasted that crop productivity, in the event of 1 to 3 °C temperature, could decline by 50 per cent in Africa and increase by 30 per cent in East Asia, decline by 30 per cent in South Asia, and the same will increase by 5 to 20 per cent in North America. William R. Cline, (2007) estimated the impact of global warming and showed that for developed countries the yield will decline by 9 per cent, for Africa by 17 per cent and for Latin America by 13 per cent.

Impact of Climate Change on Food Production

The impact of climate change on food production and agriculture depend on a range of factors, including the vulnerability of regional agricultural systems, populations and their adaptive capacities. The impacts of climate change on food production and agriculture depend on a range of factors like temperature, precipitation, CO_2 fertilization and socio-economic conditions such as market access, technology and the availability of resources needed for adaptation. Human-induced climate change is the most troubling and complex environmental problems, inter and intra-country, facing the world as a whole. Climate change is expected to have considerable impacts cn natural resource systems thereby affecting the agricultural production and productivity. The impact of global warming is visible as the yields of major crops in developing countries show a declining trend.

Table 1
Growth Rates of Yields for major Cereals in Developing Countries.

Year	*Wheat*	*Rice*	*Maize*
1960	5.8	3.0	3.2
1970	3.6	2.1	2.9
1980	4.9	3.3	3.3
1990	2.0	1.6	2.8
2000	0.9	0.8	1.4
2005	2.2	1.1	0.9

Source: Food and Agriculture Organization, 2006.

As indicated by the Table 1 the growth rate of yields for the major cereals (rice, wheat, and maize) in developing countries has slowed sharply since the 1980s. The growth rate of yield of wheat has declined from 5.8 per cent in 1960 to 2.2 per cent in 2005. Similarly the growth rate of yield in rice crop has declined thrice since 1960. The same is the case with the maize crop. These declining trends in growth rate of major cereals in developing countries are may be due to the change in climate which is more pronounced in these areas, because the addition of modern inputs is not associated with increase in productivity in these areas except some part of Africa.

Table 2
Growth Rates (decadal) of Agricultural Production in the World (1961-2007)

Crop	*Countries*	*1961-71*	*1971-81*	*1981-91*	*1991-2001*	*2001-07*
Wheat	Africa	24.86	7.02	29.60	1.32	1.64
	Asia	27.24	20.94	17.54	7.37	5.69
	India	33.58	18.58	18.13	10.16	3.65
	World	19.39	11.18	8.50	3.27	1.18
Rice/Paddy	Africa	23.26	6.60	20.42	6.72	1.23
	Asia	16.77	10.55	9.57	5.92	4.27
	India	8.19	9.22	14.69	9.64	1.42
	World	16.82	11.08	10.20	6.14	4.29
Maize	Africa	16.99	14.37	4.54	4.99	6.23
	Asia	22.95	20.07	19.45	7.71	6.48
	India	7.29	13.10	6.79	21.26	15.85
	World	18.46	15.36	11.39	9.51	10.94

Source: Compiled from FAOSTAT 1961 to 2007.

Table 2 reflects that the decadal growth rates of wheat in Africa, Asia and the World as a whole are declining decades by decades. Similarly, the growth rates of rice and maize are also declining despite the use of modern technology in larger part of the world. The world's growth rate for wheat has declined from an average of 19.39 per cent (1961-71) to 1.18 per cent (2001-07). Wheat yields in Asia grew at an average decadal rate of 27.24 per cent during the (1961-71), but it dropped to 5.69 per cent in the period (2001-07). After rapid growth of 16.77 per cent per decade during (1961-71) rice yield growth fell to 4.27

per cent in period (2001-07) in Asia. In Africa, the situation is even more dramatic, with wheat yield growth decreasing drastically from 24.86 per cent to 1.64 per cent during the period on 1961 to 2007. The same also happened with the growth rate of rice in the Africa.

Impact on India

Agriculture still continues to be backbone of Indian economy as it provides employment to 52.1 per cent of people and contributed 17.1 per cent in GDP (2008-09). The share of agriculture sector in total exports is 12.2 per cent. The total foodgrains production is estimate to be 229.85 million tones in 2008-09 as compared to 208.60 tones in 2006-07. But the total foodgrains production is below the targets.

Table 3
Growth Rates(decadal) of Agricultural Production in India (1961-2007)

Crop	*1961-71*	*1971-81*	*1981-91*	*1991-2001*	*2001-07*
Wheat	33.58	18.58	18.13	10.16	3.65
Rice	8.19	9.22	14.69	9.64	1.42
Maize	7.29	13.10	6.79	21.26	15.85

Source: Compiled from FAOSTAT 1961 to 2007.

In India the growth rates of all the principal food crops have shown a negative trend (Table 3). The decadal growth rate of wheat has declined from 33.58 per cent to just 3.65 per cent from 1961-71 to 2001-07. Similarly the growth rate of rice has also declined from 8.19 per cent to 1.42 per cent during the same period. In contrast to this, the decadal growth rate in maize production has also shown a dwindling trend after 1991-2001. As exhibited by Table 4 the net availability of cereals has been declined since 1991, when it was at maximum, i.e., 468.5 grams per person. It is interesting to note that the per capita availability of cereals was 334.2 grams in 1951 and it was 386.2 in 2001, despite the fact that the total foodgrains production has increased more than four fold. The reason for this meager increase in per capita availability of cereals is the three fold increase in population. So the declining trends in both productivity and net availability of foodgrains and increasing population on the one hand and the global warming on the other hand, poses a threat to the mass our country unless some revolution in cropping pattern, crop varieties has occurred.

Table 4
Per capita Net Availability of Cereals in India

	1951	*1961*	*1971*	*1981*	*1991*	*2001*
Population (in Millions)	363.2	442.4	551.3	688.5	851.7	1033.2
Cereals (Per Capita per day in Grams)	334.2	399.7	417.6	417.7	468.5	386.2

Source: Complied from Economic Survey, 2008-09, Govt. of India.

Indian agriculture, in particular, highly depends on the climatic conditions and thereby economic growth becomes vulnerable to climate change. Agricultural lands worldwide have become vulnerable to degradation processes such as soil erosion, nutrient depletion and loss of organic matter, and have suffered a consequent decline in soil productivity or soil fertility. Consequently, agricultural production and yield growth began to slowdown in the latter part of the 20th century worldwide despite the continued development of new and improved modern varieties, greater use of chemical fertilizers and modern plant protection measures.

Table 5 revealed that the production of wheat has increased seven fold since 1960-61, the area under wheat has increased two times since then and the yield has increased more than three times in the same period. The production of rice has increased three fold, despite the fact that area under rice cultivation just increased by 50 per cent and the productivity just by two times. The production if maize has increased four fold since 1960-61 despite the less than two times increase in area under cultivation and the productivity of the same has increased more than 2.5 times. It is interesting to note that incase of wheat and rice crops, production, area and yield has reached near saturation in 2000-01, reflecting a negligible increase in the all three variables.

Table 5
Production, Yields and Area under Major Crops in India

Crop		*1960-61*	*1970-70*	*1980-81*	*1990-91*	*2000-01*	*2008-09*
Wheat	Production	11.0	23.8	36.3	55.1	76.4	77.6
	Area	12.9	18.2	22.3	24.2	27.5	27.7
	Yield	851	1307	1630	2281	2778	2806
Rice	Production	34.6	44.2	53.6	74.3	89.7	99.4
	Area	34.1	37.6	40.1	42.7	45.2	45.6
	Yield	1013	1123	1336	1986	1901	2177
Maize	Production	4.1	7.5	7.0	9.0	11.5	18.7
	Area	4.4	5.8	6.0	5.9	6.4	8.0
	Yield	926	1229	1159	1518	1792	2302

Source: Complied from Economics Survey, 2008-09, Govt. of India.
Note: Production, Area and Yields are measured in Million Tones, Million Hectares and Kilograms per Hectares respectively.

The Economic Issues

In tropical and sub-tropical areas the crops are often close to their thermal optimum[4] levels so that regional warming may result in reductions of production and productivities. Extreme weather events are likely to cast substantial negative influence on crop production, either directly or through increase in the incidence of pests attack. However the damage due to interactions of food production with droughts, heavy rains, hail storms or pest outbreaks cannot be estimated accurately. Use of specially designed genetically modified organisms could be a way to increase the adaptive capacity of crops. The global warming will accentuate inter and intra-country difference in agricultural production. The same may also results in reduction in arable land. All theses factors cumulatively lead to a decrease in income of poor farmers resulting from loss in agriculture production on the one hand and the higher cost of cultivation on the other hand. Hence the countries in tropics would lose a major portion of their GDP which accrue from their primary sector.

The developed nations are emitting more GHG's per capita which led to future loss to the developing countries due to global warming. On the other hand these countries are also going to be the beneficiaries of this global warming as their net agricultural production sets to increase. This may cause serious economics consequences for most of the developing nations as they have to import foodstuffs to feed their growing population. The terms of trade are already against them will further deteriorate causing a pressure on their balance of payments position.

Implications for India

The forecast made by Mendelson, Dinar and Sanghi (2001) could have serious repercussion for India by the turn of the present century. The northwestern regions (Punjab, Haryana, and Uttar Pradesh) are the bread basket of India and if the production as forecasted declined by 58 per cent as these states produce 65 per cent of the wheat produced in the country[5], then how the country feed its growing population (around 1800 million by that time). This situation will be more dangerous also with the fact that the productivities of major food crops are also declining in this region. The study also forecast that the revenue from agriculture will decline by 87 per cent in this region, which is amongst the most prosper area of the country. Researches have revealed that the farmers are already in bad debts and committing suicides. The study projected a very sobering situation for the North Indian agriculture as compared to the Southern region. The question to the policy makers is that what should be done to avoid or adapt these dangerous consequences of the climatic change in India.

Conclusion

The ongoing global warming put one-third of the population at risk of hunger in Africa, one-quarter in Western Asia, and about one-sixth in Latin America. Due to decline in food production in tropics some exporting countries would gain in export value due to increased food prices in the world. The major implications of climate change are thus largely for the distribution of agricultural production. These changes have implications for food production, food security and food safety. The developing countries possess least adaptive capacity are going to be worst affected by the climate change, which ultimately results in regional disparities in terms of crop yield and income too. The widening disparities between north and the south may further tilt the terms of trade in favor on the developed nations. The regions already under water stress today, climate change will exacerbate the situation. Such a steep increase in the numbers of people under water stress in such a short time span is likely to overburden available adaptive capacities such as sea-water desalinization or long-range transport, and thus cannot be termed tolerable.

Notes

1. Pre-industrial period refers to the period from 1865 to 1890, in which the concentration of CO_2 was estimted at 350 ppm level.
2. Benchmark levels or pre-industrial levels of CO_2 indicate concentration of CO_2 at 350 ppm level.
3. Carbon fertilization means the Earth's ecosystem uses some CO_2 and converts it in the other gases such as Oxygen.

4. Thermal optimum means a maximum limit of temperature which a crop can bear without adverse effects on yield.
5. Economic survey 2008-09, Govt. of India, page A20.

References

EPA (Environmental Protection Agency), (1989), *"The Potential Effects of Global Climate Change on the United States"*, Washington.

Fischer, G., M., van Velthuizen, H. T., Shah, M. M. and Nachtergaele, F. O. (2002a), *Global Agro-ecological Assessment forAgriculture in the 21st Century: Methodology and Results*. Laxenburg.

Food and Agriculture Organization (2007), United Nations, FAOSTAT FAO 1961 to 2007.

IPCC (Intergovernmental Panel on Climate Change), (2007a), Summary for Policymakers. *Climate Change 2007: The Physical Science Basis.* Working Group I Contribution to IPCC Fourth Assessment Report: Climate Change, Geneva.

IPCC Workshop on New Emission Scenario", Laxenberg, Austria, (2005), and "Climatic Change (2001): Working group III: Mitigation".

Jorgenson, Dale W., Richard J. Goettle, Brian H. Hurd, and Joel B. Smith. (2004), U.S. *Market Consequences of Global Climate Change*. Arlington, VA: Pew Center on Global Climate Change.

Mendelsohn, Robert, Ariel Dinar, and Apurva Sanghi, (2001), The Effect of Development on the Climate Sensitivity of Agriculture. *Environment and Development Economics* 6: 85–101.

Michaelowa, A Butzengeiger, S, Jung, M and Dutschke, M (2003), *Beyond 2012 – Evolution of the Kyoto Protocol Regime. An Environmental and Development Economics.*

Nicholas Stern, "Stern Review Report on the Economics of Climate Change" (2006), Cambridge University Press.

Parry, Martin, Cynthia Rosenzweig, and Matthew Livermore, (2005), Climate Change, Global Food Supply and Risk of Hunger. *Philosophical Transactions of the Royal Society B* 360.

Reilly, J., N. Hohmann, and S. Kane, (1994), Climate Change and Agricultural Trade: Who Benefits, Who Loses? *Global Environmental Change* 4, No. 1: 24–36.

Tol, Richard S. J. (2002), Estimates of Damage Costs of Climate Change. *Environmental and Resource Economics* 21: 47–73.

WBGU (German Advisory Council on Global Change) (2003), *Charging the Use of Global Commons. Special Report*. WBGU, Berlin.

William, R. Cline, (2007), Global Warming and Agriculture: Impact Estimates by Country", Peterson Institute of International Economics.

World Bank (2003), United Nations, Washington D.C.

Some Aspects on Rules and Legislations for Solid Waste Management in Delhi

SUBHASH ANAND

Introduction

Institutional arrangements generally constitute a set of rules governed relationship among organisations to achieve a set of aims and objectives. The important thing to note is that the coordination of teamwork and resources (human, financial and physical) is crucial whenever more than one person or institution is involved. It should also be understood that institutions are the means by which people order their relationships with others to come to a productive, collaborative or compatible arrangement. To protect public health and the environment, effective laws should be passed and vigorously enforced. Legislative and regulatory instruments are designed to ensure adherence to set rules and regulations for the protection of environment. Institutional arrangements for environmental protection cannot exist without appropriate legal foundations. In general, combining governmental and legal institutes is a very cumbersome arrangements. The result is often a host of laws, rules and regulations, orders, ministries, departments, divisions, commissions, boards and councils in a complex web of interrelationship (Dwivedi, 1997).

As the Brundtland Report indicated, environmental protection and management can no longer be left to weak and under funded government departments to several departments with overlapping responsibilities. To this end, all nations have established departments or ministries of the environment as a structural response to the changing perceptions of the environmental challenge. Among the environmental challenges and issues, solid waste management is very crucial. Solid waste management (SWM) is the systematic administration of activities that provides for collection, processing and disposal of solid waste (Parker, 1980). It forms an integral part of the urban environment and the planning of the urban infrastructure to ensure a safe and healthy human environment, while considering the promotion of sustainable economic growth. The problem of waste disposal is aggravated by the lack of financial as well as human resource trained in SWM practices in the sphere of collection, transportation, processing and final disposal. Management of solid waste is a major challenge these days for administrators, engineers and planners. The policy makers and planners must examine and take into account the directly or indirectly aspects related to daily collection and disposal of garbage.

According to Constitution of India, SWM falls within the purview of the local bodies. The municipal body executes the tasks of waste disposal with its own staff, equipment and available funds. The urban local bodies are exploring various methods for processing, treatment and disposal of municipal garbage to solve their ever-increasing problem of waste management. Effective SWM depends upon an appropriate distribution of functions, responsibilities and financial position of concerned local bodies. Improved municipal solid waste management capacity normally requires innovations in the organisational structures, staffing plans and jobs descriptions of the bodies.

The Ministry of Environment and Forest, Government of India is the nodal agency at the central level for planning, promoting and coordinating the environmental schemes and programmes. Municipal Solid Waste (Management and Handling) Rules, 2000 has been notified with the intentions of streamlining the technical and administrative methods in which the solid waste is handled. These rules apply to every municipal authority responsible for processing and disposal of garbage. In Delhi, the highest responsibility is assigned to Municipal Corporation of Delhi (MCD) as it is providing its services to 97 per cent people of Delhi, while New Delhi Municipal Council (NDMC) and Delhi Contonment Board (DCB) provides its services to a very smaller area. Because of huge population, larger areas and lack of sufficient finance, MCD is feeling burden. This article deals with rules, regulations, legislations, government's initiatives, institutional arrangement and financial aspects with regard to solid waste management in Delhi. It also highlights of garbage management activities of few non-governmental organisations and Master Plan for waste processing and disposal for Delhi.

Management of Solid Waste in Class-I Cities

The writ petition was filed by Almitra H. Patel in 1996 regarding management of solid waste in Class-I cities. The petitioner alleged that practices adopted by the municipalities for disposal of waste were deficient. The management of solid waste by the municipalities had direct impact on the health of the people in the country. The petitioner had appreciated the guidelines and recommendations made by the Central Pollution Control Board (CPCB) for the management of municipal waste. In its reply, the Central Pollution Control Board submitted that the responsibilities of management of solid waste were vested with the municipal corporations of the municipalities which are under the administrative control of respective States/Union Territories. At the central level, the Ministry of Urban Affairs is the nodal Ministry to deal with the matters relating to municipal solid wastes. The Central Pollution Control Board itself has taken several initiatives for improvement, collection, transportation disposal and utilisation of municipal solid wastes. On the basis of the replies of the various departments, Central/ State Pollution Control Boards and concerned State Government, the Hon'ble Supreme Court on 16.1.1998 directed to constitute a committee to look into all aspect of solid waste management in Class-I cities (having population over one lakh) of India. The terms of reference for the committee was to look into all aspects of urban solid waste management, particularly examine and suggest ways to improve conditions in the formal and informal sector for promoting eco-friendly sorting, collection, transportation, disposal, recycling and reuse etc. The Chairman of the Committee nominated was Shri Asim Barman (Commissioner of Calcutta Municipal Corporation). The committee has submitted its report in the month of March 1999 before the Supreme Court for consideration. The committee made several recommendations including technical aspects also for the management of solid waste in class-I cities. The recommendations were further categorised under three heads (i) mandatory recommendations for citizens/ associations; (ii) mandatory recommendations for local bodies/state Governments; and (iii) Discretionary recommendations for urban local bodies.

Classification of Recommendations

The recommendations made in this report can be grouped into two categories as under:

A. Mandatory Recommendations

B. Discretionary Recommendations

Mandatory Recommendations for Citizens/ Associations

(1) Not to throw any waste on the streets, lanes, bylane, footpaths, open spaces, water bodies etc.

(2) Store the organic food and bio-degradable waste at source in personal domestic bins.

(3) Segregate/store separately recyclable waste/non-biodegradable wastes as well as domestic hazardous waste at source.

(4) Provide community bin/bins in commercial complexes.

(5) Deposit domestic, trade institutional wastes in the food cart/recycles/ community bins/vehicles as may be notified by the local body.

(6) Trim the garden waste by the days notified by the local body.

(7) Store the construction waste within the prepare permises.

(8) Hospitals nursing homes make their own arrangements for disposal of their industrial and bio-medical waste.

Mandatory Recommendations for Local Bodies/State Governments

(1) Initiate public awareness campaigns through Information Education and Communication (IEC) strategy.

(2) Primary collection of waste from doorstep/community bins with or without community participation.

(3) Street sweeping on all days in the year irrespective of Sundays and Public holidays, making adequate provision for giving statutory weekly off to the workers or compensating them for working on holidays etc.

(4) Provision of mobile/bulk community waste storage containers large size containers/tractors trolleys) at the waste storage depots.

(5) Transportation of waste at regular intervals before the containers start overflowing.

(6) Collection, transportation and disposal of Market waste, Hotel and Restaurants waste, Construction waste, Garden waste, Kalyan Mandap/ Marriage Hall waste with the participation of waste producers.

(7) Cities with population above 5 lakh should provide for weighbridge at the processing and disposal sites.

(8) Arrange for composting of all bio-degradable waste into bio-organic fertiliser.

(9) Disposal of non-compostable waste and rejects at the sanitary secured landfill site in an environmentally acceptable manner.

(10) Take remedial measures to make old dumpsites safe.

(11) Identify and procure the sites for composting and disposal of waste for next 20 to 25 years (States Government and local bodies jointly).

(12) Provide toilet facilities to prevent open defecation.

(13) Special attention to be paid in cleaning of slums and dirty areas.

(14) SWM activities to be kept under one umbrella.

(15) Management Information Systems to be introduced for strict monitoring of SWM services.

(16) Declaration of no-development zone in a buffer region of 500 meters from the boundary of such reserved processing and disposal sites.

Discretionary Recommendations for Urban Local Bodies

(1) Choice of technology, Type of bins and vehicles to be used

(2) Work norms for workers and vehicles deployed
(3) Measures for institutional strengthening
(4) Supervisory levels to be deployed
(5) Human Resources Development
(6) Allocation of funds
(7) Health check-ups for all SWM workers and staff
(8) Obligatory functions to be given priority over non-obligatory functions and allocate funds accordingly. Improve financial position through plugging the leakages and improving recovery of taxes, charges and introduce mechanism for automatic annual increase in taxes and charges to cover the inflationary trends.
(9) Introduce area based property tax reforms.
(10) Make suitable provisions in the building bye-laws to earmark place for keeping community bins for the storage of wastes in high raised buildings and commercial complexes.
(11) Propose amendments in the State laws to improve solid waste management services.
(12) Develop citizen charter and widely publish the same.
(13) Introduce the system of levy of administration charges/special cleaning charges – from those who violate SWM code of conduct.

Initiatives Undertaken by the Government

In the context of improvement of the management of municipal solid waste, a number of steps were taken by the Government of India, Government of NCT of Delhi, Municipal Corporation of Delhi etc. The outbreak of suspected plague in Surat in 1994, had caught the attention all over world, whereby it was realised that all Indian cities were equally vulnerable and the need to improve their condition should be a top priority. The Government of India had long recognised the futility of the passive policy, followed since the 1960s and 1970s, of managing municipal waste by offering loans and grants to urban local bodies to set up by composting facilities.

National Waste Management Council (NWMC) was constituted by Central Government in 1990 and one of its objective was managing municipal solid waste. The council decided that sample surveys be carried out in various cities to find out recyclable waste materials in the garbage. Financial assistance was provided to 22 municipalities to undertake such survey.

A High Power Committee on urban solid waste management was set up in 1995, under the Chairmanship of J. S. Bajaj, member, Planning Commission with 11 other experts. The committee called for the formulation of a National Policy on Solid Waste Management and has given a number of recommendations and some of them are as follows:

- Primary level of collection should be ensured from each household.
- Private agencies and non-governmental organisations may be involved in primary collection.
- Segregation of domestic garbage at household level should be encouraged and promoted.
- Monthly charging for door-to-door collection based on income groups may be implemented.
- The vehicles for transporting the waste from the transfer point to the disposal site should be of appropriate design, suiting the waste characteristics.
- Composting of municipal solid waste should be the next appropriate option after land filling.

- Participation of private sector in setting up pilot plants utilising appropriate technologies for urban solid waste management should be encouraged.

Municipal Solid Waste (Management and Handling) Rules, 2000

Ministry of Environment and Forest, Government of India has made the legislative rules for the management of municipal solid waste entitled "Municipal Solid Waste (Management and Handling) Rules 2000". These rules apply equally to every municipal authority regardless of its size and make them responsible for the collection, storage, segregation, transportation, processing and disposal of municipal solid waste. The Rules are divided into four schedules, which deal with the following issues:

(1) Implementation Schedule

(2) Management of Municipal Solid Wastes
- Collection of Municipal Solid Waste
- Segregation of Municipal Solid Waste
- Storage of Municipal Solid Waste
- Transportation of Municipal Solid Waste
- Processing of Municipal Solid Waste
- Disposal of Municipal Solid Waste

(3) Specification for Landfill Sites
- Site Selection
- Facilities at the Site
- Specification for Landfilling
- Pollution Prevention
- Water Quality Monitoring
- Ambient Air Quality Monitoring
- Plantation at Landfill Site
- Closure of Landfill Site and Post Care
- Special Provisions for Hilly areas

(4) Standards for Composting, Treated Leachates and Incineration (CPCB, 2003)

The major features of this manual as per following:

Segregation

In order to encourage the citizens, municipal authority shall organise awareness programmes for the segregation of wastes and shall promote recycling or reuse of segregated materials. The municipal authority shall undertake phased programme to ensure community participation in waste segregation. These programmes, with participation of non-governmental organisations and local resident welfare associations are to be held every three months interval with a focus on source segregation. The waste must be segregated in streams such as bio-degradable recyclable and non-recyclable.

Storage and Transportation

Municipalities are required to establish storage facilities where the waste from the households and others can be stored for short period of time before it will be collected by the municipalities for final treatment or disposal. The waste from these facilities should be transferred to the disposal site using appropriate vehicles.

Processing and Disposal

The Municipal Solid Waste Rules-2000 require that biodegradable component of the waste should be composted and recyclable materials and resource recovery must be there, so as to reduce the pressure on landfill sites. Only non-degradable waste should reach to landfills. Composting and incineration of waste should be carried out as per standards mentioned in the MSW Rules, 2000.

The most significant step taken by the Delhi Government was linked with institutionalisation of Bhagidari, the project on citizen participation by government as initiated in 2000. Many Bhagidari workshops on municipal solid waste had been held with the participation of resident welfare associations, rag pickers, MCD personnel and eco-clubs of schools. Seven priority areas that were related to waste management were taken up as part of this scheme. These were maintenance of community parks, supervision of sanitation services, information about sanitation staff, imposition of fines on littering, door to door collection of garbage, solutions on community bins and generation of public awareness on sanitation. A nodal officer from the MCD was appointed to meet regularly with the resident welfare associations and a cell in the Chief Minister office was set up to track the progress of action taken on mutually accepted decisions (Bhagidari, 2002). The Delhi Government implemented the Delhi Plastic Bag (Manufacture, Sale and Usage) and Non-Bio-degradable Garbage (Control) Act, 2000. This act prohibited the manufacture, sale and usage of recycled plastic bags and virgin plastic bags of less than 20 microns and forbade the use of recycled bags for food. The act further required that the recycled bags should be coloured and pigmented with permitted dyes and be marked as "recycled bag usage for carrying food". Violation of these provisions of the act could involve imprisonment for a period of three months to one year and a fine upto Rs. 25,000 (The Hindustan Times, 2002).

On the 16 July 2002, in a meeting held at New Delhi on the subject of "An Interface with Civic Authorities" two panellists, Additional Commissioner (Sanitation) and Director (CSE) of MCD sounded confident about the abilities of the body in meeting the challenges posed by the requirement to implement the Municipal Solid Waste Rules, 2000 from Ist January 2004. The MCD had introduced night sweeping in 40 commercial areas and it aimed to extend it to 60 commercial areas by 2007. The

MCD also planned to restructure sweeping operations where sweepers from regularised colonies were to be moved to unauthorised colonies (Dhamija, 2006). In each zone, a control room had started functioning with a Sanitary Task Force, equipped with a wireless and consisting of a Sanitary Inspector, eight *safai karamcharis*, a truck and a loader, which could be despatched for immediate removal of garbage on request.

One of the main objectives of the Tenth Five Year Plan in its finalised form was the implementation of Solid Waste Management Handling Rules, 2000. Plastic Bill and other orders were also passed by the Supreme Court of India and Delhi High Court from time to time for various public interest litigations. A sum of Rs. 600 crore as compared to Rs. 450 crore in the Ninth Five Year Plan was secured by Delhi Government for strengthening and mechanisation of conservancy and sanitation services (Dhamija, 2006).

Litigation and Court Decisions

In view of the Public Interest Litigation (PIL), various courts have viewed the matter of solid waste collection and disposal seriously. Some of the important decisions delivered by the courts are the following in writ petition (C) No. 286 of 1994, B.L. Wadhera vs. Union of India, Hon'ble Supreme Court of India given a judgement on Ist March, 1996. A 29 pages judgement touched garbage affairs of Delhi and issued several directions to the local municipal authorities to perform upto satisfaction of public which all their statutory obligations.

In Delhi Garbage case, the petitioner, B. L. Wadhera, sought directions to the Municipal Corporation of Delhi and the New Delhi Municipal Council to perform their statutory duties in particular the collection, removal and disposal of garbage and other waste. The Court observed that River Yamuna – the main source of drinking water supply is the free dumping place for untreated sewage and industrial waste. Apart from air and water pollution, the city is virtually an open dustbin. Garbage strewn all over Delhi is a common sight. It is no doubt that rapid industrial development, urbanisation and regular flow of persons from rural to urban areas have made major contribution towards environmental degradation but at the same time the authorities – entrusted with the work of pollution control – cannot be permitted to sit back with folded hands on the pretext that they have no financial or other means to control pollution and protect the environment.

In its earlier order dated December 16, 1994 the Court had directed the MCD and Delhi Development Authority (DDA) to place on record the list of all garbage dumping places and city garbage collection centres. They were also asked to state what steps were being taken by them to keep those places clean and tidy. The gravity of the garbage disposal problem can easily be understood by the affidavit filed before the Court by the Executive Engineer, MCD, dated January 30, 1995. It was mentioned that about 4,000 metric tons (MT) of garbage is collected daily by the MCD. The disposal of the garbage is done mainly by 'Land Fill Method'. The total number of garbage collection centres are 1804 (337 *dhalaos*, 1284 dustbins, 176 open sites and 7 steel bins). The garbage collection trucks collect the garbage from the collection centres and take it to the nearest Sanitary Land Fill (SFL). It was highlighted in the affidavit that "about 45 per cent of the total population of Delhi is living in slums; unauthorised colonies and clusters. There are about 4,80,000 jhuggies in Delhi. According to a rough estimate about 6 persons stay in each jhuggi. They throw their garbage on the road or nearby dustbins."

The Court passed another order in the case on September 15, 1995, where it was pointed out by the Court that the collection and disposal of garbage in the city of Delhi was causing serious problems. It was not for the Court to keep on monitoring such problems. The officers who were manning institutions like MCD and NDMC must realize their responsibilities and show the end result. Further the Court asked the petitioner and other learned counsel to assist the Court on next hearing regarding the statutory duties and functions of various authorities in regard to the sanitation in the city of Delhi. The officers concerned with the problem of sanitation were directed by the Court to consider various issues arising in the petition, at their own level and give information to the Court on the next date of hearing, i.e. October 12, 1995, regarding the final date by which they shall sort out the problem of collection and disposal of the garbage in the city. Authorities were also asked by the Court to place before the Court, the difficulties which were likely to come up in their way. But at the same time the Court made it clear that from the date which may be given by the authorities concerned, "not a drop of garbage is to be seen any where in the city of Delhi on early morning each day. The whole of the work of garbage collection must be completed over-night and the city is to be left absolutely clean of the residents for their use." The Court mentioned that MCD had a very large force of workers, it had 38,311 *safai karamcharis* and more than 1,400 Sanitary Inspectors to keep clean and tidy an area of 1399.29 sq. km. The simple arithmetic showed that there were 27 safai karamcharis and one sanitary inspector for one sq. km. of area. The NDMC was found by the Court in still better position, having 2,172 *safai karamcharis* for cleaning an area of 42.40 sq. km., i.e. 50 *safai karamcharis* for one sq. km The Court held that there was no reason whatsoever why with such a huge manpower at their command the MCD and NDMC cannot give a neat and clean Delhi to its residents.

After examining the relevant provisions of Delhi Municipal Corporation Act, 1957 (The Delhi Act) and the New Delhi Municipal Council Act, 1994 (The New Delhi Act), the Court opined that the MCD and the NDMC were under a statutory obligation to scavenge and clean the city of Delhi and that they had been wholly remiss in the performance of their statutory duties. Non-availability of funds, inadequacy or inefficiency of the staff, insufficiency of machinery, etc. cannot be pleaded as ground for non-performance of their statutory obligations. In the light of the facts and circumstances of the case and also keeping in view the suggestions made by the learned counsel assisting the Court in the petition, the Court issued following fourteen directions:

1. The experimental schemes placed by MCD and NDMC to distribute polythene bags and door to door collection of garbage and its disposal was approved by the Court.
2. Directions were issued to construct and install incinerators in all the Government administered hospitals and nursing homes, with 50 beds and above preferably within nine months.
3. The All India Institute of Medical Science (AIIMS), New Delhi was directed separately to install sufficient number of incinerators, or an equally effective alternate, to dispose of the hospital waste.
4. The MCD and NDMC were asked to issue notices to all the private hospitals and nursing homes in Delhi to make their own arrangements for the disposal of their garbage and hospital waste.
5. The Central Pollution Control Board (CPCB) and the Delhi Pollution Control Committee (DPCC) were assigned the job to inspect the different areas of Delhi to ascertain that the collection,

transportation and disposal of garbage and waste is carried out satisfactorily.

6. The Government of NCT of Delhi was directed to appoint Municipal Magistrates for the trail of offences under the DMC Act and the NDMC Act.
7. 'Doordarshan' was asked to undertake a programme of educating the residents of Delhi regarding their civic duties under the DMC Act and the NDMC Act.
8. The Ministry of Defence Production, Government of India was directed to have the already ordered Tippers supplied to the MCD as expeditiously as possible and preferably within three months.
9. The Development Commissioner, Government of NCT, Delhi was directed to hand over two sites, near Badarpur on Jaitpur pits and Mandi village near Jaunpur Quiry pits, to be used as SLF sites within three months.
10. Direction was issued to review and put into operation the compost plant at Okhla MCD was also asked to examine the construction of four additional compost plants as recommended by Jag Mohan Committee.
11. The MCD was instructed not to use the filled-up SLFs for any other purpose except forestry. MCD was directed to develop forests and gardens on all 12 such sites.
12. The MCD and NDMC were directed to construct or install additional garbage collection centres in the form of *dhalaos*, trolley and steelbins within four months.
13. The directions were issued to the Union of India and NCT, Delhi Administration to consider the request from MCD and NDMC for financial assistance, in a just and fair manner.

14. As disposal of garbage and solid waste by 'SLF' method may not be possible after some time due to non-availability of sites, the NCT Delhi Administration and also the MCD and NDMC were directed by the Court to join hands and engage an expert body like NEERI to find out alternate methods of garbage and solid waste disposal.

In order to ensure strict compliance of its directions the Court asked the concerned officer, authority or bodies to file affidavits before the Court regarding steps taken by them to fulfil the requirements of directions issued to them by the Court.

In a writ petition (C) no. 841 of 1988, B.L. Wadhera vs. Union of India and others, the Hon'able Supreme Court of India directed CPCB and the Delhi Pollution Control Committee (DPCC) to inspect different areas of Delhi and New Delhi to ascertain that the collection, transportation and disposal of garbage is carried out satisfactorily. The Court further directed CPCB and DPCC to file the reports by way of an affidavit bi-monthly. Since then, CPCB submitted twenty eight inspection reports to the Court which represented in depth analysis of waste management in Delhi and made several recommendations for consideration of MCD and NDMC to improve the current situation

The question of solid waste disposal once again came before the Court in Almitra H. Patel vs. Union of India through public interest litigation (writ petition no. 888 of 1996). In its order dated 16th January, 1998 the Court constituted a committee to look into all aspects of urban solid waste management and in particular to the following areas:

1. Examine the existing practices and to suggest hygienic processing and waste disposal practices and proven technologies on the basis of economic

feasibility and safety which the corporations/ government may directly or indirectly adopt or sponsor.

2. Examine and suggest ways to improve conditions in the formal and informal sector for promoting eco-friendly sorting, collection, transportation, disposal, recycling and reuse.
3. To review municipal bye-laws and the power of local bodies and regional planning authorities and suggest necessary modifications to ensure effective budgeting, financing, administration, monitoring and compliance.
4. Examine and formulate standards and regulations for management of urban solid waste, and set time-frame within which the authorities shall be bound to implement the same.

The local authorities and State Governments and Union Territories concerned were instructed by the Court to extend all cooperation and assistance to the committee for its proper functioning. The committee first submitted a preliminary and then the final report before the Court. Notices were issued to all the States who were required to file their responses to the report of the Committee. The Court noted that none of the States really opposed the recommendations made by the committee and their responses were in fact positive.

The Court in its order dated 15th October, 1999 indicated that it will take up the question of cleaning of four metropolitan cities, namely Mumbai, Chennai, Kolkatta and Delhi as also the city of Bangalore. During the proceedings of the case the Court came to know that the 14 directions issued by it in B.L. Wadhera's case have not been complied till date. The Court opined that keeping Delhi clean was not an easy task but then it was not an impossible one either. What was required was initiative, selfless zeal, dedication and professional pride. The Court pointed out that domestic garbage and sewage were large contributor of solid waste and the problem became more complex because of a large number of inhabitants living in unauthorised colonies and slums, having no proper means of dealing with the domestic effluence. The Court stressed upon the need to deal with the garbage and solid waste generated by slums most expeditiously and on the basis of priority. The effect of failure of the direction issued in Delhi Garbage case was clearly reflected in the judgment when the Court accepted its limitations and observed: "We believe it is not for this Court to direct as to how the municipal authorities should carry out their functions and resolve difficulties in regard to the management of solid waste. The Court, in fact, is ill-equipped to do so. Without doubt the Government agencies including the local authorities have all the powers of the State to take action and ensure that the city remains clean. They have only to wake up and act. The Court should, however, direct that the local authorities, Government and all statutory authorities must discharge their statutory duties and obligations in keeping the city at least reasonably clean."

Thus, the same Court which was averse to see "a drop of garbage any where in the city of Delhi an early morning each day" just four years ago settled for discharge of duties by local authorities and Government in keeping the city "at least reasonably clean" this time. The Court passed further directions which were in addition to and not in derogation of the orders passed by it in Delhi Garbage case. Those directions may be enumerated as follows:

1. The Municipal Corporation of Delhi, the Cantonment Board and all concerned officials were directed to ensure that the relevant provision of the DMC Act, 1957, New Delhi Municipal Council Act, 1994 and the

Cantonments Act, 1924 relating to sanitation and public health prohibiting accumulation of any rubbish, filth, garbage or other polluted obnoxious matters in any premises or prohibiting any person from depositing the same in any street or public place shall be scrupulously complied.

2. The streets, public premises such as parks, etc. shall be surface cleaned on daily basis, including on Sundays and public holidays.
3. The MCD, NDMC and other statutory authorities were authorised to levy and recover charges and costs from any person littering or violating provisions of the diverse Acts, bye-laws and regulations relating to sanitation and health, for violating the directions being issued by the Court.
4. The MCD, NDMC and other statutory authorities were directed to ensure proper and scientific disposal of waste in a manner so as to sub-serve the common good.
5. The Court directed that sites for landfills be identified bearing in mind the environmental considerations and the requirement of Delhi for the next twenty years within a period of four weeks. The sites so identified shall be handed over to the MCD or NDMC within two weeks of the identification, free from all encumbrances and without MCD or the NDMC having to make any payment in respect thereof.
6. Union of India, Government of NCT of Delhi, MCD, NDMC and other statutory authorities like DDA and Railways were directed to take appropriate steps for preventing any fresh encroachment or unauthorised occupation of public land for the purpose of devilling resulting in creation of a slum. Further direction was given

to take appropriate steps to improve the sanitation in the existing slums.

7. The Court further directed to identify and make available to the MCD and NDMC within four weeks eight sites for setting up compost plants. Such sites be handed over to the MCD and NDMC free of cost and free from all encumbrances within two weeks of identification. MCD and NDMC were asked to take appropriate steps thereafter to have the compost plants established or caused to be established and to be in operation by 30th September, 2000.
8. Direction was issued by the Court to regularly publish the names of concerned Superintendents of sanitation and such equivalent officers who are responsible for cleaning Delhi, who can be approached for any complaint by the citizens of Delhi together with their latest office and residential telephone numbers and address.
9. The Government of NCT of Delhi was directed to appoint Magistrate within a period of six week for each Board/Circle/Ward for ensuring compliance of the provisions of the MCD and NDMC Acts and to try the offence specified therefore, in relation to littering and causing nuisance, sanitation and public health (Plate 1).
10. The Court asked all the concerned authorities to file compliance reports of these directions within eight weeks. The Central Pollution Control Board was also directed to file within the same time an affidavit indicating as to what extent the directions issued have been complied with.

It is clear from aforesaid discussion that the Court has maintained it's view point that Municipalities, Nagar Mahapalikas and other similar local bodies must perform their duties to keep towns clean and tidy. The expert committees were constituted by the Court for solving the problems relating to sanitation and waste management. The Central and State Governments were also asked by

The Times of India, New Delhi
Thursday, April 14, 2005

14/4/05

TIMES

HC to MCD: City is open dustbin

1,500 Tonnes Of Garbage Lies Uncollected Every Day

TIMES NEWS NETWORK

New Delhi: The stink of the garbage heaps accumulating on the streets of Delhi has struck a chord with the Delhi High Court, with the judges rapping MCD hard on the knuckles by observing that the city has become an "open dustbin".

The court has also ordered MCD to file a status report on the general state of sanitation in the city, manpower deployment and steps taken to improve efficiency within four weeks. The civic body may not be able to duck behind procedural issues to cover up for its slackness.

A division bench comprising Chief Justice B C Patel and Justice Sanjay Kishan Kaul made these observations during the hearing of a PIL filed by an NGO. The judges were irked over MCD's failure to comply with a February 2, 2005, order to submit a status report on the matter.

While MCD has been devising various schemes for "scientific disposal and treatment of solid waste" and efficient transportation of garbage from colony *dhalaos* to municipal sanitary landfill sites (SLFs), the corporation to date manages to leave over 1000 tonnes of trash in bins overnight.

This is apart from the garbage generated in unauthorised colonies where MCD does not collect refuse. The city generates over 6,500 tonnes of garbage, and only 5,000 tonnes reaches the SLFs.

For unauthorised colonies, the corporation had devised a plan to form co-operatives for youths working as sanitary workers under contractors. These co-operatives would, in turn, have been awarded tenders for sanitation in unauthorised colonies. This plan, too, ran into political rough weather.

The civic agency recently prepared a master plan for municipal waste disposal and treatment, which has suggestions for effective garbage disposal. But getting a grip on the 50,000-strong army of mostly recalcitrant sanitary workers has been an uphill task for the MCD.

Most of these workers have also been protesting against the MCD move to involve private players and NGOs for collection and transportation of garbage in a bid to improve services.

"One of the reasons we have been pushing plans like private transportation of garbage and door-to-door collection of garbage in planned and unauthorised colonies by NGOs is that the municipal staff is unwilling to do its job. At any given time, 30% of the *safai karamcharis* are on leave," said a senior MCD official.

The privatisation bid has also been stiffly opposed by the political wing of MCD, which brands the move a ploy to outsource the civic agency's functioning completely. While MCD officials admit that both the 50,000-strong sanitation staff and 500 working garbage trucks are adequate for Delhi, they assert that major restructuring is required for any change to creep in.

The petitioners had also brought to the court's notice former Chief Justice of India B N Kirpal's order, decreed in 2000, stating that mobile magistrates should be appointed to issue challans against those found littering.

Although MCD had appointed magistrates, the agency failed to comply with the directions stating that the commissioner, inspectors and assistant sanitary inspectors have to take effective steps to prohibit accumulation of garbage on the roads.

Plate 1: High Court's Comment on Status of Solid Waste Management

the Court to help local bodies in maintaining proper sanitation and waste management facilities (Tiwari, 2006).

Conclusion

Institutional arrangements for environmental protection cannot exist without appropriate legal foundations. Lack of professionalism in administration, weak financial position and inefficient institutional arrangements are the main problems in solid waste management sector. There are number of rules, regulations, legislations and policies devised for improving the situation of solid waste management stream. But laws do not have sufficient provisions to deal with the situation effectively. Present rules and regulations are inadequate both in terms of assessing environmental impacts of waste and its economic and social implications. Supreme Court of India had to direct many times to concerned local bodies for managing and processing the waste scientifically. In spite of implementing numerous plans, regulations, initiatives, desired success has not taken place so far.

References

Agarwal, A. *et al.* (2005), "Municipal Solid Waste Recycling and Associated Markets in Delhi, India", *Resource Conservation and Recycling*, Vol. 44, pp. 73-99.

Anand, Subhash (2005), "Solid Waste Generation and Management in Delhi: A Sustainable Approach" in Singh, J. (ed.), *Environment and Development*, I.K. International Publishing House, pp. 621-639.

Anand, Subhash (2010), "Recycling of Solid Waste Materials" in Singh, J. and A.L. Ramanathan (eds.), *Solid Waste Management*, I.K. Publishing House, pp. 76-83.

Anand, Subhash (2010), "Role of Waste Pickers in Urban Solid Waste Management in Delhi", in Dubey, R.N. (ed.), *Urbanisation and Urban Planning in Delhi*, Shree Natraj Prakashan, New Delhi, pp. 303-310.

Ansari, J. H. (1999), "Solid Waste Management in Delhi – Need for Partnership Arrangements", Congress Paper at 27th National Congress, 19-22 September.

Asian and Pacific Centre for Transfer of Technology (1997), *Waste Minimisation*, Circular of Sub-Regional Training Workshop organised by APCTT, New Delhi.

Beukering, Pieter Van (1994), "The Recycling Sector in Bangalore: An Economic Analysis of Different Types of Formal and Informal Entrepreneurs, Recovering Urban Solid Waste" in Baud, I. and Sachenk, H. (eds.), *Solid Waste Management*, Manohar Publishers, New Delhi, p. 108.

Bhagidari (2002), *Delhi Government*, Delhi, pp. 4-6.

Central Pollution Control Board (2003), *A Support Manual for Municipal Solid Wastes (Management and Handling) Rules, 2000,* CPCB, Delhi, p. 1.

Central Pollution Control Board. *Management of Municipal Solid Waste*, CPCB, Delhi, pp. 8-9.

Cointreau, S. J. (1987), *Living with Garbage: Cities Learn to Recycle in Development Forum*, Jan-Feb, pp. 12-13.

COWI-Kadam (2004), *Feasibility Study and Master Plan for Optimal Waste Treatment and Disposal for the Entire State of Delhi based on Public Private Partnership Solutions*, Vol. 4, COWI-Kadam, Delhi.

Datta, M. (1996), *Waste Disposal in Engineered Landfills*, Narosa Publishing House Pvt. Ltd., India.

Dhamija, Urvashi (2006), *Sustainable Solid Waste Management: Issues, Policies and Structures*, Academic Foundation, New Delhi.

Dreze, J. and Sen, A. (1995), "Basic Education as a Political Issue", *Journal of Educational Planning and Administration*, Vol. 9, p. 2.

Dwivedi, O. P. (1997), *India's Environmental Policies, Programmes and Stewardship*, Macmillan Press Ltd., Great Britain, pp. 31-45.

Girard, James E. (2005), *Principles of Environmental Chemistry*, Jones and Bartelett Publishers, USA.

Gotosh, S. (1989), "Issues and Factors to be Considered for Improvement of Solid Waste Management in Asian Metropolies", *Regional Development Dialogue*, 10(3).

Govt. of NCT of Delhi (2001), *Delhi Urban Environment Infrastructure Improvement Project*, Delhi-21 Report, GNCTD, Delhi, p. 50.

Graham, Hilary (1984), *Women, Health and the Family*, Wheatsheaf Books Ltd., Britain, p. 39.

Indian Environmental Society. *Solid Waste Management: An Introduction*, IES, New Delhi, p. 14.

Indian Express (2006), "NDMC Reaches out for Private Hand to Clean up Capital", *Indian Express*, Daily Newspaper, September 13, Delhi.

Indian Society of Environmental Management (2000), *Development of Statistics in Environment Sector: Solid Wastes*, ISEM, New Delhi, p. 175.

Khan, M.U. (2004), "Regulations and Management of Municipal Waste: Indian Scenario", *NIST ADS News*, Vol. 6, No. 1, p. 43.

Krishna, D. and Sangeetha, T. L. (1994), Approaches to Urban Solid Waste Management: Linkages between Formal and Informal Systems of Source Separation and Recycling – A Profile of Hyderabad, Discussion Paper for the Workshop on "Linkages in Urban Solid Waste Management", Bangalore, pp. 39-41.

Mohan, S. (2006), "Technologies for Sustainable Urban Development – A Study of National Capital Region" in Singh, R. B. (ed.), *Sustainable Urban Development*, Concept Publishing Company, Delhi, p. 176.

Municipal Corporation of Delhi (2004), *Bhagidari Workshop on Municipal Solid Waste Segregation*, Report 8-10 Jan., MCD, Delhi, pp. 13-15.

Nagar Palika Network (2002), *Database on Municipal Governance in Some Major States*, All India Institute of Local Self Government, New Delhi, October.

Naresh, Gautam (1996), "The Municipal Corporation of Delhi: A Fiscal Analysis", *Urban India*, Vol. XVI, No. 1, Jan-June, pp. 49-50.

NISTADS (2004), "National Institute of Science Technology and Development Studies", *Official Newsletter*, Vol. 6, No. 1, April, New Delhi.

Pal, Surinder and Dass, Ravi (2004), *Feasibility Study and Master Plan for Optimal Waste Treatment and Disposal for the Entire State of Delhi based on Public Private Partnership Solutions – A Review*, MCD, Delhi.

Parker, S. P. (1980), *Encyclopedia of Environmental Sciences*, Second Edition, McGraw Hill Books, New York.

Planning Department (2002), *Economic Survey of Delhi, 2001-02*, Govt. of NCT of Delhi, Delhi.

Sarkar, Papiya (2003), "Solid Waste Management in Delhi – A Social Vulnerability Study in Martin", J. Brunch *et al.* (eds.), *Proceedings of the Third International Conference on Environment and Health*,

Chennai, 15-17 Dec., Department of Geography, University of Madras and Faculty of Environmental Studies, New York, pp. 451-464.

Schubeler, Peter (1996), Conceptual Framework for Municipal Solid Waste in Low-Income Countries, Working Paper No. 9, SKAT, Switzerland, pp. 47-48.

Singh, R. B. and Subhash Anand (2008), "Current Status of Municipal Solid Waste Management Practices in Delhi", *Annals of NAGI*, Vol. XXVIII, No. 1, pp. 36-50.

Singh, Satpal (2006), *Solid Waste Management in Resettlement Colonies of Delhi*, Bookwell, New Delhi, pp. 203-207.

Tchobanoglous, G. *et al.* (1993), *Integrated Solid Waste Management: Engineering Principles and Management Issues*, McGraw Hill, Inc., New York.

The Hindustan Times (2002), "The Battle Against Waste", *The Hindustan Times*, Daily Newspaper, October 17, Delhi.

The Hindustan Times (2004), "Garbage can Earn Millions", *The Hindustan Times*, Daily Newspaper, October 30, Delhi.

The Hindustan Times (2005), "In the Wrong Bin", *The Hindustan Times*, Daily Newspaper, January 1, Delhi.

The Hindustan Times (2006), "Garbage Outsourcing Plan Falls Flat", *The Hindustan Times*, Daily Newspaper, July 23, Delhi.

The Hindustan Times (2007), "Delhi Ranks Poorly in Solid Waste Management", *The Hindustan Times*, Daily Newspaper, Feb. 21, Delhi.

The Hindustan Times (2007), "Swipe at Plastic", *The Hindustan Times*, Daily Newspaper, February 22, Delhi.

The Times of India (2006), "Waste Management in a Complete Mess", *The Times of India*, Daily Newspaper, Aug. 18, Delhi.

Tiwari, A. K. (2006), *Environmental Laws in India*, Deep & Deep Publication Pvt. Ltd., New Delhi, pp. 324-333.

United Nations Environment Programme (1999), *International Source Book on Environmental Sound Technologies for Municipal Solid Waste Management*, accessed at *http://www.unep.or.jp/ietc/EST dir/pub/MSW.*

US Environmental Protection Agency (2004), *Municipal Solid Waste Facts*, USEPA, accessed at *http://www.epa.gov/epao/swer/nonhw/muncpl/facts. htm.*

Venkatesan, R. (2006), "A Menance on Islands – Disposing Plastics", *Science Reporter*, Vol. 43, No. 6, June, pp. 34-36.

Vogler, J. (1998), *Work from Waste: Recycling Wastes to Great Employment*, Oxfam.

White, R. and Whiteny, J. (1992), "Cities and the Environment: An Overview" in Stren, R. E. *et al.* (eds.). *Sustainable Cities Urbanisation and the Environment in International Perspective*, Westwise Press, Boulder, pp. 23-24.

Anthropogenic Stimulation to Climate Change: Integration of Cellular Automata, GIS and Remote Sensing in Village Information System (VIS)

SWARNIMA SINGH

Introduction

The history of development is long and diverse series of steps by which man has preceded his needs as well as greed at the cost of environment. Human societies over the ages have depleted natural resources and degraded their local environments. Population have also modified their local climates by cutting down trees, occupying cultivable land and converting it into towns and cities. It is not a new phenomenon, it is occurring through ages but the concern is being felt to reduce this threat to create a sustainable life. It is now apparent that human activities are deranging the climate system at the global scale. Climate change is likely to have wide-ranging and potentially serious consequences resulting from disturbances to complex ecological processes.

Global climate change is caused by the accumulation of greenhouse gases in the lower atmosphere. The global concentration of these gases is increasing, mainly due to human activities, such as the combustion of fossil fuels (releases carbon dioxide), deforestation (because forests remove carbon from the atmosphere). The atmospheric concentration of carbon dioxide, the main greenhouse gas, has increased by 30 per cent since preindustrial times. Projections of future climate change are derived from global climate model or General Circulation Model (GCM) and regional circulation model (RCM) experiments. Climatologists of the Intergovernmental Panel on Climate Change (IPCC) review the results of these experiments for global and regional assessments. It is estimated that global mean surface temperature will rise by 1.5° to 3.5° C by 2100. This rate of warming is significant. Large changes in precipitation, both increases and decreases, are forecast largely in the tropical countries. Climate change is very likely to affect the frequency and intensity of weather events, such as storms and floods, around the world and will also cause sea level rise due to the thermal expansion of the oceans and the melting of the mountain glaciers. Global mean sea level is anticipated to rise by 15 to 95 centimetres by 2100. Sea level rise will increase vulnerability to coastal flooding and storm surges. The faster the climate changes, the greater will be the risk of damage to the environment.

Objectives

The chief objective of the current work is to develop integrated village information system by providing complete information for decision-making based on their existing resources and capabilities and to create a dynamic data management system for mitigating micro level factors which are likely to induce climate change.

Methodology

This is the high time to think globally and act locally. Therefore, micro level information and its integration with regional level planning can play a major role in global hue and cry across the world. Planning is panoramically accepted as a way to handle complex problems at local and regional level. It is indispensable to bring information technology in order to provide a more effective and better planning and development. Hence the need for a suitable information system is increasingly being felt in all levels of planning and developmental activities in rural areas. Planners and decision-makers at micro level have to depend upon spatial and non-spatial data for optimization of the location of basic infrastructure. This paper tries to justify the idea of incorporating cellular automata, Geographic Information System into micro-level planning and usage of spatial database, which will enhance the effectiveness and efficiency of such plans (Figure 1).

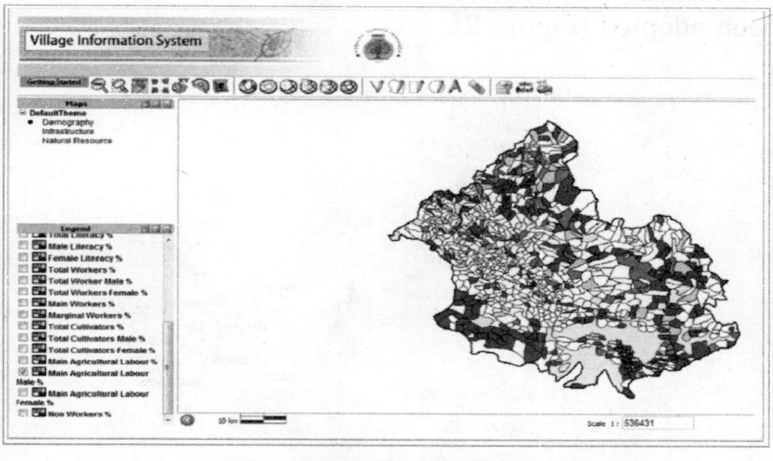

Figure 1: A Detailed Outline of Village Information System (VISs)

This also attempts at encompassing and addressing all the factors necessary to develop village information system (VIS). More than 62 per cent people in India still lives in villages, mapping and monitoring the information at micro/village level is still in amorphous stage. VIS is a Geographical Information System (GIS) based application, which provides detailed information pertaining to demography, infrastructure and natural resources for every village, district and state. It displays geo-referenced maps of virtually any political boundary in India – from village to a district, a state or the entire country with data of about 175 parameters from census of 1991 and 2001. For this kind of analysis, single criterion or multicriterion procedures are involved.

There are two approaches on which a micro level study can be conducted; one is Agro-Climatic Zone (ACZ) and the other one is Agro-Ecological Zone (AEZ). India has been divided in to 20 AEZ and these are uniform in terms of physiography, climate and soil type. Therefore, AEZ approach of selecting representative villages for monitoring, mapping and tracking land use changes has been adopted (Figure 2).

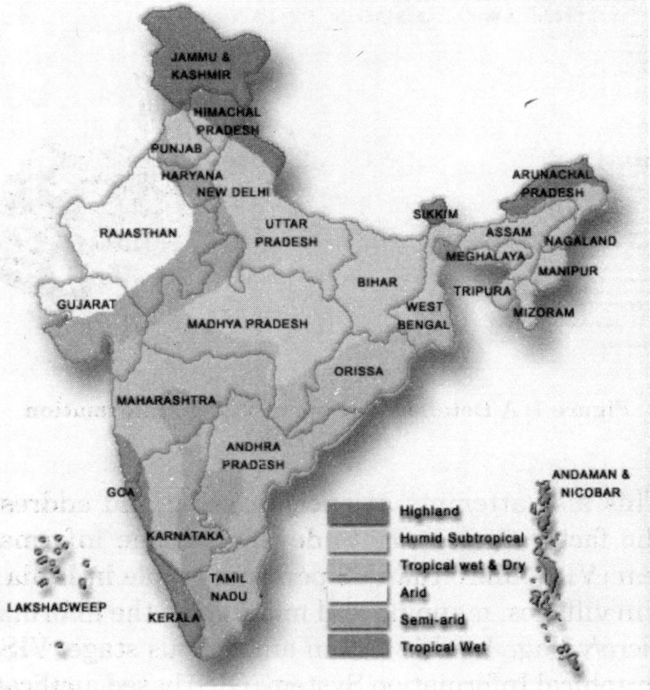

Figure 2: Agro-Ecological Zones of India

The offline data management package, Village Natural Resource Data Management System (VNRDMS) is the product of rigorous research involving identification of indicators, development and field testing of methods across the states. After identification, the second stage is incorporation of Remote Sensing and Geographic Information System. The emergence of Remote Sensing and Geographic Information System has acted as a powerful tool for spatial analysis and storage. It has alleviated the problem by computerization of the large spatial data in less time. Thus, the main objective of the study is to develop an integrated village information system through Cellular Automata (CA) and GIS interlinking. For providing complete information to the decision making based on their existing resources capabilities and changes. CA has been taken in to consideration. Although information system has been developed in order to generate maps, but CA gives an added advantage to create a dynamic analysis. A suitability-based cellular automata model has been developed to simulate land use change dynamics through the concepts dynamic phenomenon. Various climatic problems caused by fossil fuels burning, land degradation, land use change are dynamic phenomenon therefore only dynamic solution can solve the problem. The flow chart in figure 3 shows the elaborated as well as crisp outline of the micro climate modeling in the study. It can be used as a useful planning tool to test the effects of different land use change scenarios at micro level.

Geographic Information Systems (GIS) provide rich spatial databases but have been traditionally static. The coupling of dynamic models to GIS provides an insight to the evolution of spatial phenomena as discussed by Grossman and Eberhardt (1992).

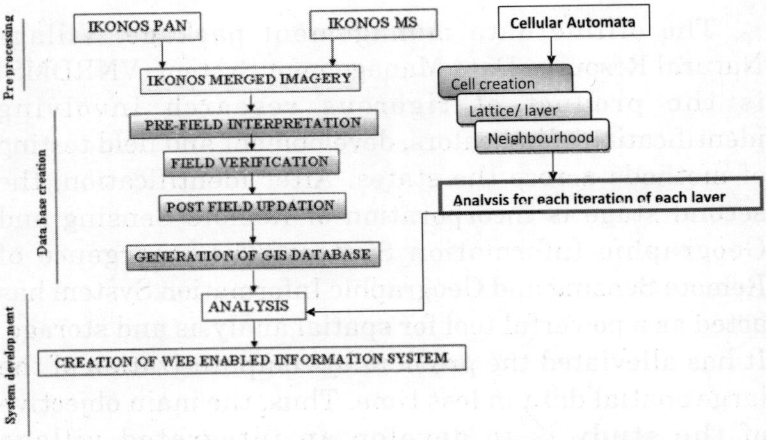

Figure 3: Methodological Flow Chart Describing the Integrated Output of Remote Sensing, GIS and Cellular Automata

Village Level Information System and Role of GIS

Research in science has paved a path to many brilliant inventions and discoveries for the sustainable development of the community. With a firm belief in this principle, Indian Space Research Organisation (ISRO) envisage to extend its expertise and achievements in the spheres of space science and information technology for the rural development and initiated the Village Resource Centres (VRCs) concept. This step has been put forward to set up of the Village Information Centres (VICs) with a view to integrate its capabilities in satellite communications and satellite based earth observations to propagate a variety of services. The information centre growth and development at micro level planning is definitely a stride for reaching to the unreached. The VIC is a totally interactive VSAT (Very Small Aperture Terminal) based network. These nodes can be further extended using other technologies like Wi-Fi, Wireless and Optical Fibre. These extensions may serve as local clusters around the areas where the VRC is located. Originating from the remote system technology to address the changing and critical needs of rural communities and salubrious sustainable development is definitely an aid for integrated/entwined multi level planning.

There are few major steps involved in creation of village information system. For example to make information system more interactive and elaborative Dehradun district of Uttarkhand has been taken as a case study to show the major steps for developing VIS, these are;

(i) **Generation of Spatial Village Maps:** District and taluka boundaries were drawn from the SOI topographical maps. These were brought in as ArcView or ArcInfo coverage after following the standard procedure: digitization, geo-referencing, etc. Taluka maps, collected from the local government offices in the district, contain village boundaries with contour lines and other topographical information. These taluka maps, which do not show geographical co-ordinates, were traced, marking GCPs with respect to the SOI base map, digitized and brought to the real world coordinate system after projecting the maps and carrying out the editing functions such as appending, edge-matching, etc. in Arc/Info. Each village in this map was assigned unique ids (user-defined) in a regular sequence. Thus, taluka map with village boundaries with in-built table having aerial extent, village-ids, etc., was generated for selected Dehradun district and its villages (Figure 4). The village map was opened in ArcView 3.2 for further processing.

(ii) **Transformation of Non-spatial Census Database into GIS:** 2001 Census data collected, census information can be displayed using the customized NIC software. This file contains vast

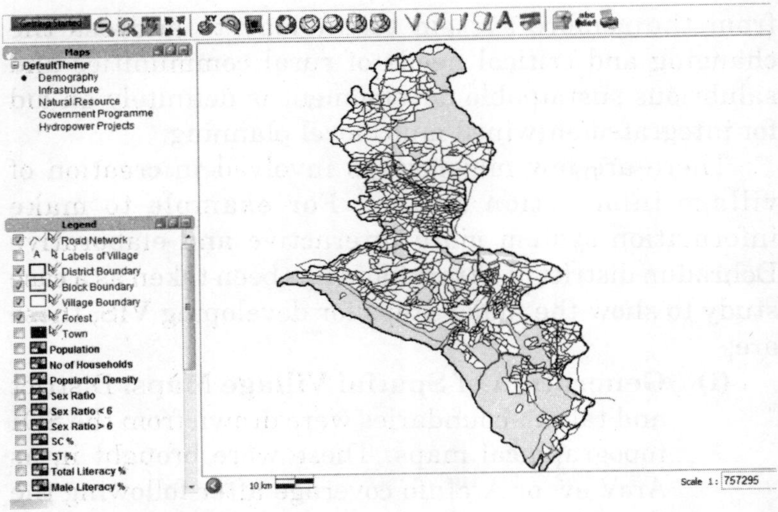

Figure 4: GIS Interactive Map of Demographic theme for Dehradun District of Uttarakhand

information on physical, social and economic characteristics of each village, including census-codes on region, district, taluka and village. The dbf file was later displayed in MS Excel sheets. Village-ids, same as given in the spatial village map, were fed in the Excel sheets for all the villages in the district. Thus, spatial and non-spatial data contain similar column in their respective Tables with uniform village ids. These ids will help in joining the spatial and non-spatial data in the system.

(iii) **Integration of Spatial and Non-spatial Database:** The MS Excel file was converted into text format to enable to open in ArcView. The Tables of both spatial village map and non-spatial census information were opened in ArcView and joined together, with the help of user-defined ids, using table-join function.

Thus, an information system has been generated for the district showing the village map with its boundaries and the relevant census information containing physical, social and economic dimensions. Generalized processing flow for developing VLIS model is depicted (Figure 5).

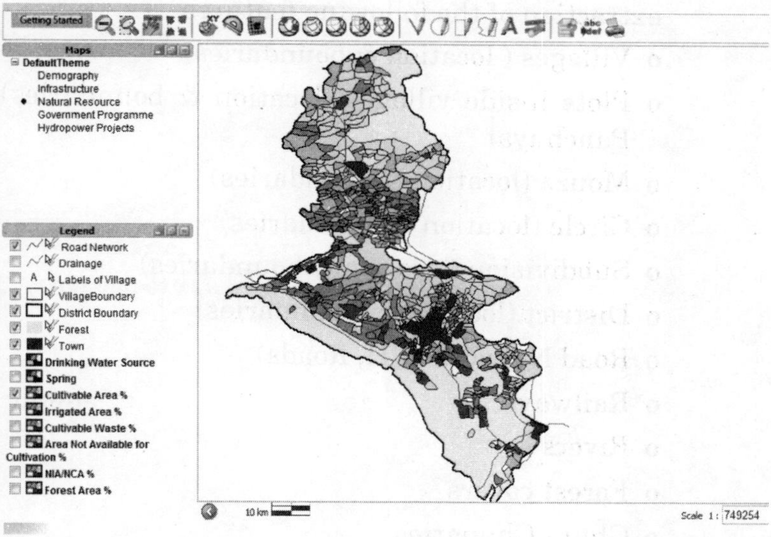

Figure 5: GIS Interactive Map under Natural Theme for Dehradun District of Uttarakhand

(iv) **Generation of Views/Scenarios for Decentralized Planning:** In general, the rural development schemes are on watershed basis. Each scheme is bounded by certain government policies, which have social, economic and biophysical dimensions. Policy is enshrined in their directives that establish the scheme and these commonly lay down the criteria for site selection.

(v) **Methodology used for Spatial Analysis:** Micro level planning involves creation of spatial database and then the establishment of linkage between the spatial database and its related attributes. Customized application software is designed and developed to use this linkage and present the planners the output as desired by them. For this purpose, various base maps with village and plot level details are digitized for extraction of the following features.

- o Villages (location & boundaries)
- o Plots inside villages (location & boundaries) Panchayat
- o Mouza (location & boundaries)
- o Circle (location & boundaries)
- o Subdivision (location & boundaries)
- o District (location & boundaries)
- o Road Network (NH, Roads)
- o Railway
- o Rivers
- o Forest covers
- o Char / Chaparies
- o Land use
- o Agriculture

Micro-level modelling attempts to analyse patterns of land use change as the aggregate outcome of many disparate individual land use decisions across space. In this case space may be represented as networks or cellular automata (White and Egelen, 2000). The advantage of micro-level modelling is that it provides sufficient resolution to represent land use decisions (by agents or individuals) in terms of discrete choices of land usage and location selection (Schotten *et al.*, 2001).

The difficulty is that there are potentially an explosive number of actors and interactions needed to account for any land use decision. The inter-play between the two levels reveals an interesting and complex relationship: macro-level patterns emerge from micro-level processes and behaviours which are highly controlled by macro-level constraints. Some developmental activities like establishment of new schools, hospitals and other amenities could be effective done through GIS so that the public can avail the optimum path of road, railways, canals etc. Further, incorporation of VIS would amplify its scope manifold. This new technology can reduce the time and cost to the planners in organising the data in arriving at precise conclusion and decisions at their doorstep. Land records and facilities available with that area of land can be put in digital form which can be quickly updated. These can be assessed by an user and the latest information can be given in a shortest time. These will surely reduce the litigations pertaining to land and related issues. This paper presents a case study of Web-GIS working as a catalyst in the resource planning in Dehradun with the defined objectives as follows:

1. To build dynamic Model for Micro level decentralized planning for sustainable development.
2. To provide a web-based platform for collaboration and data sharing between specialists, planning agencies, citizens, and private entities
3. To provide a better tool for understanding micro-level climate change
4. Provide scientific approach to facility planning
5. To provide better management of resources and facilities.
6. Enable faster response to changing ground realities and prompt corrective action.
7. To plan and monitor various Govt. Schemes and their covered span of area.

8. To provide faster way of extraction of plotwise detailed information of the land for land related issues.

A few physical, social and economic views or scenarios for identifying selected villages for different rural watershed management schemes were generated for sub-division of Dehradun district.

1. Total population
2. SC/ST population
3. Total geographical area
4. Bank facilities
5. Land utilization →
    - Irrigated areas
    - Net area sown
    - Forests
    - Fallow lands

Likewise, the user can generate various views by themselves depending upon their needs, using the VLIS module, to make plans and take appropriate rural development decisions. VLIS could be customized to make it friendlier interface using the scripts available in Arc/View. Integrated CA-GIS approaches can enhance the current poor spatial dynamic modeling capability of GIS (Li *et al.*, 2001). This can act as a good platform to model micro level climate change.

Merits of Interactive GIS and CA in VIS

1. Spatial resource database with infrastructure attributes is created for the Dehradun for improved decision making in an interactive way.
2. It can facilitate decentralized micro level planning for sustainable development of the region.
3. It can provide a way scientific approach to facility/ infrastructure planning and enable faster response to

ganging ground realities thereby making planners to take prompt proactive actions.

4. The study shows that Web-GIS is a very user friendly and low cost solution for making Spatial Analysis at micro level for sustainable development of the region.

Conclusion

VIS tool has demonstrated its potential for grass-root level development planning taking into consideration the local needs and constraints. It has also established its usefulness to the decision makers in the district to generate views or scenarios for decision-making at local level. Integrated CA-GIS approaches can enhance the current poor spatial dynamic modeling capability of GIS. CA models can be completely developed within GIS for easily accessing the information stored in the GIS database during the modeling processes. Constraints for modeling can be defined using GIS and remote sensing data. Remote sensing can be used to obtain land use and other land related data, and this data can be transformed, so that it can be used in GIS for analysis and modeling. Therefore, the development of CA within GIS greatly enhances the ability of dynamic spatial modeling within GIS.

Achieving prediction accuracies of the order of 85 per cent is a significant task, as dynamic terrain-related processes are complicated and there could be a possible omission of some of the driver variables, which could not be recorded or monitored. The integrated CA-GIS framework proves to be a promising environment, wherein a variety of spatial-dynamic phenomena can be modelled. This interactive GIS-CA tool will serve as a first step towards the development of Decision Support System for decentralized planning at village, district or sub-district level. With a better database, we can provide a better service to the user organizations to assist in their own decision-making process for developmental planning. A few possible future development programmes for better service to the users for decentralized planning. The model developed for simulating the spatial dynamic process can be used as a planning tool to test the effects of different land use change scenarios. Cellular Automata are seen not only as a framework for dynamic spatial modeling, but also as a paradigm for thinking about complex spatial-temporal phenomena and an experimental laboratory for testing ideas.

References

Ali, A. S. M., and Sato, Y. (2001), How Can GIS and RS Support the Decision Making for Rural Development Programs in Egypt; Paper Presented on the 22nd Asian Conference on Remote Sensing.

Batty, M. and Xie, Y. (1994), From Cells to Cities. Environment and Planning B: Planning and Design, 21, 531-548.

Sarma,B., Reddy, D.Y., Rao, V. B., and Gohain, M. (2003), Development of a Web-based Land Information System using integrated GIS and Remote Sensing Technology for Guwahati City, India.

Burrough, P. A. (1986), Principles of Geographic Information System for Land Resource Assessment; Oxford University Press; New York.

Couclelis, H. (1997), From Cellular Automata to Urban Models: New Principles for Model Development and Implementation. Environment and Planning B: Planning and Design, 24, 165-174.

Deadman, P., Brown, R. D., and Gimblett, H. R. (1993), Modeling Rural Residential Ettlement Patterns with Cellular Automata. *Journal of Environmental Management*, Vol. 37, 147-160.

Deekshatulu, B. L., Krishnan, R., Novaline, J. (1999), Spatial Analysis and Modelling Techniques – A Review. Proceedings of Geoinformatics - Beyond 2000 Conference, Published by Indian Institute of Remote Sensing (1999), DehraDun, India, 268-275.

Grossman, W. D. and Eberhardt, S. (1992), Geographical Information Systems and Dynamic Modeling. The Annals of Regional Science, Vol. 26, 53-66.

Howarth, P. J. (1986), Landsat Digital Enhancements for Change Detection in Urban Environment. *Remote Sensing of Environment*, Vol. 13, 149-160.

Jain, S., and Jain, R. K. (2003), An Integrated Information System for Urban and Regional Planning: Some Suggestions. Spatio-Economic Development Record (SDR), Vol. 10, No. 2, pp. 16-21.

Li, X., and Yeh, A.G.O. (2001), Calibration of Cellular Automata by using Neural Networks for the Simulation of Complex Urban Systems. *Environment and Planning A*, 33, 1445-1462.

Li, X., and Yeh, A.G.O. (2002), Neural-network-based Cellular Automata for Simulating Multiple Land use Changes using GIS. *International Journal of Geographical Information Science*, Vol. 16, No. 4, 323-343.

Sokhi, B.S.,and Ramesh, B. (2003), Physical Improvement of Jharpada Village; Jawahar Publishers and Distributers, pp. 111.

Schotten, K., Goetgeluk, R., Hilferink, M., Rietveld P., and Scholten, H. (2001), Residential Construction, Land Use and the Environment- Simulations for the Netherlands Using a GIS-based Land Use Model, *Environmental Modeling and Assessment*, 6, 133-143.

Star, J., and Estes, J. (1990), Geographic Informations System, an Introduction; Prentice-Hall;Inc.; New Jersey.

Subudhi, A. P., Sokhi, B. S. and Roy, P.S. (1999), Remote Sensing and GIS Application for Urban Studies; IIRS Publication.

Tahir, M., Hussain T. (2003), Environment, Ecology and Natural Resource Management; Jawahar Publishers and Distributers.

White, R., and Engelen, G. (2000), High-resolution Integrated Modelling of the Spatial Dynamics of Urban and Regional Systems. Computers, *Environment and Urban Systems*, Vol. 24, 383-400.

Wu, F. (1998), An Experiment on the Generic Polycentricity of Urban Growth in a Cellular Automata City, Environment and Planning B: Planning and Design, Vol. 25, 103-126.

Environmental Degradation: A Case Study of Water Quality of River Yamuna in Delhi

VANEETA CHANDNA

Introduction

Water has no alternative – it is known as life which is essential for the sustenance of living organisms including plants, animals and plants. It is regarded as one of the most exploited natural resources since the dawn of human civilization. Demand for water has been increasing steadily over the years from various sectors, and supplies are decreasing, thus creating ever increasing mismatch between these two. With the vertiginous growth of population followed by industrial revolution and concomitant urbanization particularly in river basins, the demand of water has tremendously gone up on the one hand and indiscriminate discharge of domestic and industrial effluents on the other. However, due to uneven distribution of water resource, increase in demand and unplanned use, the planning and management of scarce water resource has become a key issue in this century. Water is plentiful, but pure water is scarce.

Study Area

River Yamuna is the largest tributary of river Ganga and the main source of fresh water in north India. Originating from the Himalayas, its total length is 1376 Kilometers covering catchment area of 3,66,220 square kilometer. The river Yamuna occupies a unique position in the cultural ethos of India. It is one of the most sacred rivers in our country after river Ganga. It is also a life supporting system for the people. Most of the civilizations flourish along this river. The density of population is dense along this river.

The river Yamuna after originating from the lower Himalaya enters the plains after flowing through seven states of north India. It enters Delhi about a mile north of Palla village at an altitude of 690 feet above sea level. Delhi is situated between the Himalaya and the Aravalli ranges in the heart of the Indian sub-continent. Delhi is situated on the banks of river Yamuna between latitude 28° 23'17" North to 28° 53'00" North and 76° 50'24" East to 77°20'37" East.

As it reaches Delhi, it is located between the two barrages, Wazirabad and Okhla barrage. It traverses a distance of about 32 kms in the NCT of Delhi. It leaves at a point east of Jaitpur, south of Okhla at an altitude of 650 feet. At Okhla the progress is checked by the head works of the Agra Canal. Some 19 drains open into the Yamuna polluting the entire stretch in the NCT of Delhi. Out of the total area in Yamuna River catchment, 1,485 sq. km. or 0.04 per cent of the area falls in the National Capital Territory of Delhi.

Water Quality of River Yamuna: Causes

Conceptually, water quality refers to the characteristics of a water supply that will influence its suitability for a specific use, i.e. how well the quality meets the needs of the user. It is defined by certain physical, chemical and biological characteristics. It is highly variable, from place to place and from time to time, even within a particular river system. It is dependent on many factors, both natural and as a consequence of human activities. According to Central Pollution Control Board, it is categorized as point and non point sources. The point source of pollution is measurable and includes sewage discharge into the river due to generation of domestic and industrial waste. The non point sources are the non measurable ones which include agricultural runoff, mass bathing and floral offerings during festivals, open defecation and dead bodies etc. Any change in the dynamic equilibrium in aquatic ecosystem disturbs the normal function and properties of pure water and gives rise to the phenomena of water pollution or deteriorating water quality. The main factors influencing the water quality are as follows:

Minimum Flow

The quantum of flow in rivers including the river Yamuna and its quality are inter-related. Impact on river water quality resulting from discharges of treated or untreated waste water into the river will depend upon the dilution offered by the quantum of flow in the river. A minimum flow of 10 cum/sec in the recipient river is required in order to maintain the desired water quality. The in stream uses of water specially in our country, such as religious and mass bathing, regular bathing and washing also require adequate flow to be maintained so that the pollution of the river caused by such uses can be kept within acceptable limits.

Agricultural Runoff

Agricultural water pollution is caused by use of fertilizers, insecticides and pesticides, farm animal wastes and sediments. The chemicals used in fertilizers enter the ground water by leaching and the surface waters by run-off. Large quantities of commercial fertilizers are applied during agriculture in the Yamuna basin, especially in the northern region. According to Central Pollution Control

Board the use of fertilizers has gradually increased from approximately 12 kg/ha in the 1970's to 61 kg/ha in the 1990's. The transport of fertilizers and pesticides' residues from agricultural fields to the river occur through the transport medium, which is mainly water available through rain and flooding the fields.

Domestic Sewage

Domestic sewage refers to waste water that is discarded from households. It is also referred to as sanitary sewage containing a wide variety of dissolved and suspended impurities. The main organic materials are food and vegetable waste, plant nutrient from chemical soap, washing powder etc. Domestic sewage is also very likely to contain disease causing microbes. Besides these, total salts, chlorides, nutrients, detergents, oil and grease etc. are also contributed by the domestic sources. Due to the presence of unauthorized colonies along the river, the dumping of untreated waste water is more significant and has become a major non-point source of pollution. When sewage is discharged into a stream of water the organic nutrients in the sewage help bacteria to grow very fast. Due to this enhanced activity, air dissolved in water becomes poorer and poorer in oxygen. Hence the aquatic life like fish cannot survive.

The quality of water at the point where the river leaves Delhi has deteriorated drastically with large amounts of untreated sewage falling into the river. Only 55 per cent of the 15 million Delhi residents are connected to the city's sewage system which empties itself into the river.

The River Yamuna enters Delhi at Palla village 15 km upstream of Wazirabad barrage, which acts as a reservoir for Delhi. Delhi generates 3,296 million litres per day (mld) of sewage, against an installed waste water treatment capacity of 2,330 mld. Thus, 966 mld of untreated and a significant amount of partially treated sewage enter the river every day.The capacity of Delhi's sewage system is inadequate. Only 46.4 per cent of waste gets partial or full treatment and the rest gets back into the river Yamuna through the 17 open drains.

About 150 tonnes of BOD, 300 suspended solids (SS) are added to the river Yamuna everyday through the 17 drains in the city. Seven drains, Najafgarh, Trans-Yamuna, Sen Nursing Home, Barapula, Maharani Bagh, Kalkaji and Tughlakabad contribute about 95 per cent of this pollution load. The outfall at the Najafgarh Drain alone is about 50 per cent of the total discharge and 25 per cent of the total BOD load that joins the river Yamuna and canals at Delhi (Table 1).

Table 1
Discharge and Load Contribution of Various Drains to River Yamuna at Delhi (2006)

Name of the drain	*Discharge contribution (in per cent)*	*BOD contribution (in per cent)*
Najafgarh drain	54.62	33.98
Sen Nursing Home	2.79	7.51
Shahdara	17.31	21.55
Sarita Vihar Bridge	14.12	9.74
Other drains	11.15	20.79
Power House drain	-	6.43

Source: CPCB, 2007.

Also, the sewage treatment plants are under-utilised. Only 68 per cent of the city's sewage treatment capacity is utilized. It is due to the fact that the sewage has to be transported over long distances for treatment, through largely defunct conveyance systems. In 2001, only 15 per cent of Delhi's sewerage system was functional. On top of this, almost 45 per cent of Delhi lives in unauthorized colonies, generating 'illegal' sewage, which is unaccounted.

Industrial Effluents

All the Indian rivers have been polluted by industrial effluents. Industries generate water pollution loads, which is toxic and varied in nature, highly concentrated in terms of space and time. The River Yamuna has become a highly polluted river due to various types of industrial discharges. In Delhi there are about 1.29 lakh industries, out of which major polluting industries are thermal power plants, electroplating, dying, silicate manufacturing, rolling mills etc. Of these about 20,000 units are in 28 approved industrial areas. From Delhi industrial area alone, more than 8 lakh tonnes of industrial waste is discharged into the River Yamuna.

The total BOD load discharged by the industrial areas in Delhi has been estimated using discharge and concentration data for the five sewerage zones (Table 2).

Table 2
Sewerage Zonewise Estimated Generation of Industrial Pollution Loads

Sewerage Zone	*Estimated Industrial Discharge (MLD)*	*Estimated Industrial BOD load (MT/day)*
Okhla	10.00	9.41
Keshopur	28.53	80.65
Rithala	30.39	33.04
Coronation Pillar	2.48	0.68
Shahdara	10.00	3.54
Total	81.40	127.32

Source: CPCB, 2000.

The Status of Pollution Control in 17 categories of industrial units in NCT of Delhi shows a total of 3 thermal power plants, one pesticide and one fertilizer industry each of which discharge waste water into the River Yamuna including pulp and paper, sugar etc. The ash ponds are located close to the River Yamuna in case of

Rajghat and Indraprastha Power Plants. The River Yamuna is vulnerable to overflows from the ash ponds, particularly during the monsoon. Besides, ground water contamination may take place due to leaching of heavy metals present in the fly ash. The Government has already initiated action to issue direction for closure of unit number 5 of Indraprastha Thermal Power Station due to excessive pollution from this unit. In order to compliance to the environment laws, it is compulsory for these industries to treat the effluent to achieve prescribed standards before discharging effluent into the river.

Festivals

It seems rather incongruous that status of water pollution has deteriorated in a nation where water is generally revered by all the religions in some form or the other. The practice of mass bathing during the festivals and immersion of idols during festivals have adverse impact on water quality parameters.

It is not only idols, but the puja material like flowers, garlands, oily substances, polythene bags, eatables and decoration material which increases suspended matter and trace metals like zinc, lead, iron, chromium, arsenic and mercury. During Durga Puja the year 2005, alone 300 idols were immersed in the Yamuna in Delhi stretch, 36 tonnes of paints and other chemicals and the offerings had 15 tonnes of flowers, leaves, coconut husk, clothes and other puja material.

Spatial Analysis of Water Quality of the River Yamuna

The water quality of the river at any point depends upon a number of factors like drainage inflows, agricultural runoffs, industrial effluents which are the source of pollutants discarded into the river. When toxic substances enter streams they get dissolved or lie suspended in water or get deposited on the bed. This results in the pollution of water whereby the quality of water deteriorates affecting aquatic ecosystems.

Location of Sample Sites in Various Zones

The River Yamuna enters the NCT of Delhi at Palla and after traversing a distance of approximately 48 kms it leaves Delhi at village Jaitpur below Okhla. The distance between the two barrages at Wazirabad and Okhla is around 22 kms. This stretch between the two barrages receives the maximum amount of pollution. The study analyses the problem of the intensity of water pollution at various locations during the same time period.

To study this, the entire stretch of the river was divided into various zones. For this purpose, the longitudinal profile of the entire stretch of river Yamuna in NCT Delhi was made. Starting from the Uttar Pradesh border in the north of Delhi, 5 zones of approximately 5 kms length each were identified (Figure 1). To assess the spatial extent of water quality in River Yamuna in Delhi, ten potential point sources of pollution were identified along the river by ground truthing and frequent site visits and accessibility to the river site. Out of these, eight sites are situated in the area between the two barrages and one site each above the Wazirabad barrage and below the Okhla barrage. All the sample sites are located on the western bank except for site no. 5 and 6 which are on the eastern bank of the river. The number of sample sites with the number of drains falling into the river in each zone is depicted in table 3.

Methodology of Selection of Parameters and Sample Collection

The parameters selected for testing water quality from various sites along the river Yamuna in Delhi is strictly on the basis of water quality kit provided by the Central

Table 3
Number of Sample Sites and Drains in Various Zones

Zone	*No. of Sample Sites*	*No. of drains*
1	1	0
2	3	7
3	3	4
4	2	6
5	1	0
Total	10	17

Source: Primary Survey

Pollution Control Board. It is a portable laboratory provided with the apparatus and reagents needed for the field testing to assess the physical, chemical, bacteriological and biological quality of water in the field.

The selection of parameters was done by taking into account the most common pollutants affecting the human, animal or plant life through polluted water. Also, care was taken that the tests can be easily performed by the researcher. The tests performed were categorized broadly into three; namely

1. Physical tests: include, (i) colour, (ii) odour (iii) temperature (iv) stream velocity
2. Chemical tests: include, (v) pH (vi) dissolved oxygen (vii) total hardness (viii) calcium (ix) total dissolved solids (TDS) (x) magnesium (xi) chloride (xii) nitrate (xiii) fluoride
3. Biological test: (xiv) total coliform

The task of planning an effective and optimal structure monitoring network, involves a decision on location of sites, methods and analysis, frequency of sampling is a complex phenomena. It requires the collection of water quality samples from a large number of sites over a short time. In this case the survey was carried out in June 2007

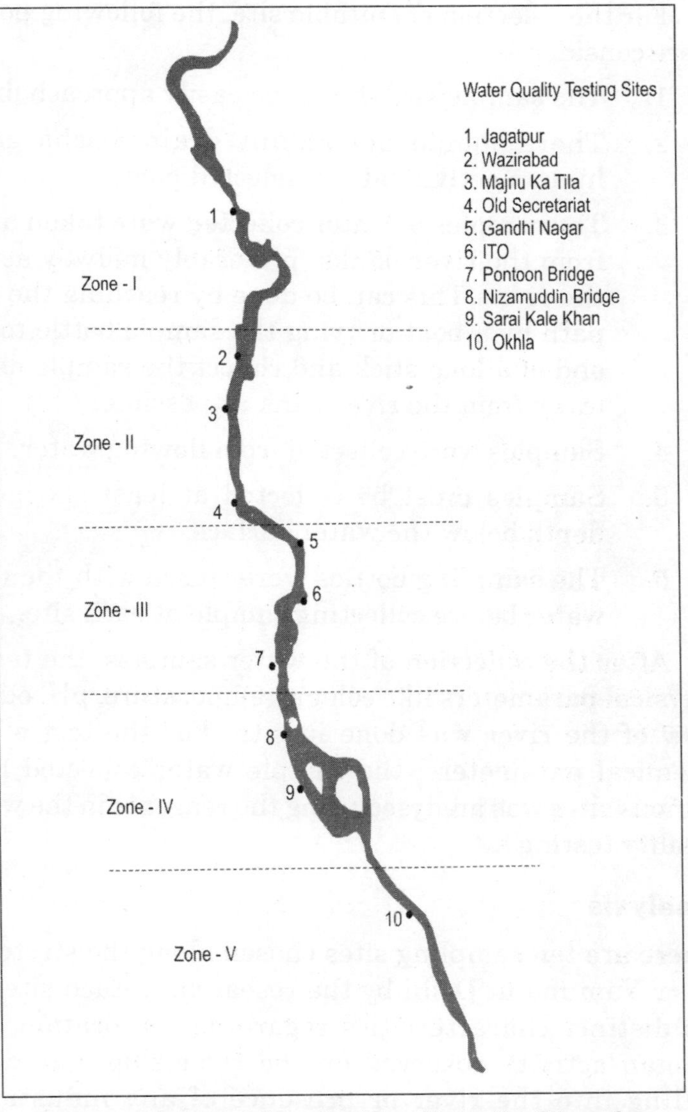

Figure 1: Selected Sample Sites Along the River for Determination of Water Quality

and is subjected to physical, chemical and biological analysis as per standard methods of Central Pollution Control Board.

For the selection of suitable site, the following points were considered:

1. The sample site should be easily approachable.
2. There should not be any drain discharge or human activity at the selected point.
3. The samples of water collected were taken away from the river banks- preferably midway across the river. This can be done by reaching the mid path by a boat or tying the sample bottle to the end of a long stick and collect the sample as far away from the river bank as possible.
4. Samples were collected from flowing water.
5. Samples must be collected at least six inches depth below the water surface.
6. The sampling bottles were rinsed with the river water before collecting sample at each site.

After the collection of the water samples, the test of physical parameters like colour, temperature, pH, odour, flow of the river was done in situ. For the test of the chemical parameters, the sample water collected from various sites was analysed using the reagents in the water quality testing kit.

Analysis

There are ten sampling sites chosen along the stretch in river Yamuna in Delhi by the researcher. Each site has its distinct characteristics regarding its location, any human activity observed on the river side, any drain falling into the river or presence of any industry or agriculture practiced near the river, presence of littered material like paper, utensils, garbage, plastics, flowers etc. on the river side. The physical parameters selected such as colour, odour, stream velocity were observed and noted. These are discussed in detail as below.

Physico-Chemical Characteristics of Water Quality

Of the thirteen physico-chemical parameters selected for the analysis of water quality along the ten sites of River Yamuna, 04 parameters are physical, 08 parameters are chemical and 01 parameter falls in the biological category.

Physical Parameters

The physical parameters selected to study the water quality for the 10 sites along the river include colour, odour, temperature and stream velocity. Colour and odour have been analysed on the basis of personal observation and temperature is noted using a thermometer (Table 4).

The observation of the physical parameters from site 1 to site 10 indicates that the colour of the river water changes from clear to dark black. The odour also becomes pungent as we move downstream till site 10. The temperature shows an increase from 30°C to 32.5°C and the stream velocity show a remarkable decrease in the values from 9.5 m/10 sec at site 1 to less than 2 m/10 sec at site 10. The gradual change in colour and odour depicts deterioration in water quality from site 1 to site 10. Also, increase in temperature and decrease in stream velocity are indicative of declining water quality.

Chemical Parameters

The chemical analysis of the water samples collected for the ten sites mentioned above have been depicted in the tabular form (Table 5). There are eight parameters chosen for chemical analysis and 01 parameter for biological analysis. These have been tested for their chemical and biological constituents. The detailed analysis of the chosen nine parameters have been discussed below.

The pH value is indicative of the acidity or alkalinity of the water sample. The value 0 to 7 is acidic and greater than 7 is basic. The pH value is 7.7 at site 1 and then it gradually decreases till site 4. At site 5 is near Gandhi

Table 4
Physical Parameters of Water Quality Along Chosen Sites

Parameter	*Site 1*	*Site 2*	*Site 3*	*Site 4*	*Site 5*	*Site 6*	*Site 7*	*Site 8*	*Site 9*	*Site 10*
Colour	clear	Light brown	Grey blue	Grey blue	Grey	Greyish	Grey black	Dirty black	Dirty black	Dark black
Odour	No odour	No odour	Slight odour	Unple-asant	Stink-ing	Rotten egg	Rotten egg	Pungent	Pungent	Pungent
Temperature (in ° C)	30	30	30	30.5	31.5	31.5	31.5	32	32	32.5
Stream velocity (m/10 sec)	9.5	9.5	7-8	6.5	6	5.5	4	3	2-3	< 2

Source: Primary Survey

Table 5
Chemical Parameters of Water Quality along Chosen Sites

	Sites									
Parameters	*1*	*2*	*3*	*4*	*5*	*6*	*7*	*8*	*9*	*10*
pH	7.7	7.4	7.5	7.2	7.9	7.8	7.7	7.4	7.5	7.4
DO	11.8	9.5	4.3	2.1	1.5	1.2	1.0	0.8	0.2	0.0
Total Hardness	200	280	300	400	400	450	480	500	500	600
TDS	80	90	200	600	850	100	1300	1450	1680	1950
Calcium	75	150	240	200	210	210	280	260	250	260
Magnesium	40	42	45	75	75	85	85	85	90	100
Nitrate	50	50	75	90	100	120	100	100	110	125
Fluoride	0.5	0.5	1.0	1.25	1.25	1.25	1.25	1.50	1.50	1.50
Total Coliform	8000	40540	62950	103000	603×10^4	8×10^6	84×10^6	107×10^6	197×10^6	2×10^7

Source: Primary Survey

Nagar, it increases drastically to 7.9, may be due to the basic content discharged by the drains carrying industrial and domestic effluents. Then from site 6 till site 8 the pH value decreases to 7.4. There are not many variations in the pH value along the entire stretch of Delhi, only the values are fluctuating between 7 and 8 indicating the basic nature of water.

The dissolved oxygen is the amount of oxygen present in the water which goes on decreasing from site 1 to site 10. The decrease is drastic from site 1 to site 4 i.e. 11.8 to 2.1 and then it fluctuates between 1.5 to 0 from site 5 to site 10. The permissible limit of DO is 4.0 mg/l. Hardness of water is the property of water which prevents lather formation with soap and increases the boiling point of water. It is mainly due to the presence of calcium and magnesium ions in water. From the table, it is evident that hardness of water has increased gradually from site 1 to site 10. The permissible limit of total hardness is 200 mg/l.

The total solids are the ones which are dissolved or suspended in one litre of water. It is measured in mg/l of water. A limit of 1000 mg/l of total solids has been imposed by World Health Organization as standards for drinking water. It is observed that there is an increase in values of total dissolved solids from site 1 to site 10. The calcium content has increased from site 1 to site 3 and then decreased to site 6. It shows a drastic increase at site 7 and then decrease to site 10 which is much beyond the permissible limit of 75-200 mg/l. Magnesium is one of the important contributors to the hardness of water. Its concentration in water may vary from zero to several hundred milligrams per litre depending upon the source of water. Chemical softening or ion exchange reduces the magnesium and associated hardness to acceptable levels However, Magnesium values have increased from site 1 to site 10 i.e. from value of 40 mg/l to value of 100 mg/l. The permissible limit of magnesium is 30 mg/l.

Cultivation of crops causes chemicals from fertilizers i.e. nitrates and pesticides to seep into the water from irrigation runoff causing significant amount of water pollution. The nitrates when ingested by living organisms beneficial to man, e.g fish and other marine life, changes into nitrites and kills them. Similarly, phosphorous based fertilizers cause eutrophication and kill plants and animal life. The presence of nitrates in the sample water in the present study can be observed to be increasing from site 1 to site 10.

Fluoride may occur naturally in water. Fluoride concentration of approximately 1.0 mg/l in water effectively reduces dental carries without harmful effects on health. Some fluorosis may occur when the fluoride levels exceeds the recommended limits of 1.0 mg/l.. The values of fluoride concentration from site 1 to site 10 show a steady increase from values 0.5 to 1.5 mg/l.

The coliforms are group of bacteria which are rod shaped, lactose fermenting, non spore forming and gram negative Whenever there is no coliforms, chances of occurrence of pathogens is very rate. The drinking water should be free from coliform. The water quality of the Yamuna River with respect to total and faecal coliform is in the deteriorated condition from site 1 to site 10. The total coliform count ranging from 8000/100 ml at site 1 to $2x10^9$/100 ml at site 10 which is extraordinary high for most of the beneficial uses.

Conclusion

The increase in population has resulted in unplanned habitations and increase in industries leading to waste water discharge without any treatment into the rivers/ streams downgrading the quality of natural waters causing water pollution. The spatial analyses of River Yamuna in Delhi stretch at various points indicate that the river maintains a reasonably good quality in upstream of Delhi. As soon as it enters the city Delhi, it receives pollution loads from various sources as waste water, industrial effluents from the city. This input of waste water sets a progressive series of chemical and biological events in the downstream water. The water quality values are much above or below the prescribed water quality standards for each of the parameters. Delhi has spent over 15 billion rupees on cleaning the Yamuna till the year 2006 but in spite of this investment, the river runs dirtier. What is required is the the control by water quality management planning, by the enforcement of national standards, policing and licensing of discharges and good environmental awareness on the part of the public.This can be achieved only by co-ordination of various sectors accompanied by educational awareness and active participation of the local people.

References

Central Pollution Control Board (2000), Water Quality Issues and Status in India, National River Conservation Directorate, Ministry of Environment and Forest, New Delhi.

Central Pollution Control Board (2007), Annual Report, Ministry of Environment and Forest, New Delhi.

Datta, M. C. and Bhatia, S. K., (1986), Minimum Flows in Yamuna at Delhi for Acceptable Water Quality-a Case Study in Post Session Proceedings Seminar on water quality and its management, Volume II, Central Board of Irrigation and Powers, New Delhi.

Helmer, R. and Hespanhol, I., (eds.) (1997), Water Pollution Control: A Guide to the Use of Water Quality Management Principles, E and FN Spon, London.

James, A. (1979), *"The Value of Biological Indicators in Relation to other Parameters of Water Quality"* in Biological Indicators of Water Quality, A. James and L. Erison (eds.), John Wiley and Sons, Great Britain, pp. 1-16.

Jain, J. L. *et al.* (2005), Fundamentals of Biochemistry, Chand Publications, Meerut.

Kataria, H. C., *et al.* (1996), *"Assessment of Water Quality of Kolar Reservoir in Bhopal (M. P)"*, Pollution Research, Volume 2, (15), pp.191-193.

Reddy, M. V. and Kumar, A. V. (2001), *"Effects of Ganesh idol Immersion on Some Water Quality Parameters of Hussainsagar Lake"*, Current Science, Volume 81, No. 11.

Schindler, D. W. (1971), "Carbon, Nitrogen, Phcsphorus and Eutrophication of Freshwater Lakes", *Journal Physiology*, Volume 7, pp.321-329.

Sykes, G. and Skinner, E. A. (1961), Microbial Aspects of Pollution, Academic Press, New York.

Tearle, K., (ed.) (1973), Industrial Pollution Control - The Practical Implications, Business Books Limited, London.

Trivedy, R. K. and Goel, P. K. (1999), Chemical and Biological Methods for Water Pollution Studies, Ashish Publications, Delhi.

Livestock Jeopardizing the life on Earth

B. S. YADAV, SACHI RANA AND
SHALINI SHEEL

Introduction

Livestock have a substantial impact on the world's water, land and biodiversity resources and contribute significantly to climate change. Directly and indirectly, through grazing and through feed crop production, the livestock sector occupies about 30 per cent of the ice-free terrestrial surface on the planet. In many situations, livestock are a major source of land based pollution, emitting nutrients and organic matter, pathogens and drug residues into rivers, lakes and coastal seas. Animals and their wastes emit gases, some of which contribute to climate change, as do land-use changes caused by demand for feed grains and grazing land. Livestock shape entire landscapes and their demands on land for pasture and feed crop production modify and reduce natural habitats. In the varied contexts of the livestock sector, environmental issues require an integrated approach, combining policy measures and technology changes, within a framework of multiple objectives. The livelihood concerns of hundreds of millions of poor livestock holders, who often engage in livestock production because they have no alternative, must be taken into account. The demands of the emerging middle class, who are consuming growing amounts of meat, milk and eggs, cannot be ignored either. Attempts to curb the booming demand for these products have generally proved ineffective. Using animals for food and other products and services is only one of many human activities that depend on natural resources.

Urbanization and expansion of cropping have led to an unprecedented loss and fragmentation of habitats, including valuable ones such as forests and wetlands may greatly increase the number of hungry people, most severely in Sub-Saharan Africa (FAO, 2005a). According to FAO, developing countries may lose about 280 million tonnes of potential cereal production as a result of climate change. Because of habitat losses, unsustainable forms of exploitation and climate change, the loss of biodiversity continues to accelerate and will have serious implications for the future.

Need of the Study

The Millennium Ecosystem Assessment (MEA, 2005a), in a comprehensive assessment of the environmental health of the planet, estimates that species are disappearing at 100 to 1000 times the background levels seen in fossil records. The MEA gauged that one-third of all amphibians, a fifth of mammals and an eighth of all birds are now threatened to extinction. This assessment is based on known species and it is estimated that 90 per cent or more of all existing species have not been catalogued yet. While some species provide obvious services such as food, timber or clothing, most species' services are more difficult to observe and, therefore, less appreciated. They include recycling of nutrients, pollination and seed dispersal, climate control and purification of air and water. Water availability is becoming a serious constraint to the expansion of agriculture and to meeting other growing human needs. Agriculture is the largest user of water, accounting for 70 per cent of total freshwater use. While there are different views on the extent of climate change and its effect on the environment, it is now firmly established that anthropogenic climate change is indeed occurring. The most important gas associated with climate change is carbon dioxide (CO_2) while other greenhouse gases, including methane, nitrous oxide, ozone and sulphur hexafluoride also contribute. Carbon dioxide levels have increased by over 40 per cent over the past 200 years, from 270 parts per million (ppm) to 382 ppm (NOAA, 2006). Today, CO_2 concentrations are higher than at any time during the last 650 000 years (Siegenthaler *et al.*,2005). Methane concentrations today are more than twice the pre-industrial level (Spahni et al.,2005). Average temperatures have increased by 0.8°C over the past century (NASA, 2005). Combustion of fossil fuels is a major contributor to these changes. Additional land available for cultivation is limited. Therefore, most of the increase in agricultural production has come, and will come, from intensification of land that is already cropped or grazed.

Environmental degradation is often associated with war and other forms of conflict. Throughout history, peoples and nations have fought over natural resources such as land and water. By increasing the scarcity of these resources, environmental degradation increases the likelihood of violent conflict, particularly when there is a lack of governing institutions. In recent years, public attention has been drawn to the prospect that future wars will be fought over increasingly scarce natural resources (see, for example, Klare, 2001, or Renner, 2002). A Pentagon report (Schwartz and Randall, 2003) suggested that global warming could prove a greater risk to the world than terrorism and could lead to catastrophic droughts, famines and riots.

Environmental degradation at its current scale and pace is clearly a serious threat for the sustainability of natural resources. The functioning of ecosystems, both at local and global levels, is already seriously compromised. Ultimately, if left unchecked, environmental degradation may threaten not only economic growth and stability but the very survival of humans on the planet.

The Causes and Effects

Environmental degradation significantly affects human health, both directly and indirectly. Direct effects on human health include contact with pollutants. Indirect effects include increased exposure of humans and of animals to infectious diseases because of climate change. The geographic range and seasonality of a number of important diseases, including malaria and dengue fever, are very sensitive to changes in climatic conditions (UNEP 2005a). These secular changes of population, economies, diets, and technology and land use drive changes in the global livestock sector while, to some extent, the sector itself shapes these forces. Sketching these broad developments helps to understand the context within which the livestock sector operates.

The World Bank (2006) projects that GDP growth in developing countries will accelerate in coming decades. Growing incomes boost demand for livestock products. Over recent decades, the global economy has experienced an unparalleled expansion. Population growth, technological and science breakthroughs, political changes, and economic and trade liberalization have all contributed to economic growth.

In developing countries, this growth has translated into rising per capita incomes, and an emerging middle class that has purchasing power beyond their basic needs. There is a high income elasticity of demand for meat and other livestock products (Delgado *et al.*, 1999) – that is, as incomes grow, expenditure on livestock products grows rapidly. Therefore, growing per capita incomes will translate into growing demand for these products. This will close much of the gap in average consumption figures of meat, milk and eggs that currently exists between developed and developing countries. The effect of increased income on diets is greatest among lower- and middle-income populations. This observation is true at individual level as well as at the national level (Devine, 2003).Over the decade 1991 to 2001, per capita GDP grew at more than 1.4 per cent a year for the world as a whole. Developing countries grew at 2.3 per cent on average compared to 1.8 per cent for developed countries (World Bank, 2006). Growth has been particularly pronounced in East Asia with an annual growth rate of close to seven per cent.

Worldwide shifts in dietary preferences the advent of agriculture and the sedentarization of hunter/gatherers enabled increasing populations to be fed – but it also led to a narrowing of the human diet. Prior to agriculture, animal products played a much larger role in human nutrition, and intake levels were similar to, if not higher than, current consumption levels in developed countries. Increases in income and advances in agriculture enabled developed countries to enrich and diversify their diets over the last 150 years. Developing countries are currently engaged in a catching-up process, which has been termed the "nutrition transition" by Popkins, Horton and Kim, (2001). The transition is characterized by an accelerated shift from widespread undernourishment to richer and more varied diets and often to over nutrition. In contrast to the more secular nutrition transition that occurred in developed countries, this shift now occurs within a single generation in rapidly growing developing countries (Table 1).

Table 1
Changes in food Consumption in Developing Countries

	1962	*1970*	*1980*	*1990*	*2000*	*2003*
			Consumption Kg / person / year			
Cereals	132	145	159	170	161	156
Roots and Tubers	18	19	17	14	15	15
Starchy Roots	70	73	63	53	61	61
Meat	10	11	14	19	27	29
Milk	28	29	34	38	45	48

Prices have been declining in real terms since the 1950s. Currently they allow much higher consumption levels of high-value food items than was the case for developed countries at comparable levels of income in the past (Schmidhuber and Shetty, 2005). Food products or opt for certified products, such as free range or organic foods (Krystallis and Arvanitoyannis, 2006; King, et al., 2000). The growing trend towards vegetarianism, albeit still at a very low level in most societies, is another manifestation of this trend. Government promotion campaigns are also identified as potential drivers of consumption trends (Morrison *et al.*, 2003).

The livestock sector has been affected by pro-found technological change on three different fronts:

- In livestock production, the widespread application of advanced breeding and feeding technology has spurred impressive productivity growth in most parts of the world.
- In crop agriculture, irrigation and fertilization-additional factors determining consumption, as it shapes the relative costs of different food commodities. Access to marine resources, on the one hand, and to natural resources for livestock production, on the other, have drawn consumption trends in opposite directions.

Lactose intolerance, found particularly in East Asia, has limited milk consumption. Cultural reasons have further influenced consumption habits. This, for example, is the case in South Asia, where consumption per capita of meat is lower than income alone would explain. Other examples are the exclusion of pork from the diet by Muslims. Socio-cultural patterns have created a rich diversity of consumer preferences, but have also influenced consumers' views about the quality of animal products (Krystallis and Arvanitoyannis, 2006) combined with the use of improved varieties and mechanization, continue to translate into growing yields and improved nutrient composition in pasture and major crops used for feed (Table 2).

Structural Changes and their Impact

The livestock sector is undergoing a complex process of technical and geographical change, which is shifting the balance of environmental problems caused by the sector.

Extensive grazing still occupies and degrades vast areas of land; though there is an increasing trend towards intensification and industrialization. Livestock production is shifting geographically, first from rural areas to urban and periurban, to get closer to consumers, and then towards the sources of feedstuff, whether these are feed crop areas or transport and trade hubs where feed is imported. There is also a shift of species, with production of monogastric species (pigs and poultry, mostly produced in industrial units) growing rapidly, while the growth of ruminant production (cattle, sheep and goats, often raised extensively) slows. Through these shifts, the livestock sector enters into more and direct competition for scarce land, water and other natural resources.

Table 2
Past and Projected Trends in Consumption of Meat and Milk in Developing and Developed Countries

Food Demand	*Developing Countries*					*Developed Countries*				
	1980	*1990*	*2002*	*2015*	*2030*	*1980*	*1990*	*2002*	*2015*	*2030*
Annual per capita meat consumption (kg)	14	18	28	32	37	73	80	78	83	89
Annual per capita milk consumption (kg)	34	38	46	55	66	195	200	202	203	209
Total meat consumption (million tons)	47	73	137	184	252	86	100	102	112	121
Total milk consumption (million tons)	114	152	222	323	452	228	251	265	273	284

Land Degradation

The livestock sector is by far the single largest anthropogenic user of land. The total area occupied by grazing is equivalent to 26 per cent of the ice-free terrestrial surface of the planet. In addition, the total area dedicated to feed crop production amounts to 33 per cent of total arable land. In all, livestock production accounts for 70 per cent of all agricultural land and 30 per cent of the land surface of the planet. Expansion of livestock production is a key factor in deforestation, especially in Latin America where the greatest amount of deforestation is occurring – 70 per cent of previous forested land in the Amazon is occupied by pastures, and feed crops cover a large part of the remainder. About 20 per cent of the world's pastures and rangelands, with 73 per cent of range lands in dry areas, have been degraded to some extent, mostly through overgrazing, compaction and erosion created by livestock action. The dry lands in particular are affected by these trends, as livestock are often the only source of livelihoods for the people living in these areas.

Atmosphere and Climate

With rising temperatures, rising sea levels, melting icecaps and glaciers, shifting ocean currents and weather patterns, climate change is the most serious challenge facing the human race. The livestock sector is a major player, responsible for 18 per cent of greenhouse gas emissions measured in CO_2 equivalent. This is a higher share than transport.

Intensification – in terms of increased productivity both in livestock production and in feed crop agriculture – can reduce greenhouse gas emissions from deforestation and pasture degradation. In addition, restoring historical losses of soil carbon through conservation tillage, cover crops, agro forestry and other measures could sequester up to 1.3 tons of carbon per hectare per year.

Water

The world is moving towards increasing problems of freshwater shortage, scarcity and depletion, with 64 per cent of the world's population expected to live in water-stressed basins by 2025.

The livestock sector is a key player in increasing water use, accounting for over 8 per cent of global human water use, mostly for the irrigation of feed crops. It is probably the largest sectoral source of water pollution, contributing to eutrophication, "dead" zones in coastal areas, degradation of coral reefs, human health problems, emergence of antibiotic resistance and many others. The major sources of pollution are from animal wastes, antibiotics and hormones, chemicals from tanneries, fertilizers and pesticides used for feed crops, and sediments from eroded pastures.

Biodiversity

Livestock now account for about 20 per cent of the total terrestrial animal biomass, and the 30 per cent of the earth's land surface that they now pre-empt was once habitat for wildlife. Indeed, the livestock sector may well be the leading player in the reduction of biodiversity, since it is the major driver of deforestation, as well as one of the leading drivers of land degradation, pollution, climate change, overfishing, and sedimentation of coastal areas and facilitation of invasions by alien species. In addition, resource conflicts with pastoralists threaten species of wild predators and also protected areas close to pastures. Meanwhile in developed regions, especially Europe, pastures had become a location of diverse long-established types of ecosystem, many of which are now threatened by pasture abandonment.

The Livestock- GHG Relationship

Three perspectives can be framed out of the analysis of the live stock–GHG relationship and give rise to the following specific questions:

- Second order impacts: To gain a broader sense of livestock's impacts, an additional set of framing perspectives is needed, which are referred to as follows: the second order impacts of livestock production and consumption; the opportunity cost of land and resource use; and the ultimate question of needs. Opportunity cost: We have to eat—would plant based substitutes be more GHG efficient?
- Needs: Is it possible to define how much livestock production human society 'needs' and does production fall short of, or exceed this level?

The main input to livestock rearing is land. This can be subdivided into: land for cereals, for oilseeds (and other proteins), and for pasture. Byproducts from other food and agricultural sectors constitute an additional input. Energy (fossil fuel-derived for industrialized systems) is also needed but is not discussed here since impacts relative to other emission sources are low. Through the use of renewable such as wind and waste-derived biogas they may also be more straightforward to tackle.

Cereals are a major source of nutrition for pigs, poultry, dairy cows and for cattle in intensive beef systems. More than half of the UK's cereal output is used to feed farm animals (Defra, 2006). Globally livestock have been estimated to consume a third (FAO, 2002) or more (37%) of world cereal output (WRI, 2004). While livestock in the developing world generally consume far fewer cereals and rely more on foraging and byproducts, this situation may change (Keyzer *et al.*, 2005), as production systems intensify and as we see a growth in chicken and pig production where grains feature heavily in their diets.

Various authors have calculated how much plant food is required to produce an equivalent quantity of animal food. This 'feed conversion efficiency' has a major bearing on GHGs since losses of nutritional energy through the production chain – from plant to animal nutrients – means that more GHGs are emitted for a given quantity of nutritional output.

Described in terms of its energetic value (calories available for consumption) too, the conversion efficiency of animal based foods is significantly lower than that of plant foods, as Wirsenius (2003) shows. Nevertheless, it is of course the case that if people did not consume livestock products, fewer cereals would be required for livestock but more for direct human consumption. It is possible then that a reduction in livestock consumption might lead to an increase in the land area required to grow cereals, perhaps leading to additional land use change derived CO_2 emissions. This is also the case when considering the GHG impacts arising from livestock production. Studies attribute all emissions to the edible outputs of production. Properly speaking, however, these should be divided among the various outputs, such as leather or wool. The research shows that 7.7% of all agricultural impacts are attributable to the leather itself. This, if the logic were carried to its natural conclusion, would reduce emissions from the edible output by 7.7%. If allocations were made to all the other non-food outputs of livestock production, the GHG emissions attributable to meat and dairy products will be lower still. Research finds (Allard *et al.*, 2007) that on temperate unfertilized grasslands where cattle are reared without the use of feed inputs or additional fertilizer, the carbon sequestering role of livestock outweighs their CH_4 and N_2O emissions.

When land is overgrazed the combination of vegetative loss and soil trampling can lead to soil carbon losses and the release of CO_2 (Abril and Bucher, 2001; Abril et al., 2005). Overgrazing is a significant concern in the developing world and it has been estimated that 20% of land globally is degraded—up to 73% in dry lands (FAO, 2006). Animals provide us not only with food, but with non-food goods such as leather, manure, and wool; and with benefits such as soil quality, species diversity and landscape aesthetics. The questions that arise in relation to these outputs are: to what extent does society need these goods? To what extent are substitutes available at a lower GHG cost? These questions are discussed in relation to three key outputs—food, leather and manure.

Food and Livestock Farming

Clearly, food is the major and most important output from livestock farming. Meat, eggs and dairy products provide a range of essential nutrients, including protein, iron, calcium, vitamin B_{12} and fat, and if we did not consume animal foods we would have to obtain these nutrients from somewhere else.

There may also be an opportunity cost in using byproducts actually need animal source foods to obtain key nutrients; and if not, whether their substitution with plant foods would produce more or fewer GHG emissions.

The second question – the environmental cost of substitutes – then arises. It was noted that the GHG burden of plant based foods tends to be lower but also emphasized that comparisons on a CO_2 e/kg of food consumed basis are not meaningful. A kg of broccoli is not comparable to a kg of sugar or beef. Nevertheless, some studies have sought to compare the GHG emissions of vegetarian and meat-based whole meals, both balanced nutritionally. (Carlsson - Kanyama 1998).

Manure and GHG Emissions

Livestock manure is a mixed blessing. On the one hand, it contributes vital nutrients to the soil; it has been estimated that globally, around 22% of total nitrogen and 38% of phosphate applied is of animal origin, over half of which comes from beef cattle. Manure can improve the quality and fertility of soil and it has been shown that soil fertilized with manure is more biologically active and fertile than soil fertilized by mineral fertilizers alone (Fließbach *et al.*, 2007). Manure can also build the carbon storage potential of the soil and so help remove carbon from the atmosphere (FAO, 2001).

As with artificial fertilizers, however, manure emits N_2O and CH_4 as it breaks down in the soil. Manure-derived emissions contribute to more than 5% of total anthropogenic GHG emissions, with N_2O the main culprit (FAO, 2006) and manure is particularly problematic when overly abundant, as can be the case in intensive livestock systems. Moreover, as already observed, a third of all cereals grown worldwide are used to feed animals, so in effect, some of their manure is used to sustain their own existence.

Leather and GHG

Leather, is another key output of livestock production. As for food, its use raises the question both of the opportunity cost (what is the GHG of producing a substitute?) and more fundamentally that of need—do we need leather in the quantities available to us? While there are some LCAs of alternative materials (Kalliala and Nousiainen, 1999; Laursen *et al.*, 1997).

The amount of substitute materials required will clearly depend on how far we judge there to be a need for a durable material such leather, and which leather products are essential. Such judgments will always be arbitrary. Traditionally footwear has been the main output of leather production: the FAO (2003) gives footwear and nothing else as an indicator of trends in the production and trade of manufactured leather goods. Using this as a very crude marker of need it appears that the proportion of light bovine leather going into shoe uppers.

Suggestions and Discussion

Animal foods are susceptible to pathogens and often carry chemical residues. Food safety requirements must be met, and are generally a prerequisite in formal markets. The previous assessments of the Livestock Environment and Development (LEAD) Initiative (de Haan, Steinfeld and Blackburn, 1997; Stein-feld, de Haan and Blackburn, 1997) emphasized the livestock sector perspective and analyzed livestock-environment interactions from the perspective of a livestock production system. A major motivation for this assessment is that the environmental issues linked to livestock have not generally received an adequate institutional response neither in developing nor in developed countries. Livestock sector growth in some places, and stagnation with poverty in others, go largely uncontrolled. Although usually considered part of agriculture, in many places livestock production has grown in the same way as industry, and is no longer directly tied to land or to specific locations.

Conclusion

By 2050, the global population is projected to top 9 billion. Demand for land, food and energy will grow. If land is used for livestock, there will be less available to grow other food or biomass. In parts of the world with access to fertilizers and other inputs, the response will be to intensify production, leading to a range of environmental problems including soil and water pollution. Elsewhere,

where resources are lacking, the consequences will be soil degradation and dwindling yields (FAO, 2006).

A great deal of research is clearly needed, so that policy can move in the right direction. A number of research questions have been raised above and this paper concludes by drawing them out as recommendations for further policy-oriented research.

- Land use: Quantify CO_2 emissions arising from land use change attributable to livestock consumption in the UK and examine how this affects the GHG 'hierarchy' of different livestock products. Broader research looking at all the land use change impacts of all food groups is also needed.
- Ecological leftovers: Define what land areas of the UK can be classed as suitable only for grazing and quantify the volume of byproducts available for livestock consumption that does not compete with other uses. Assess how many and what kind of livestock could be supported and quantify the subsequent volume of meat and dairy products available for national consumption.
- Breeding: Examine what breeding and other strategies are needed to increase the resilience and productivity of grazing livestock reared on upland and marginal areas; adopt comparable breeding strategies for monogastrics.
- Bio fuels and biomass: Examine the potential long term relationship between bio fuels and animal production; assess whether first generation bio fuels help or hinder certain types of livestock production and what the effects on GHG emissions might be. Consider what scope there might be for biomass production in upland areas.
- International development: Consider, for the developing world, the role and utility of livestock

rearing for peoples practicing subsistence agriculture; and in assess the contribution livestock make to household economic and food security. Examine how international development strategies could be reoriented to maximize food security at minimum GHG cost while safeguarding the benefits that vulnerable peoples derive from livestock keeping.

References

Websites

http://www.claremontclimatereport.com/2010/01/livestock-related-greenhouse-gas.html

http://www.fao.org/ag/magazine/0612sp1.htm

www.sciencedirect.com

www.elsevier.com/locate/envsci

www.virtualcentre.org

References

Abril, A., Bucher, E. H. (2001), Overgrazing and Soil Carbon Dynamics in the Western Chaco of Argentina. *Applied Soil Ecology* 16 (3), 243–249.

Abril, A., Barttfeld, P., Bucher, E. H. (2005), The Effect of Fire and Overgrazing Disturbances on Soil Carbon Balance in the Dry Chaco Forest. *Forest Ecology and Management* 206 (1–3), 399–405.

Allard, V., Soussana, J. F., Falcimagne, R., Berbigier, P., Bonnefond, J. M., Ceschia, E., D'hour, P., He´nault, C., Laville, P., Martin, C., Pinare's-Patino, C. (2007), The Role of Grazing Management for the Net biome Productivity and Greenhouse Gas Budget (CO2, N2O and CH4) of Seminatural Grassland. Agriculture, Ecosystems and Environment 121, 47–58. American Dietetic Association, 2003. Position of the American Dietetic Association and Dieticians of Canada: vegetarian diets. *Journal of the American Dietetic Association*.

Appleby, P. N., Thorogood, M., Mann, J. I., Key, T. J. (1999), The Oxford Vegetarian Study: An Overview. *American Journal of Clinical Nutrition* 70 (Suppl. 3), 525S–531S.

Baer, P., Athanasiou, T., Kartha, S. (2007), The Right to Development in a Climate Constrained World. Heinrich Bo¨ ll Foundation,

Christian Aid, Eco Equity and the Stockholm Environment Institute, Berlin.

Basset-Mens, C., van der Werf, H.M.G. (2005), Scenario-based Environmental Assessment of Farming Systems: The Case of Pig Production in France. *Agriculture, Ecosystems and Environment* 105, 127–144.

Berlin, J. (2002), Environmental Life Cycle Assessment (LCA) of Swedish Semi-hard Cheese. *International Dairy Journal* 12, 939–953.

Berlin, D., Uhlin, H. E. (2004), Opportunity Cost Principles for Life Cycle Assessment: Towards Strategic Decision Making in Agriculture. *Progress in Industrial Ecology* 1 (1/2/3), 187.

Bhogal, A., Chambers, B. J., Whitmore, A. P., Powlson, D. S. (2007), The Effects of Reduced Tillage Practices and Organic Material Additions on the Carbon Content of Arable Soils. Summary Report for Defra Project SP0561 by ADAS and Rothamsted Research, Defra, UK.

Biffaward (2006), Poultry UK. Biffaward, UK.

British Pig Executive (2007), Pig Year Book. Milton Keynes, UK.

British Pig Executive (2008), Personal Communication.

British Standards Institution (2008), PAS 2050:2008— Specification for the Assessment of the Life Cycle Greenhouse Gas Emissions of Goods and Services. London.

Carlsson-Kanyama, A. (1998), Climate Change and Dietary Choices—how can Emissions of Greenhouse Gases from Food Consumption be Reduced? *Food Policy* 23 (3/4), 277–293.

Casey, J. W., Holden, N. M. (2005), The Relationship between Greenhouse Gas Emissions and the Intensity of Milk Production in Ireland. *Journal of Environmental Quality* 34, 429–436.

Casey, J. W., Holden, N. M. (2006), Quantification of Greenhouse Gas Emissions from Suckler-beef Production in Ireland. *Agricultural Systems* 90, 79–98.

Cederberg, C., Mattson, B. (2000), Life Cycle Assessment of Milk Production — A Comparison of Conventional and Organic Farming. *Journal of Cleaner Production* 8, 49–60.

Cederberg, C., Stadig, M. (2003), System Expansion and Allocation in Life Cycle Assessment of Milk and Beef Production. *International Journal of Life Cycle Assessment* 8 (6), 350–356.

Chen, Y. C., Nakthong, C., Chen, T. C. (2005), Improvement of Laying hen Performance by Dietary Prebiotic Chicory Oligofructose and Inulin. *International Journal of Poultry Science* 4 (2), 103–108.

Committee on Climate Change (2008), Building a Low-carbon Economy—the UK's Contribution to Tackling Climate Change. The First Report of the Committee on Climate Change. The Stationary Office, London, UK.

Cottrill, B., Smith, C., Berry, P., Weightman, R., Wiseman, J., White, G., Temple, M. (2007), Opportunities and Implications of using the Co-products from Bio Fuel Production as Feeds for Livestock. Report prepared by ADAS and Nottingham University for the Home-Grown Cereals Authority, English Beef and Lamb Executive and British Pig Executive.

Dalgaard, R., Schmidt, J. H., Halberg, N., Christensen, P., Thrane, M., Pengue, W.A. (2007), LCA of Soybean Meal. *International Journal of Life Cycle Assessment.*

Davis, J., Sonesson, U. (2008), Environmental Potential of Grain Legumes in Meals. Life Cycle Assessment of Meals with Varying Content of Peas. SIK Report 771. Swedish Institute for Food and Biotechnology, Go¨ teborg, Sweden.

Defra (2006), Agriculture in the UK - 2005 Figures: Figures Derived from Tables 5.1, 5.2, 5.3, 5.4.

Defra (2007), Personal Communication, Work in Progress.

Defra (2007), Table B.10 The British Survey of Fertiliser Practice: Fertiliser Use on Farm Crops for Crop Year 2007, Defra.

Defra (2008), The Milk Road Map. Produced by the Dairy Supply Chain Forum's Sustainable Consumption & Production Taskforce, London.

Druckman, A., Bradley, P., Papathanasopoulou, E., Jackson T. (2007), Measuring Progress Towards Carbon Reduction in the UK. *Ecological Economics* 66 (4), 594–604.

Edwards-Jones, G., Plassmann, K., York, E. H., Hounsome, B., Jones, D.L., Mila' i Canals (2008), Vulnerability of Exporting Nations to the Development of a Carbon Label in the United Kingdom.

Environ. Sci. Policy (available on line December 2008).

Elferink, E.V., Nonhebel, S., Schoot Uiterkamp, A. J. M. (2007), Does the Amazon Suffer from BSE Prevention? Agriculture, Ecosystems and Environment 120, 467–469. English Nature, 2005.

European Environment Agency (2007), Estimating the Environmentally Compatible Bio Energy Potential from Agriculture. EEA Technical Report No12/2007. Luxembourg.

European Commission (2006), Environmental Impact of Products (EIPRO): Analysis of the Life Cycle Environmental Impacts

Related to the Total Final Consumption of the EU25. European Commission Technical Report EUR 22284 EN.

Fadel, J. (1999), Quantitative Analyses of Selected Plant Byproduct Feedstuffs, a Global Perspective. *Animal Feed Science and Technology* 79, 255–268.

FAO (2001), Soil Carbon Sequestration for Improved Land Management. World Soil Resources Reports 96. Food and Agriculture Organization, Rome, Italy.

FAO (2002), World Agriculture: Towards 2015/2030. Summary Report. Food and Agriculture Organization, Rome, Italy.

FAO (2003), World Statistical Compendium for Raw Hides and Skins, Leather and Leather Footwear 1984–2002, Rome, Italy.

FAO (2006), Livestock's Long Shadow—Environmental Issues and Options. Food and Agriculture Organization, Rome, Italy.

FAO (2008), Food Outlook: Global Market Analysis. Statistical Appendix Table A24—Selected International Prices.

Henning Steinfeld, Pierre Gerber, Torn Wassenaar, Vincent Castel, Mauricio Rosales, Cees De Haan, Lead, FAO, Fiat Paris – Livestock's Long Shadow – Environmental Issues and Options.

Levy, Mark (1995), 'Is the Environment a National Security Issue?', International Security 20(2): 35–62.

Matthew, Richard A. (2002), 'Environment, Population and Conflict', *Journal of International Affairs* 56(1): 235–254.

Miguel, Edward; Shanker Satyanath & Ernest Sergenti (2004), 'Economic Shocks and Civil Conflict: An Instrumental Variables Approach', *Journal of Political Economy* 112(4): 725–753.

Pervis, Nigel & Joshua Busby (2004), 'The Security Implications of Climate Change for the UN System', Environmental Change and Security Project Report 10: 67–73.

Suhrke, Astri (1997), 'Environmental Degradation, Migration and the Potential for Violent Conflict', in Nils Petter Gleditsch *et al.*, eds, Conflict and the Environment.

Tänzler, Dennis & Alexander Carius (2002), 'Climate Change and Conflict Prevention', in Federal Ministry for the Environment, ed., Climate Change and Conflict Berlin: Federal Ministry for the Environment (3–8).

Tara Garnett (Food Climate Research Network, Centre for Environmental Strategy, University of Surrey, UK) – Livestock Related Greenhouse Gas Emissions: Impacts and Options for Policy Makers.

Urdal, Henrik (2005), 'People vs. Malthus: Population Pressure, Environmental Degradation, and Armed Conflict Revisited', *Journal of Peace Research* 42(4): 417–434.

Wolf, Aaron T.; Annika Kramer; Alexander Carius & Geoffrey D. Dabelko (2005), 'Managing Water Conflict and Cooperation', in Michael Renner *et al.*, eds, State of the World 2005: Redefining Global Security.

High Risk to an Earthquake Disaster and Its Possible Management in Delhi, India

JAGBIR SINGH

Introduction

Disasters can strike at any time, at any place. Nearly three million people worldwide may have been killed in the past 20 years due to natural disasters such as landslides, earthquakes, floods, snow avalanches, cyclones etc. Ninety per cent of the natural disasters and ninety five percent of the total disaster related deaths worldwide occur in developing countries in which India has the second largest share. India with diverse hypsographic and climatologically conditions has 70 per cent of the cultivable land is prone to drought, 60 per cent of the land area is prone to earthquake, 12 per cent to floods, 8 per cent to cyclones, 85 per cent of the land area is vulnerable to number of natural hazards and 22 states are categorized as multi hazards states. Tens of thousands of people are affected by these natural and man made disasters. Disasters come in many forms, both natural and manmade earthquakes, floods, hurricanes, locust swarms, livestock plagues, war and the breakdown of law and order. About 300 people killed by a massive earthquake in Chile on 27 Feb. 2010 and scientists believe that it is the effect of climate change. The 90 per cent of the people killed on 30th Sep. 2009 due to a massive Tsunami in Pago American

Samoa in Pacific Ocean were women. It was the same in USA, when hurricane Katrina hit New Orleans African American women were among the worst affected. India is one of the most densely populated countries in the world with over one billion people. During the earthquake occurrences in the last century in India and the neighboring countries as well as the most recent earthquakes in 2001 in Gujarat and 2005 in Jammu & Kashmir, the most extensive damage has occurred in masonry buildings of different types, which has resulted in massive loss of lives besides the loss of dwellings. Currently, as per Census of housing 2001, India has a total of 249 million housing units which by wall material consist of 111 million brick houses, 25.48 million stone houses and 73.79 million claymud/unburnt brick houses. These masonry constructions account for more than 85% of the total dwelling units, which are spread over the rural as well as the urban areas. Thus, in any earthquake of magnitude 6.0 or higher on the Richter scale or earthquake Intensities of VII or higher on the MSK

Intensity scale, these buildings will be vulnerable to various damage categories from 'moderate' to 'total collapse'. These Intensities are likely to occur in Seismic Zone III (Intensity VII), Seismic Zone IV (Intensity VIII) and Seismic Zone V (Intensity IX or higher). These three zones cover 59% of the total land area of the country which include most of the mega cities, medium cities as well as the rural areas some time or the other in future. Delhi is one of the mega-cities of the world with population of 10 million and has been projected to become the second largest city in India by the year 2025. Both politically and economically, Delhi is one of the most important cities in our country and its ability to with stand disasters is of critical importance to the entire country. Delhi faces severe earthquake hazard due to its vicinity to the tectionically active Himalayan region. The population of

Delhi is just over 17 million people, making it the second largest city in India. An earthquake disaster in Delhi has the potential to go well which also happens to be a major commercial and industrial centre will have huge economical and political implications which will affect the entire country and not just the population of Delhi. This adds an extra dimension to an earthquake problem for Delhi.

The below table 1 shows the status of whole India regarding different zones.

Table 1
56% of India are prone to Quake

Zone	*Area*	*Parts of the state*
V	**12%**	J&K, N.E States, H.P, Gujarat, Bihar, Uttaranchal, A&N
IV	**18%**	J&K, HP, Uttaranchal, Bihar, Delhi, W. Bengal, Haryana, Punjab, Sikkim, Gujarat, Maharashtra, Lakshdweep
III	**26%**	Punjab, Haryana,Uttaranchal, Bihar, W. Bengal, Gujarat, Rajasthan, M. P, Orissa, A. P, Goa, Kerala, Tamil Nadu, Maharashtra

The below table 2 shows some major earthquakes in India and abroad-from 1819 to 2010.

Table 2

Date	*Earthquake*	*Time*	*Magnitude*	*Max. Intensity*	*Deaths*
16 June 1819	Great Kachchh	11:00	8.3	IX	1,500
12 June 1897	Great Shillong	17:11	8.7	XII	1,500
8 February 1900	Coimbatore	03:11	6.0	VII	Not Known
4 April 1905	Great Kangra	06:20	8.0	X	19,000
15 January 1934	Great Bihar-Nepal	14:13	8.3	X	11,000
26 June 1941	Great Andaman	??	8.1	X	Thousands

table 2 contd.

Date	*Earthquake*	*Time*	*Magnitude*	*Max. Intensity*	*Deaths*
15 August 1950	Great Assam	19:31	8.6	XII	1,530
21 July 1956	Anjar	21:02	6.1	IX	115
10 October 1956	Bulandshahar	??	6.7	VIII	Many
28 December 1958	Kapkote	??	6.3	VIII	Many
2 September 1963	Badgam	07:04	5.5	VII	Hundreds
10 December 1967	Koyna	04:30	6.5	VIII	200
23 March 1970	Bharuch	20:56	5.2	VII	30
19 January 1975	Kinnaur	??	6.5	VIII	Not Known
29 July 1980	Pithoragarh	18:28	6.8	??	>150
31 December 1984	Silchar	04:53	5.6	??	20
26 April 1986	Dharamshala	13:05	5.5	??	6
??	Shillong	??	??	??	??
21 August 1988	Bihar-Nepal	04:39	6.6	IX	1,004
20 October 1991	Uttarkashi	02:53	6.4	IX	768
30 September 1993	Killari (Latur)	03:53	6.2	VIII	7,928
22 May 1997	Jabalpur	04:22	6.0	VIII	38
29 March 1999	Chamoli	00:35	6.6	VIII	63
26 January 2001	Bhuj	08:46	7.7	X	13,805
14 September 2002	Diglipur	03:58	6.0	VII	-
26 December 2004	Great Sumatra	06:28	9.3	XII	10,749
08 October 2005	Kashmir	09:20	7.4	X	1,308
14 February 2006	Sikkim	06:25	5.7	VII	2
27 February 2010	Chili	08:23	8.8	X	300

Source: IMD & New Papers

As per the Indian seismic zone information, Delhi and its surroundings are placed in what is termed seismic zone IV. As per the Vulnerability Atlas of India (1997), for a shaking intensity of VIII, 6.5% houses in Delhi have a high damage risk, and 85.5% houses have a moderate phase with a large number of bridges and flyovers and together with the metro project which is under construction. After a severe earthquake, this transport infrastructure is earthquake resistant but the old one must be seismically retrofitted. Damage could be slight in specially designed structures, considerable in ordinary substantial buildings with partial collapse and very heavy in poorly built structures.

Methodology and Data Bases

The assessment of disaster preparedness of certain areas is to a large extent liable to subjectivity. For a modern society, however, objectivity is a prerequisite. A methodology is presented for calculating the disaster preparedness of NCR. Standardization of this methodology could allow comparison of disaster preparedness between and among the national capital region. Under the Disaster Mitigation Act, all state and local entities must have approved mitigation plans, and these plans must be based on empirically derived hazard vulnerability assessments. Therefore, GIS-based hazard assessment methodology is standard and would definitely play a major role in insinuating disaster preparedness in NCR. This project proposal presents tools and methodologies to analyse women's role in disaster preparedness in NCR through multi-modal approach.

Secondary Data Sources

The table below shows the types and sources of secondary information that may be available and helpful.

Primary Data Collection

Secondary data is unlikely to provide the detailed disaggregated information about people's roles, needs and problems required for gender analysis. Primary data collection methods would be employed. The main methods of data collection for use at this stage are rapid appraisal (RA) methods. The basic aim of rapid appraisal techniques is to obtain quantitative and qualitative information adequate for use in the assessment process. RA methods use a combination of techniques, including semi-structured interviewing; focus group interviewing and participant observation. The aims and objective would be reached by means of integrating geo-technologies.

Delhi Profile

- City population exceeds 14 million.
- Around 60% population lives in informal housing and slums.
- Building codes/standards have been poorly implemented.
- Many building and other structure are very old and weak.

Administrative Set-up

- Delhi is the capital of India as well as a State in itself.
- Both national and state governments are based in Delhi.
- Different agencies are controlled by State or National governments/administrators.
- Lt.Governor and Chief Minister are the highest authorities.
- Disaster management strategies must consider interaction between different agencies.

Risk Management

- Disaster management plan should consider the following factors;

- Hazard
- Vulnerability
- Exposure
- External context (Economic impact)
- Earthquake risk of Delhi is the intersection of all these factors.
- Disaster management should consider cost benefit of different options for risk reduction.

As we know due to high migration from all over India, the capital city is under population pressure and due to the lack of preparedness we can face a big challenge from disaster by earthquake.

History of Earthquakes around Delhi

A detailed seism tectonic discussion of the area is beyond the scope of this paper. Yet, it will be of considerable interest to see if Delhi has sustained strong earthquake motions in the past. As poer Iyengar (2000) damaging earthquakes have occurred around Delhi since ancient times. He points out that the great epic, Mahabharata mentions about earthquakes during the war at Kurukshetra (Circa 3000 BC). More recently, damage to Delhi in the 1720 earthquakes (intensity IX in Delhi) is well discussed by Kafi Khan (Iyengar, 2000). Tandon (1953) mentions of damage to the Qutab Minar during the 2803 earthquake near Mathura. Srivastava and Roy (1982) discuss several more earthquakes in Delhi region. These include: (a) earthquake of year 893 or 894 (Intensity XI/XII) which took place not far from Delhi in which many persons died; (b) earthquake of 22 March 1825 near Delhi Intensity VII; earthquake of 17 July 1830 near Delhi (Intensity VIII); and (d) earthquake of 24 October 1831 near Delhi (Intensity VI).

Delhi has also sustained earthquake damage in more recent times. For instance, Srivastava and Somayajuluy

(1966) mention of (a) Khurja earthquake (M6.7) of 10 October 1956 in which 23 persons were killed in Bulandshahr and some injured in Delhi; (b) M6.0 earthquake of 27 August 1960 near Delhi wherein about 50 persons in Delhi were injurred; and (c) an earthquake near Moradabad on 15 August 1966 that killed 14 persons in Delhi. Iyengar (2000) also mentions about damage to one of the minarets of Delhi's Jama Masjid during the M4.0 earquakes on 28 July 1994.

Most recently, the 1999 Chamoli earthquake (M6.5) took place about 280 km from Delhi. Such a moderate earthquake does not normally cause damage at such large distance. And yet, several buildings in Delhi sustained non-structural damage possibility due to peculiar geological and geotechnical features if this area. In 1985, an earthquake about 400 km from Mexico city caused very considerable damage and deaths in Mexico city, primarily due to the peculiar site conditions there. The Chamoli earthquake effects in Delhi indicate that there is real possibility of a large earthquake in the Himalaya causing considerable damage to Delhi.

Infrastructure and other Implications

Delhi is currently passing through a major infrastructure development phase with a large number of bridges, flyovers and the metro project under construction. After a severe earthquake, the transport infrastructure is earthquake resistant and the old one is seismically retrofitted. Indian seismic code (IS:1893-1984) is not applicable for major projects which require special studies on seismic design criteria. Moreover, the Indian seismic codal provisions on bridges as these exist today are obsolete and inadequate (e.g Jain and Murty, 1998). Earthquake disaster in Delhi has the potential to go well beyond the statistics of deaths and injuries. Such a disaster in the country's capital, which also happens to be a major commercial and industrial centre, will have huge economic and political implications which will affect the entire country and not just the population of Delhi. This adds an extra dimension to the earthquakes problem for Delhi.

The dangers facing an earthquake and possible management in Delhi are as follow:

1. Delhi suffers from an over population dilemma.
2. The roads are already congested and even one ambulance cannot get through.
3. There is a complete lack of infrastructure.
4. There is also poor quality in the housing construction and there is no public awareness of what to do in an earthquake.
5. Also, there is no knowledge of what possible food to have on hand in a disaster.
6. There are questions to be asked. How can the general public help the injured? The danger from fire is very high as many gas bottles are used in India.
7. Helipads to withstand an earthquake need to be constructed as they will provide the only help possible.
8. New Buildings are to have a limited height and be able to "Shake" as in Japan.
9. Awareness is needed for communities as well as individual Apartments to be organized in earthquake Disaster Management.
10. There is a lack of locally trained people everywhere.
11. Many people have to many glass picture frames on the walls and objects near their beds.
12. Single brick walls have been built in slum areas without cement. This will cause injuries and be an obstacle to any rescue operation.

13. There is also a lack of emergency hospital tents and their location.
14. Every house should have a disaster management domestic kit-whistle, small torch, mask, bandages, water bottle and phone numbers for the hospital, police and ambulance.
15. A Helicopter Disaster Team should be trained and on hand for any disaster.

Future Visions

In the disaster situations, a quick rescue and relief measures are inevitable to save the life and livelihood sources without much damage and destruction. However, considerable damage can be minimized if adequate preparedness levels can achieved. Indeed, it has been noticed in the past, that as when attention has not been paid to adequate preparedness measures, the loss to life and property has been considerably increased. The project will operate at grassroots level hence; it would provide a constant support to the women through mobilization, participation in all disaster preparedness activities and other rescue and relief measures. Therefore, a concerted effort through this project is to bring women through mainstreaming and to organize as well as to strengthen them on the lines of disaster preparedness and protection in NCR.

Conclusion

It is fact that Delhi is unlikely to be at the epicentre of a moderate to large earthquake. The highest earthquake magnitude experienced in Delhi in about a century was on July 27, 1960. It registered 5.6 on the Richter scale. Some buildings in the New Delhi area were partially damaged during that quake. A seismic damage survey, by the Central Public Works Department (CPWD), put the damage at about Rs 5 lakh. Delhi has had many damaging earthquakes in the past and is placed in a high seismic zone (zone IV). Delhi is prone not only to damaging earthquakes in or near Delhi, but due to its peculiar geological setting it could also sustain strong shaking due to a large earthquake in the Himalayas. Unfortunately, most buildings in Delhi many not meet Indian standards on seismic construction and may be considered deficient from a seismic safely view point. Thus, there is a real potential for a great earthquake disaster in Delhi, the implications of which go well beyond causalities because of their political and commercial significance. There is an urgent need for healthy debates on the seismic risk aspects of Delhi be for held a reasonable assessment of the problem. Studies are needed on seismic hazard evaluation for different types of construction. Using these, seismic risk scenarios must be developed and implementation strategies chalked out for new and old constructions. The efforts required are truly multidisciplinary and should include components on technical training, institutional development, development of technical manuals, legal and enforcement aspects and public awareness programmes. Most importantly, we need the political will to handle this problem and the biggest challenge perhaps lies in drawing the attention of political leadership to this problem while the city faces many other urgent problems as well.

References

Bhandani, R. K. An overview on Natural & Man made Disaster & their Reduction, CSIR, New Delhi.

Gupta, H. K. (Ed), (2003), Disaster Management, Universities Press, India.

Gupta, M. C. (2009), Manuals on Natural Disaster Management in India, National Centre for Disaster Management, IIPA, New Delhi, 2001.

Kates, B. I. & White, G. F. (1978), The Environment as Hazards, Oxford, New York.

Satender, (2003), Disaster Management in Hills, Concept Publishing Co., New Delhi.

Singh, Jagbir (2009), "Tsunami Disaster and its Mitigation" Publisher- I. K. International Pvt. Ltd. S-25, Green Park Extension, Uphaar Cinema Market, New Delhi, India (www.ikbooks.com).

Singh, Jagbir (2007), Society, Sustainability and Environment", 2007. Publisher-Shivalik Publisher & Distributor , 27/16, Shakti Nagar, Near Nagia Park, New Delhi-110007, India.

Singh, Jagbir (2008), "Disaster Management: Future Challenges and Opportunities" Published by- I.K. International Pvt. Ltd. S-25, Green Park Extension, Uphaar Cinema Market, New Delhi, India (*www.ikbooks.com*).

Singh, R. B. (1994), Space Technology for Disaster Mitigation in India (INCED), University of Tokyo.

Singh, R. B. (Ed), (2000), Disaster Management, Rawat Publication, New Delhi.

Tandon, A. N. (1953), The Very Great Earthquake of August 15, 1950', A Compilation of Papers on the Assam Earthquake of August 15, (Compiled by M.B. Ramachandra Rao), The Central Board of Geophysics, Govt. of India, pp. 80-89.

Municipal Solid Waste Management in Delhi

Binu Sangwan and Ashwani Kumar

Introduction

The concept of sustainable development has created more concern on environment among people in recent years. In the urban cities, higher disposable income and changed pattern of life has resulted in high rate of consumption. In 20th century, handling of waste is typical problem for urban authorities. Now, it has become more critical due to rapid population growth and economic development in countries. The enormity of the situation is exemplified at the global level in case of Tokyo city which has grown so quickly, quardrupling its population since 1950, that municipal authorities have devised ways to build islands of waste in Tokyo city (UN Publication, 1997).

As a natural corollary, management of solid waste is one of the crucial aspect of urban management. It is an issue of day to day life and involves each and every citizen. The solid waste management process includes all the functions from generation to disposal of waste. Considering various functional elements, waste management may be conceptualized as the discipline associated with the identification, reduction, storage, collection, transfer and transport, reuse, recycling and effective treatment, disposal of waste in a manner that is in accordance with the best principles of management, economics, engineering, public health, conservation, aesthetics and other environmental conditions, including all administrative, financial, legal planning and engineering functions involved in the whole spectrum of solutions to the problem of waste. This involve interdisciplinary relationships among various fields such as political science, management, demography, geography, sociology, public health conservation, city and regional planning, ecology, environmental sciences, material sciences, engineering etc (Subhash, 2007).

According to World Health Organization (WHO) five million people die due to disease related to improper disposal and poor collection practices of waste over the years. Twenty two diseases are directly related to improper management of municipal solid waste. Rodent and vector insects transmit various diseases like dysentery, cholera, plague, typhoid, infective hepatitis and others (Singh, 2004).

In India, cities and towns generate on an average 300-400 grams of solid waste per capita per day. This translates into approximately 80,000 metric tones/day or 30 million metric tones per annum. So this huge amount of waste is a major problem. In India, municipal solid waste management (MSWM) is one of the prime responsibilities of Urban Local Bodies. But the municipalities of Indian Cities work in a traditional manner and they have neither adequate financial resource nor the trained staff for dealing with the increased difficulties of garbage collection and disposal.

Available infrastructure and present practice of solid waste management is not in a position to keep pace with increasing garbage quantity and requirement leading to low level of waste collection and disposal efficiency. Lack of financial resources, organizational weakness, infrequent waste removal, obsolete tools and equipment and unscientific management practices are major problems faced by the civic bodies in India. The present study strives to examine and analyze the status of municipal solid waste management in Delhi. It explains and discusses the issues like collection, transportation, disposal and services which are provided by MCD. It also traces out various polices of Govt. of Delhi to improve the existing MSW Management System.

Study Area

The area selected for research study has a unique position with regard to its location. National Capital Territory of Delhi (NCT) is the second–largest metropolis in India and it is the sixth metropolis in the world by population. It has 13.8 million populations, 7.6 million male and 6.24 million females population and Delhi has been divided into 9 district 27 tehsils and 165 villages, as per 2001 Census.

The area of the National Capital Territory of Delhi is 1483 sq. kms. (0.04 per cent of total geographical area of India). It is encircled by Utter Pardesh in the East, Haryana on the North, West and South.

Aims and Objectives of Study

The purpose of the present research study is to understand and analyze the municipal solid waste management in Delhi. The current paper has the following objectives:

- To study the quantity and characteristics of solid waste in Delhi.
- To understand the spatial pattern of waste management.
- To trace the initiatives of Government of Delhi and give some suggestions for better management of (MSW) in Delhi.

Data – Base and Research Methodology

The study is based on secondary statistics. The data has been accumulated from Census of India for the year 2001. The other data collected from Municipal Corporation of Delhi, Central Pollution Control Board, Directory of Economic and Statistics of Delhi and National Environment Engineering Research Institute (NEERI).

The data has been analyzed keeping Zones as unit of observation. The simple percentage, mean and other statistical techniques has been used. Statistical Method like Z –score has been used for the waste service index. The formula for waste service index given as below:

$$z = \frac{x - \overline{x}}{S.D.}$$

Different cartographic technique has been used, in order to show the spatial and temporal pattern various tables and graphs has been drawn.

Characteristics of Municipal Solid Waste

The term Municipal Solid Waste (MSW) consist of household waste, construction and demolition debris, sanitation residue and waste from streets. This garbage is mainly generated from residential and commercial complexes. Municipal waste stream should be characterized by their source, by types of waste produced as well as the generation rates and composition. The physical and chemical composition of MSW is given below.

Physical Composition

The physical composition of the waste depends on a wide range of factors such as food, food habits, cultural traditions, life styles, climate and income etc. In Delhi, the composition of municipal waste changing with 3-5 years because of industrialization, urbanization, highly growth of population and changing life style. The composition of municipal waste i.e. paper (6.29%), plastic (0.85%), metal (1.21%), glass (0.57%) ash and earth (36.00%), degradable material (35.00%) and inert material (20.08%). In Delhi, ash and earth and degradable waste is most generated in quantity (Table 1, Figure 1). This shows a huge amount of raw material for composting plants in Delhi. In this composition paper, plastic and metal is also in great quantity. These are recyclable matter, so it can be said that a huge amount of raw material for recycling industry.

Table 1
The Physical Composition of Municipal Solid Waste in Delhi

Characteristics of MSW	*In per cent*
Paper	6.29
Plastic	0.85
Metal	1.21
Glass	0.57
Ash and Earth	36.00
Degradable	35.00
Inert Material	20.08

Source: http://www.mospi.nic.in

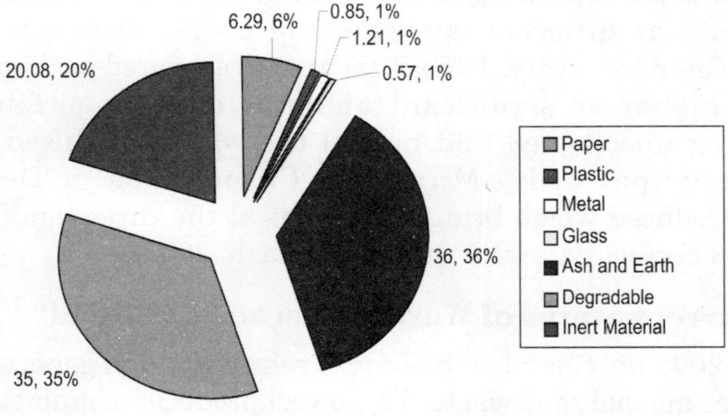

Figure 1: Physical Composition of Municipal Solid Waste

Chemical Composition

The chemical composition has shown the chemical characteristics of municipal waste at the landfill sites. In Delhi, the chemical characteristics of solid waste: moisture (43.65%), silt-inert (34%) organic carbon (20.47%), nitrogen (0.85%), potassium (0.69%) and phosphorous (0.34%). This chemical composition is changing with the time because of changes in the taste of residents.

Waste Generation

In the system of waste management, generation is basic fundamental aspects which change with location, socio-economic conditions and season to season. As per Central Pollution Control Board (CPCB) 2006, the highest quantity of MSW is being generated by Delhi among the top ten metro cities of India followed by Greater Mumbai, Chennai and Kolkata. Hydrabad, Bangalore and Ahmedabad occupy fifth, sixth and seventh position respectively in terms of generating maximum waste quantity. The MSW produced in Delhi has been rising over the year. It is estimated that Delhi has generated about 6000-8000 metric tones waste per day. But there is confusion regarding the amount of waste generation in Delhi as different agencies and bodies are showing different quantity. The actual waste generated per capita is higher as significant quantity of garbage found lying uncollected and not all the waste are lifted by municipal body. Municipal Corporation of Delhi introduced weigh bridge machines at the three landfills and correct data now exists (Subhash, 2007).

Current Status of Waste Generation in Delhi

In 2008-09, Shahdra (S) Zone is relatively high generator of municipal solid waste. The lowest producer of municipal solid waste is Narela i.e. 1.65 per cent of the total waste production on the basis of Figure 2, the state can be divided into following classification.

1. Zones with relatively high proportion of municipal solid waste (above 10 per cent).

   Three Zones namely Shahdra (S) Rohini and West Zone included in this category. Shahdra (S) Zone is the outer Zone of MCD. Here the quantity of the municipal waste is really high because of unauthorized colonies and migrants living in this Zone. Rohini and West Zone are the residential area and these are occupied by middle and high income groups people, here the waste management facilities are relatively high.

2. Zones with moderate proportion of municipal solid waste (5-10 per cent)

   There are six Zones of MCD, NDMC and DCB lies in this category. The six zones of MCD are Shahdra (N), Civil Line, Karol Bagh, Central, South and Sadar Paharganj. The Sadar Paharganj, Karol Bagh, Civil Line and Central Zones are densly populated and commercial centres and most of the Zones as under the Public Private Partnership. So the collection efficiency is really better than the other Zones of MCD. The NDMC and DCB are as also highly developed region of Delhi. These are moderate producer with 5-10% of the total waste of Delhi.

3. Zones with lower proportion of municipal solid waste (below 5 per cent)

   In this category, three Zones namely Najafgarh, Narela and City Zone are included. Najafgarh and Narela Zones are the geographically biggest zone in MCD area. These two Zones are mainly rural kind of culture here waste management practices are managed by villagers in the villages but in town the waste management responsibility of the MCD. Here the collection efficiency is very low and other management services are not up to level.

Storage and Collection of Municipal Solid Waste

Storage, collection and transportation of waste is an important activity of municipal solid waste management practices. As per Delhi Municipal Corporation Act, 1953 it is the duty of residents to deposits waste from their houses into municipal receptacles. There are about 2200 of dhalaos, dustbins and open sites located at convenient places for the storage and collection of the municipal waste. The community bin system is being followed in the Delhi. In MCD controlled territory, the waste is collected by community bin system in the roofless masonary structure which is called dustbin. Large size roofed structure is called dhalaos. Table 2 clearly shows that core areas are having more receptacles than the outer areas. Karol Bagh, Sadar Paharganj and City Zones are having equally receptacles, and other core areas are also in better condition. Because these Zones comes under the PPP project. But outer Zones like Narela, Najafgarh, Shahdra (N) are not in better position. The maximum number of receptacles comes under the South Zone (399) and minimum is Sadar Paharganj Zone (56). In the outer Zones the placement of receptacles are not as the definite standard which is set by the MCD. So, here a poor kind of facilities in respect to collection receptacles. A lot of dustbins and dhalaos are in broken condition, so the waste is spread on outside the receptacles. It is one of the major cause behind the environmental pollution.

Transportation of Waste

After the storage and collection of waste in the receptacles it is transported to landfill sites. Transportation is a link between the collection and disposal of garbage. The transportation of waste is capital intensive task. The

Table 2
Distribution of Dhalaos/ Dustibins and Open Sites in MCD Zones in Delhi

Zone	*Dhalaos*	*Dustbins*	*Open Sites*	*Total*
Shahdra (S)	66	118	20	204
Shahhdra (N)	107	17	5	129
Rohini	251	32	10	293
Civil Line	99	46	23	168
Najafgarh	72	0	13	85
Narela	34	7	62	103
City	28	44	0	72
Karol Bagh	57	21	6	84
Central	204	35	51	290
South	231	124	44	399
Sadar Paharganj	52	4	0	56
West	182	5	50	237
Total	1383	453	284	2120

Source: MCD (2008)

transportation cost is around 20-45% of the total SWM budget. The main types of modern vehicles used in Delhi are Refuse Removal Trucks (RRT), Loaders, Mini Dumpers and Tractors, Trailers. As per MCD records, it has 777 transport fleet engaged in collecting and transporting the refuse material. There is high unequal distribution of waste transportation in all the MCD Zones. Six Zone like City, South, West, Central, Karol Bagh and Sadar Paharganj are under the Public Private Partnership (PPP) scheme. So here the situation is better than the peripheral area. Under utilization of the transportation fleet is a major problem that results in delay in transportation of the garbage from the collection points of dumping grounds. The major problem is poor maintenance as a result of which a number of vehicles are not functional. Over 50-60 per cent of which are on the roads at one time. Transporting vehicles are really very old, so everyday a lot of trucks, loading machines are standing in workshops. Financial problem is a major problem in maintenance of transporting vehicles (Figure 2).

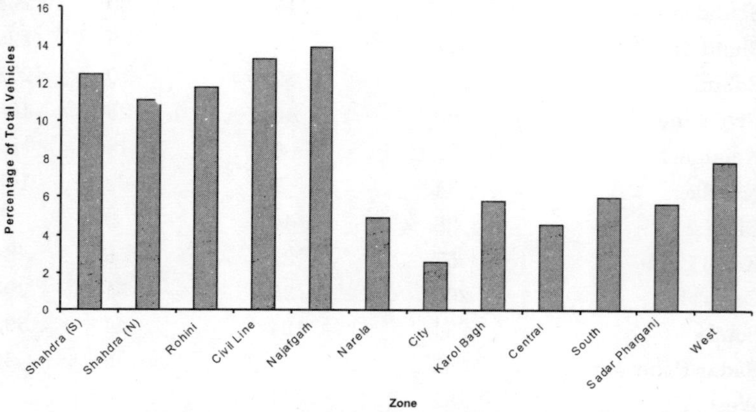

Figure 2: Zone-Wise Waste Transport

Disposal of Waste

The waste collected from various dustbins or dhalaos is disposed off by various methods such as dumping, sanitary landfill and composting. Only open dumping and sanitary landfills are in more attractive practices in developing countries like India. Delhi generates 6000-8000 metric tones of municipal solid waste every day and it will be increased 18, 000 metric tones a day by the year 2020, as per the MCD survey. The main waste generated in Delhi is from the vegetable and fruit markets, retail and commercial centers, slaughter houses and construction and demolition activities. In Delhi about, 93 per cent of municipal solid waste is disposed on land through land filling and rest 7 per cent goes to composting. As per the information provided by the MCD, presently Bhalswa, Gazipur and Okhla landfill sites are in operation. According to the Table 3, Bhalswa landfill site occupies the 21.06 hacters area and it is serving the 50.3%

population of the total population. Every year 44.48% waste deposited in the landfill site of the total waste generation. Bhalswa landfill site serving the Civil Line, Karol Bagh, Rohini, Narela, Najafgarh and West Zone. After the Bhalswa landfill site Gazipur landfill site which is serving the Shahdra (N), Shahdra (S) City and Sadar Paharganj zone. Gazipur landfill site having an area of 29.16 hectare and it is serving the 30.6 per cent of the total MCD Zones population. Every year 36.55 per cent waste deposited in the Gazipur landfill site of the total waste generation in the MCD Zones .

Table 3
Landfill Sites Serving MCD Zones

Landfill Site	*Area (Ha)*	*Population (in %)*	*Waste Dumped (in %)*	*Zones Supplying Waste*
Bhalswa	21.06	50.3	44.48	Civil Line, Karol Bagh, Rohini, Narela, Najafgarh and West
Gazipur	29.16	30.6	36.55	Shahdra (S), Shahdra (N), City, Sadar Pharganj
Okhla	16.20	18.9	18.90	Central, Najafgarh South

Source: Information Provided by MCD.

Okhla landfill site acquires the 16.20 hectare area and it is serving the 18.9 per cent population of the total population and it received 18.90 per cent waste of the total waste. Okhla landfill site serving the Central, Najafgarh and South Zone.

These three landfill sites are running at overflowing level. So it is an urgent need to stop the garbage disposal into existing landfill sites. But due to lack of an alternative site the MCD has continued to dump here. Delhi's situation is bad because neighboring states like Haryana, U.P which fall in National Capital Region have refused to handover any land to dispose Delhi garbage (Khan 2004).

MCD has proposed six new landfill sites for future land filling. Among these, the Narela, Bawana, Deorala, Kair, Put Khurd, Jaitpur and Bawana Kanjhawala. But these proposed landfill sites are not handed over to the MCD till this year because MCD is dependent on Delhi Development Authority (DDA) and Land Building Department for acquisition of land in the area of Delhi.

Processing of Waste

The objective of processing of the MSW is to reduce the weight and volume of waste that has to be disposed off in a landfill. Processing and treatment processes for MSW include physical processes for separation of different fraction, chemical processes like combustion and biological processes like composting and bio-fuel generation.

Composting is to provide much awaited answer to growing problem of municipal solid waste but operational and other problems began to appear. In Delhi, only 7 per cent of the total waste generated goes to composting plants. The first composting plant was set up in Okhla in 1981. It was semi-mechanized plant of 150 tonnes per day capacity. It was closed in 1992. The second plant was established in Okhla by NDMC in 1985. The plant capacity is 200 tonnes per day. The third composting plant was set up in Bhalswa under the private sector named Exnora Private Limited Company. The plant capacity is 500 tonnes per day. The fourth plant is also established in Tikri Khurd with a capacity of 125 tonnes per day. Presently all the composting plants are working below than the full capacity. The main causes behind the low capacity are (I) old machinery and unskilled work force (II) poor quality of waste (III) no any marketing committee or market for selling the final product (IV) poor quality of final product.

So these problems faced by the MCD. Government of Delhi has also tried to treat the waste using incineration cum power generation plant of 3.7 megawatt at Timarpur in 1989. But its operation had to be discontinued with in a short period of time due to low calorific value of waste received at the plant. It can be said that in future composting, incineration and gasification will be the answer of huge municipal solid waste, but in these processes a huge cost will be involved.

Waste Service Index

The waste service index derived through composite index and categorized into three categories of Highly Served Zone (HSZ), Moderate Served Zone (MSZ) and Low Served Zone (LSZ). Three variable i.e. no. of dhalaos/ dustbins in per ten sq. kms, no. of safai karamcharies (SK) in per ten sq. kms area, and no. of vehicles serving in per ten sq. kms area MCD Zones were taken into account for assessing waste service index. Data given by MCD on waste services have been converted into per ten sq. kms of each and every Zone (Table 4, Figure 5).

This waste service index is mainly with the collection and transportation aspects of Municipal Solid Waste Management Practices. Mainly this index shows the services which are provided by the MCD in the respected zones. The zones can be divided into following classification:

I. **Highly Served Zones:** The Waste Service index shows that three zones namely City Zone, Karol Bagh Zone and Sadar Paharganj Zones are highly served Zones. It means that here, waste management services are highly concentrated. The vehicles facility, dustibins/ dhalaos and no. of safai karamcharies are in highly concentration. These three Zones have maximum facilities than the other areas of MCD, and one interesting point is that these are core and important area of Delhi.

II. **Moderately Served Zones:** There are three Zones namely Shahdra (S), Shahdra (N) and Rohini Zone comes under the moderate service category. Shahdra (S) and Shahdra (N) Zones are peripheral areas. The main causes behind the moderate kind of services are unauthorized colonies, migration from the neighbouring states and slums. Rohini Zone is mainly divided in sectors and it is the residence of medium class peoples. So things are changing with the time and facilities are also increasing with time.

III. **Low Served Zone:** In this category six Zones namely Civil Line Zone, Najafgarh Zone, Narela Zone, Central South and West Zones are included of these, some Zones comes under the peripheral areas of Delhi. So the main cause behind the low services is village background less educated population and migration from the nieghbouring states. Because high migration is also a major cause, the population is increasing due to migration, so the facilities also short down to these areas.

The Govt. initiative also the major cause because Govt. gives the best possible facility to core and important areas. These are commercial historical and other institutional areas of Delhi. The given Table 4. shows the waste serviced index of MCD area.

Public Private Partnership

The MCD aimed at increasing efficiency and effectiveness of waste management activities, MCD planned the public private partnership project in six Zones city, South, West, Central, Karol Bagh and Sadar Paharganj. These selected Zones are highly dense populated, relatively developed

Table 4
Waste Service Index Based on Composite Score

Zone	*No. of Dustbins / Dhaloas in per 10 sq. km*	*No. of S.ks in per 10 sq. kms*	*No. of vehicles in per 10 sq. km*	*Total Score*	*Index*
Shahdra (S)	-0.06	0.22	0.02	0.18	MSZ
Shahdra (N)	-0.59	0.04	-0.14	-0.69	MSZ
Rohini	0.29	-0.28	-0.22	-0.21	MSZ
Civil Line	-0.53	-0.73	-0.06	-1.32	LSZ
Najafgarh	-1.30	-1.01	-0.7	-3.01	LSZ
Narela	-1.30	-1.06	-0.78	-3.14	LSZ
City	1.23	1.09	0.02	0.34	HSZ
Karol Bagh	0.88	1.25	0.82	2.95	HSZ
Central	-0.47	-0.68	-0.78	-1.93	LSZ
South	0.23	-0.79	-0.62	-1.18	LSZ
Sadar Pharganj	2.23	2.29	2.98	7.5	HSZ
West	-0.59	-0.33	-0.54	-1.46	LSZ

Source: Compiled by author on the basis of data provided by MCD

HSZ - Highly Served Zone

MSZ - Moderate Served Zone

LSZ - Low Served Zone

than the other Zones of MCD and are generating about 40-50 per cent of municipal waste of MCD area. Under this project 398 sq. km area of Delhi from 61 MCD wards is covered which around 62 lakh population of Delhi (Table 5).

Three agencies with which this civic body signed the 'Concession Agreement' for the project are Delhi Waste Management Private Limited for South, Central and City Zones. Noida based Ag. Environment Infrastructure. Projects (P) Ltd., for Karol Bagh and Sadar Paharganj Zones and Delhi Based Metro Waste Handling (P) Ltd. (MWH) for West Zone only. One of the main objectives into ensure clean and hygienic disposal of municipal waste

Table 5
Volume of Work in Selected MCD Zones under Public Private Partnership

Total Area (aq. Km)	*398*
No. of MCD Wards	61
Estimated Population (Lakh)	62
No. of Dhalaos	1069
No. of Steel Corner Binds	1780
No. of Operated Vehicles	150
Landfill Sites	2

Source: MCD

in the selected MCD Zones. The scope of work includes collection, storage, segregation, transportation and disposal of waste, public awareness campaign. The cost of collection and transportation vary from Rs. 597 to Rs. 693 metric tonne of waste in all the selected Zones (Table 6). The total amount of Rs. 53, 44, 23, 875 is to be paid by MCD to the private companies for this contract of services.

Government Initiatives for MSWM

Some initiatives are undertaken by the Government, these are as follows:

- Daily door-to-door garbage collection.
- Municipal authority shall organize awareness programme for segregating the waste at source as biodegradable and non biodegradable and promote recycling or reuse of segregated materials.
- Vehicles used for transportation of waste shall be covered and waste should not be visible to public.
- Monthly charging for door-to-door collection based on income groups may be implemented.

Table 6
Quantity and Cost of Waste Transportation in PPP project

Quantity & Cost of Waste Transported	*Central*	*South*	*City*	*West*	*Karol Bagh*	*Sadar Paharganj*
Daily collected at Landfills (Metric Tonne)	382	368	452	477	331	279
Collection and transportation costs (Rs./MT)	597	597	597	693	682	682
Total cost /day (Rs./MT)	228054	219696	269844	330561	225742	190278
Total cost per year (Rs. / MT)	83239710	80189040	98493060	120654765	82395830	69451470

Source: Subhash, 2007

- The non-governmental organizations may be involved in the waste management practices.
- Municipal authorities shall adopt suitable technology or combination of such technologies to make use of wastes so as to minimize burden on landfill.
- Composting and Energy Generation from municipal solid waste should be the appropriate option after land filling.
- Partnership of private sector in making up the plans for appropriate management of solid waste should be ensured.
- Up gradation of the existing dump sites and disposal of inert wastes in sanitary landfills.

Conclusion and Suggestions

Management of municipal solid waste in the city is far from satisfactory. There are problems in the solid waste management practices prevailing in the city at every level, i.e., collection, transportation, processing and disposal. Mismanagement of solid waste is a matter of serious concern for public health and environment.

In MCD zones there is no clear cut standard for the placement of waste receptacles. Present design and location of waste receptacles is not accepted to the people. The distribution of transportation vehicles in MCD zones in uneven. Presently, all three landfill sites are fully packed and overflowing Government must encourage composting, vermi-composting, incineration, refused derived fuel etc. processing and treatment methods for reducing the solid waste disposal problems because the processing of the waste is only the answer of municipal solid waste. Some suggestions would be helpful in improving SWM system in Delhi these are as follow:

- Open waste storage sites and other unhygienic street bins should not be allowed.
- The placement of waste receptacles should be correct.
- Segregation of household waste at the source would reduce the burden of solid waste significantly while at the same time improve the supply of composite serving the nutrient poor farmer near Delhi.
- Proper maintenance of vehicles and other equipments.
- Government should adopt 4R's (Reduce, Reuse, Recycle and Resource Recovery) principle.
- Government should increase the number of composting and energy generation plant.
- Public awareness programmes should be arranged by the government regularly.

References

Aggrawal, V. (2008), "Solid Waste Management: Issues and Perspectives" Prashanshika, Vol. XXXV, No. 2, pp. 102-116.

Ali, Sabir (1998), "Environmental Scenario of Delhi Slums" Gyan Sagar Publication, Delhi.

Anand, Subhash (2005), "Solid Waste Generation and Management in Delhi: A Sustainable Approach" IK International Publishers, New Delhi, pp. 621-639.

Ansari Abdul Sameem (2006), "Best Practices in Solid Waste Management: A Case Study of Suryapet" Nagarlok, Vol. XXVIII, No. 4, pp. 12–19.

Bansal S. P. (1998), "Planning for Solid Waste Management in Delhi" In Spatio Economic Development Record, Vol. 5, No. 3, pp. 28–31.

Basu, S. (1998), "Solid Waste Disposal and Sanitation of Durgapur" *Geographical Review of India*, Vol. 62, No. 3, pp. 263–268.

Bhattacharya, A. K. (1977), "Solid Waste Management" *Yojana*, Vol. 41, No. 8, p. 28–34.

Bhattacharyya, J. K. (2008), "Treatment and Disposal Options in Municipal Solid Waste Management" *Journal of IAEM*, Vol. 35, No. 1, pp. 29–33.

Bhide, A. D. (2002), "Solid Waste Management in Urban Areas" Nagarlok, pp. 17-24.

Census of India (2001), "Administrative Atlas" Govt. of India, New Delhi.

Central Pollution Control Board (2005), "Twenty–Seventh Inspection Report on Collection, Storage, Transportation and Disposal of Municipal Solid Waste in Delhi" CPCB, New Delhi.

Central Pollution Control Board (2006), "Assessment of Status of Municipal Solid Waste Management in Metro Cities and States Capitals," CPCB, New Delhi.

Choudhary, B. K. (2005), "Solid Waste Management: A Case Study of Waste Pickers in Delhi, M. Phil. Dissertation, JNU, New Delhi.

Heinen, J. T. (1995), "Review of and Research Suggestions for Solid Waste Management Issues: The Predicated Role of Incentives in Promoting Conservation Behavior, "*Journal of Environmental Conservation*, Vol. 25, No. 2.

Hina Zia and V. Devadas (2006), "Municipal Solid Waste Management in Kanpur City" Nagarlok, Vol. XXVIII, No. 4, pp. 20-31.

Jayasheela, U. Arabi and M.Shoukat Ali (2007), "Urban Solid Waste Management in Karnataka: A Case Study of a City Corporation" Nagarlok, Vol. XXXIX, No. 1, pp. 50–55.

Kappagantula, S. R. (2006), "Solid Waste Management – A Study of Municipalities in West Godavari District, Andhra Pradesh" Nagalok, Vol. XXXVIII, No. 1, pp. 31–39.

Karnad, T. G. (2006), "Planning for Solid Waste Management in Mysore" Nagarlok, Vol. XXXVIII, No. 1, pp. 8–21.

Kausal, A. (2002), "Solid Waste Management Strategies for India" *Indian Journal of Environmental Protection*, Vol. 22, No. 41.

Singh, R. B. and Anand, Subhash (2008), "Current Status of Municipal *Solid Waste Management Practices in Delhi*" NAGI, Annals, Vol. XXVIII, No. 1, pp. 36-49.

Singh, S. (2006), "Solid Waste Management in Resettlement Colonies of Delhi" Bookwell Publications, New Delhi.

Subhash (2007), "Generation and Management of Municipal Solid Waste in Urban Delhi: A Geographical Analysis" Ph. D. Thesis, Department of Geography, Delhi University, New Delhi.

Climate Change and Agriculture in Uttar Pradesh

JITENDRA DWIVEDI AND K. K.SINGH

India is a large developing country with nearly 700 million rural population directly depended on climate sensitive sectors (agriculture, forests and fisheries) and natural resources (such as water, biodiversity, mangroves, coastal zones, grasslands) for their subsistence and livelihoods. Further, the adaptive capacity of dry land farmers, forest dwellers, fisher folk and nomadic shepherds is very low. Climate change is likely to impact all the natural ecosystems as well as socio-economic systems as per the National Communications Report of India to the UNFCCC. The Intergovernmental Panel on Climate Change, in its 2007 report, predicts that global temperatures will rise by 2-4.5.0 °C by the end of this century, with a 2.7-4.3 °C increase over India by the 2080s. The panel also predicted an increase in rainfall over the Indian sub-continent by 6-8 per cent and that the sea level would rise by 88 centimeters by 2100.

India is an agriculture dependent nation and agriculture is still the back bone of Indian economy. Agriculture is still the base of livelihood of around two-third population of our country. There is geographical and climate diversity in India that's why it has been divided into 15 agriculture climate zones. The agriculture and farmer community has been marginalized due to the government policy on one hand and the climate change has worsened the already fragile situation on the other. Today the whole world is concerned about the prevailing situation created by climate change and facing it in different level. All are afraid of the situation of denudation of low land and melting of glaciers of the arctic. Developing countries are more fearful. According to Dr R. K. Pachauri, there is possibility that food production of many countries including India would be badly affected by the climate change, particularly wheat, paddy and other crops would be directly affected and these are the primary food of common Indian people. We can only estimate how the climate change would impact upon the livelihood of majority of people.

According to the recently issued status report on disaster management, around 85 per cent part of India is under disaster prone zone, 40 million hectare of land is under drought situation, 4 per cent is under violent storm and 68 per cent of land is under drought. Not only that the graph of natural calamity have gone up during the last 100 years which has affected human and physical resources. According to the report of World Bank there was loss of 2.25 per cent GDP between period from 1996 to 2000 and the government has to bear the loss of 12.5 per cent.

We have no accurate estimate of the impact of climate change on Indian agriculture but looking at the third report of IPCC it can be said that few crops, which grow in tropical region; have reached at the high level and after that these crops cannot bear (face) the hot situation so after this level the production of crops would be badly affected. For example in 2006 production of wheat was reported to be reduced of 10-15% due to increase in 4.6 centigrade temperature. Production of wheat crop is very important for food security point of view. In such situation production of food would be badly affected if the temperature of the atmosphere increases and it would also affect the quality of the food. Paddy crop can bear the heat more than wheat so there is a possibility that our farmers expand the area of paddy cultivation which is not appropriate from the food production point view.

Uttar Pradesh: An Example

Uttar Pradesh is home to 166.2 million people living in high density- 689 person per sq km. The population is spread over 70 districts, with 300 tehsils and 813 blocks across 4 geographic regions western, eastern, central and Bundelkhan. It is the third poorest state in India with a per capita annual income of US $ 200. About 80 per cent of the people in UP live in rural areas and 66 per cent depend on agriculture for their livelihood, agriculture for their livelihood agriculture accounted for 38 per cent of the GSDP in 2001-02. Six per cent of the population is involved in household industries and 28 per cent in other services.

The social composition of its people is as diverse as its regions. As per the 2001 census, Hindus constitute 80.6 per cent, Muslims 18.5 per cent, SCs 21.5 per cent and STs 0.7 per cent of the population, while other religious minorities constitute 0.9 per cent. Large disparities exist amongst the social groups despite the affirmative action from the government to bridge the gaps of exclusion through affirmative action policies for SCs/STs, besides other policy interventions for addressing alleviation of poverty and deprivation of marginalized social groups.

According to the UP Human Development Report (UPHDR) 2007, UP ranked 15th in terms of per capita income out of the 18 major Indian states considered for the study. Its poverty ranking amongst the states was 11. 32.8 per cent of the population was below the poverty line in 2004-5. Every 6th malnourished child in the country lives in UP. The Infant Mortality Rate (IMR) is 75 per 1000 much higher than the National average of 58 per 1000 (2000-2005). Almost 40 per cent of its population remains uneducated, seriously affecting their ability to make any demands on the state for a better quality of life, access to development initiatives or healthcare or even a livelihood.

Land Holdings, Land Distribution and Agricultural Production

Of the state's total geographical area of 24.2 million hectares (mha), 16.8 mha (69.42 per cent) is under cultivation. Agriculture contributes about 40 per cent to the State Gross Domestic Product (SGDP) as against 25 per cent at the national level. According to the 2001 Census, 62.12 per cent of the state's total workers are engaged in agriculture. UP contributes, on an average, 21 per cent to the national production of food grains. With the average food grain production of about 42.7 million tonnes and per capita production of 234 kg per year, third highest among major states, UP is considered to be a food grain surplus state.

UP's agriculture is characterized by very small sized land holding; around 90 per cent of the farmers in the state are small and marginal farmers. Some 73.8 per cent of the total operational holdings in the state are marginal (below 1.0 ha.) and another 15.5 per cent holding are small (between 1 and 2 ha.). Due to the preponderance of the small holdings, UP agriculture is still largely subsistence oriented.

Landholding in UP is becoming more fragmented over the time. Thus, between 1985-86 and 1995-96, while the total size of the holding increased from 18.98 million to 21.53 million ha, the proportion of the area under medium and large holding declined from 8.3 per cent to 7.4 per cent and from 3.22 per cent to 2.5 per cent respectively, and the share of the area under marginal holdings increased from 72.6 per cent to 75.4 per cent, the estimated average holding of marginal farmers and small farmers are 0.35 ha and 1.38 ha respectively. The average size of land holding in UP cutting across all categories of farmers is 0.85 ha.

Health

Although UP has a fairly large public sector health infrastructure which had an outlay of 4.6 per cent till the 10th plan, it received a boost to 7.3 per cent outlay for the 11th plan. However, only 9 per cent people actually make use of this facility for treatment of ordinary ailments. The Primary Health Centers (PHCs) at the village level in most cases function minimally. Doctors absent themselves regularly as most of them live in urban areas, far from the centers. The (ANMs) (auxiliary nurse and midwife) are not available on a daily basis as they have their own issues regarding commuting to work areas.

Patients with serious diseases or child birth complications have no choice but to resort to District Civil Hospital. Here again, lack of ready transportation and financial poverty impair recovery. In UP, only half of all pregnant women receive antenatal care services and only 18 per cent women have institutional deliveries. More than 80 per cent women in the rural areas are dependent on local traditional birth attendants (dais) for deliveries.

The Total Fertility Rate (TFR) for rural women in UP remains at 3.8, much higher than the national average of 2.8. It could be assumed that women in rural areas have more children as immunization is low, diseases cannot be contained and hygiene is of low priority, and thus more children die at infancy, resulting in more pregnancies to replace the loss. The proportion of public resources the state commits to health services is one of the lowest in the country. UP had a low increase in per capita expenditure on health after 1998 even though the decadal population increase has been 16.6, UP from 15.2 from 1991-2001. Per capita expenditure for Family Welfare in 2001-02 rose to only Rs 18.5 from Rs 15.3 in 1995-96.

Japanese encephalitis has emerged as a serious problem in rural areas of eastern UP. The scourge of the disease is most severe in Gorakhpur District. The disease is now gradually spreading to other parts of the state. This was recently evident by the death of more than four dozen children infected with the disease in Sarahanpur District of western UP. According to State Government sources, Japanese encephalitis has claimed more than 2700 lives since 2001 in UP. Medical surveys in the state reveal that of the total diseased persons, children represent 80 per cent. Thus, compared to adults, children are more susceptible to the disease. Children surviving the disease often develop complex problem relating to the brain. Water-logging and pig farming are the two most important predisposing factors favoring the development and spread of the disease.

Impact of Climate Change in UP

Increasingly, the impact of climate change and global warming is being felt in the state, as in other parts of the country. On the whole, there has been a greater frequency of droughts and floods, one of the hallmarks of climate change. By the first half of the four-month monsoon season in 2007, according to the World Meteorological Organization, the region experienced double the number of monsoon depressions, causing heavy rainfall and flooding across South Asia. Nearly a third of India's meteorological districts had received higher than average rainfall by the first week of August 2007.

U.P. has been divided in 9 Agro climatic zones. Around 80% of the population of the state is engaged in agriculture activities; out of 91% population is under the category of small farmers who are solely dependent on agriculture for their livelihood. Condition of agriculture has worsened. There is prevailing situation of drought and flood due to its physical formation and natural calamities of last 10 years are indicating any big crisis. Uncertainty in monsoon, changing pattern of flood, increasing duration of drought and flood, increase in water crisis, increase in violent storm, etc. can be seen in the region.

Thus, in eastern UP, chronically flood prone, the very nature of the flooding has changed, with a greater intensity of flash floods. Embankment along the various rivers, unable to withstand the heavy flooding, often breaks, washing away croplands and mud homes. There has been change in flooding patterns too. The flood here are frequently flash and accidental (that is, sudden and unexpected). Crop damage is on an increasing trend. Cropping patterns are changing and pulses (once a major crop in the area and a major source of protein) are not grown due to longer periods of water logging, which disrupts the whole crop cycle and production, even in the rabi season, is severely affected. Many become victims of water borne diseases like diarrhea, cholera, dengue and Japanese encephalitis as the flood waters stagnate, the natural lines of drainage being disrupted due to embankments, roads and other encroachments. As Action Aid's Hunger Monitoring Project found in 2007, the last four to five years have seen an accentuation of adverse weather conditions with farmers being affected by prolonged drought, floods and hailstorms and decrease in annual rainfall.

In the Bundelkhand area, with its high levels of poverty, many small and marginal farmers are indebted, both to moneylenders and government banks. As the weather gets hotter, the chances of paying back loans become difficult, leading to stress and, in some extreme cases, suicides. The growing water scarcity poses further problems of survival to people and animals alike. In recent years, the level of the water table has gone down significantly and there are reports from the region of cattle dying due to water shortage. While climate change is affecting everybody, regardless of caste, ethnicity, sex, race or level of income, women in the poorer households suffer the most. Their unequal position in society means that women have less access to money, land, and food protection from violence, education and healthcare. They are more dependent on natural resources for subsistence. Hence, they are more exposed to climate shocks and have fewer resources to protect their own lives, assets and livelihoods while looking after their families.

Flood and Water logging

The impact of climate change is reflected in the aggravating flood situation; particularly the area of eastern Uttar Pradesh is badly affected where many small and big rivers exist. Most of the rivers are flown from Nepal which is affected by the aggravating flood situation. Gradually, the change in frequency, nature and impact of flood is seen in the regions which are as follow:

- There is a change in frequency, timing and quantity of rainfall.
- Days of monsoon have been decreased.
- Heavy flood situation and dams are collapsed due to heavy flow of water which increases the chances of flood.
- The problem of water logging is aggravating even from the small rivers because silt is formed on the river bed. As a result there is no depth in river and even low rainfall creates the situation of flood.

There is very less gap between two rivers and the aggravating situation of flood has increased the change of water logging. There is very low slope (20 cm per km) in the region which delays the flow of water from the region. There are many crops which are badly affected as a result of this and many of the crops have been disappeared from the region. Cases of water born diseases are increasing. Some of the areas of eastern U.P. and Bihar are badly affected by Japanese encephalitis which aggravate primarily during monsoon and the reason of its spread is water logging.

Drought

One region of the state is facing situation of flood and water logging and the other southern part is facing situation of drought. In some of the region, the situation of drought has become a permanent feature. The reasons of this situation are:

- Low and untimely rainfall.
- Increase of process of vaporization due to increase in the temperature.
- Decrease of moisture in the soil due to drought.
- Decrease in carrying capacity of water in soil.

Change in Temperature and Flowing of Hot Wind

- Rabi crops have been affected due to rise in temperature during winter.
- Untimely increase in the temperature destroys the growing flower of crops which resulted in decrease in the agriculture production.
- Before time flow of wind from west reduces the size and number of food grain which resulted in low production.
- The rise in temperature is also increasing the cases of many new diseases, particularly skin disease.

Water Crisis

If we assess the ground water then we find that ground water table is going down. The main reason is that farmers are not getting rain water in time of their need because of heavy rainfall in short span of time and farmers are dependent on ground water for irrigating their agriculture land. Increase in temperature is also increasing the demand of irrigation.

Women in UP

In 2000,UP's Department of women and child development, with the help of the Mahila Samakhya and in consultation with multiple stakeholder government, NGO's, planners, academicians, activists and many other committed people drafted a policy for Women. The draft Women's Policy has been vetted, revised and is finally ready, but has to be publicly announced by UP government, a reflection of the priorities and commitment of the state administration.

Women in UP face a considerable gender gap in human capital, due to low levels of education, which is reflected in the literacy levels, eastern UP is a case in point. Their contribution, though immense in reproductive and unpaid work, is undervalued due to gender biases and this is accompanied by lack of educational opportunities for women as well as lack of access to health care. Consequently, population density is high in certain regions due to large family size as reflected in a high TFR.

Gender wage gaps are prevalent in the rural sector with women's low bargaining power due to lack of other work opportunities. The differential wage rate continues to be prevalent in the unorganized sector. Though women's participation in the labour market formal and informal has increased, they are burdened with domestic responsibilities, suffer from job segregation choices for work/jobs due to economic constraints. Work and job insecurity is high due to low level jobs because of lack of education. Women experience poor working conditions and other work place-related discriminations and harassment. Labour market discrimination against women, especially in managerial positions, is evident across sectors and levels. Women's presence in the work arena is impacted equally by economic development and caste dimensions. The work wages for different types of work for agricultural operations varies for males and females. This is evident in the table 1:

Table 1
Differential Wages in Agriculture Work

Activity	*Amount in Rs. (Female)*	*Amount in Rs. (Male)*
Sowing	45.29	51.12
Harvesting	47.43	52.03
Threshing	47.83	51.95
Other	45.47	60.48

Source: Government of UP, 11th Plan Document, 2007-08

When mechanization replaces manual labour, it is the women who get displaced, leaving them hardly any opportunity to find them any opportunity to find alternate jobs. Men often migrate to other areas in search of work, a possibility that is closed to women due to social and cultural factors, as well as the necessities of reproductive work. Paradoxically, more than 50 per cent of the total work force comprises women but only 13 per cent are seen as workers since much of the work (87 per cent) is not economically valued as it entails reproductive work. However, in the urban areas, the proportion of women workers in secondary and tertiary sectors is roughly the same as men. Work in the household industry and informal sector entails long working hours, hardly any social interaction and low monetary compensation. So, it is not surprising that women with low bargaining power are seen here in large numbers. The organized sector employed only 9 per cent women as against 91 per cent males (1998-99). Women were most visible in the casual labour sector (Table 2).

Table 2
Percentage of Gender Distribution in the Labour Market

Work	*Male*	*Female*
Cultivations	41.0	20.0
Labourers	20.0	41.22
Household Industry	4.35	8.24
Other work	33.51	12.0

Source: Government of UP, 11th Plan Document, 2007-08

Micro-finance initiative for economic empowerment of poor women, for making accessible funds for emergencies, consumption, productive enterprises and personal saving, are yet to become a visible force in UP. This strategy for poverty alleviation and market action has largely been taken up NGOs.

Government initiatives to set up self help groups (SHGs) through Swarn Jayanti Gramin Swarojgar Yojana (SJGSY) have not taken off because of lack of concerted efforts to work with women around their economic needs, their inherent skills and capacities, and connect them with existing resources and markets. UP has been way behind in this movement.

Rural women continue to be denied rights so they are unable to access finances and agricultural inputs (technical and non technical) to raise productivity of small holdings (typical of poor SC/OBC families), particularly where men migrate or when they are engaged in share cropping. Consequently, productivity is affected, providing low yields through low inputs and inefficient practices, which ultimately makes small scale marginal farming unviable and cost intensive. As a result, women farmers, who from the bulk of the small and marginal sector in agriculture, continue to face challenges of breaking even and earning substantive profits from farming. Even agriculture extension and low cost/no cost technology and inputs are not available to them through the government extension network.

What is also becoming increasingly visible is the gap in rural-urban incomes. Agricultural wages have not increased at the pace of other wages or salaries. If the majority of people continue to be located in the agricultural/ rural sector, inequalities are likely to rise further between rural urban income and expenditures. Here too, women are more adversely affected than men because of low wages, poor work opportunities and low bargaining power. They also face vulnerability due to climatic vagaries such as floods/drought that impact work availability in rural areas.

Poor SC/ST and OBC women face violence and sexual harassment at work in all sectors. Female work force participation rates (FWPR) for them are higher than for other castes, also, The gap between males and females is much lower than in other castes, in both urban and rural areas. UP also has a lower WPR for SCs/STs than the national average. In the organized sector, only 9 per cent participate as workers compared to the national average of 16 per cent with the highest rates being in Kanpur, Lucknow and Ghazibad. The highest proportion of regular salaried workers and the lowest proportion of casual work are seen among the upper castes. The SC/ST and OBC have a low per centage of regular and salaried jobs 7.9 per cent SCs and STs constitute the bulk of casual labourers. Muslim women's labour participation is also the same, at 20.7 per cent, though salaried work is low, but higher than SCs.

Impact of Climate Change on Women

Impact of climate change is more on women particularly on poor women because they are more dependent on natural resource for their livelihood and climate change impact more on these resources. Agriculture is affected by the climate change and as a result male counter of the family have to migrate in search of their livelihood. It is the women who alone have to bear the double burden at home. Their responsibilities have been increased, Wealthy related problems have increased with the increase of water born and temperature related diseases.

Climate change affected their livelihood and as a result their coping capacities to deal with the day to problems of life have been reduced. Women farmers have to work as a labourer and to get credit from landlord on high interest rate and many times they have to sleep empty stomach. During the flood situation personally and publically stored grain, seeds and other food item is destroyed and rates of all these things go up due to its scarcity in the market. Women are mainly affected with all that the situation worsens for the women working as wage labourer.

It can be said that impact of climate change is more on the women and in future their vulnerability will only increase to deal with the changing climate. So, it is the challenge for the development professionals to increase the capacity of women to deal with the deteriorating climate condition.

Under prevailing conditions of climate change it can be said that we need to conduct a study on short and long term impact of climate change. There is no scarcity of agriculture scientist but the problems are that agriculture scientists are not accepting climate change. That's why no research is being conducted in this field. Whereas we need to do two things in this direction. First, how the climates change is affecting the agriculture cycle. Second, is it possible to grow some alternative crops to deal with the alternative crops to deal with the situation? At the same time we need to develop the crop variety which could bear the risk of climate change. For example, we need to develop such varieties of crop which could bear hotter situation and heavy and low rainfall. It is true that climate change is affecting agriculture so we cannot limit its impact. No country can escape from its impact. So it is necessary that we discuss very seriously at the ground level.

Case Study

Kamlawati Devi is a small woman farmer of her village, Vanbhagalpur is situated at block Compiergang in Gorakhpur District. The total population of the village is around 1100. The agriculture land is plain but land is some parts of the area are uneven. Around 80 per cent of the farmers in the area are small or marginalized farmers who have less than one acre of land. Paddy, wheat and oilseed are the main crops of area but vegetables are also produced in the field. The livelihood of around 80-85 per cent of people are dependent on agriculture. Kamlawati primarily grow vegetables, paddy, wheat and corn and groundnut. But do not understand the balance between nature-summer-winter and monsoon. People of the region have neglected the agriculture activities due to the impact of climate change and those engaged in daily basis work or working in the field are not able to survive in the prevailing condition and so they are migrating from the region in search of job. The change in climate has also affected our health. People of villages are suffering from unknown disease. In these circumstances, now we have changed our agricultural practices in the face of climate change, today we cultivate mixed agriculture due to drought situation. We sow together corn, Groundnut, Arhar, Nanua and other vegetable because if due to heavy rainfall the crops of Arhar and groundnut are destroyed even than we have other option of producing corn and nanua crops and in case of drought situation we produce all crops. Similarly we have increased the area producing vegetable whereas earlier we used to produce crops in the 90 per cent of our agriculture land and only 10 per cent of the land was used for vegetables. Today around 80 per cent of people of the village have started sowing vegetables.Now whatever the scheme have been implemented by the government needs to be changed and updated as per the climate change.

Local Policy and Institutional Context of Disaster Management

The natural disasters that is significant for the state floods, drought, fires and earthquakes. Loss of life and property from these disasters, especially the former three, amount to hundreds of crores of rupees, annually. The recurring floods and drought are manifestations of increased vulnerability and the inadequacy of the sporadic mitigation measures attempted.

Floods are the most commonly occurring disaster in UP, affecting approximately 2.7 mha of the area and causing losses up to Rs 432 crore. Out of the 240 mha of the total area of UP, approximately 7.3 mha is flood prone. As per the Irrigation Department's estimate, only 5.87 mha of this area can be protected and the protection provided so far is merely for 1.6 mha. The areas worst affected by floods are eastern UP as well as those situated in the terai region bordering Nepal. The Bundelkhan area is chronically drought affected.

The traditional framework for disaster management in UP has been oriented towards natural hazards and civil disturbances, with revenue, police and fire services as primary emergency responders. The focus has been more on immediate relief and recovery, with the State Relief Commissioner, District Magistrates and Superintendent of Police as crisis managers. With the increase in frequency of disasters, the emerging context is one of rising levels of vulnerability and escalating costs of disasters, significantly narrowing the differences between natural and human made disasters. There has been paradigm shift in the approach to disaster management, from reactive relief and rehabilitation to proactive mitigation of disasters and pre-disaster preparedness. UP has been progressive on disaster management measures, having enacted the Disaster Management Act 2005 (the third state to do so after Gujarat and MP). The act provides legal backing to all preparatory and post disaster measures and responses and allocates major responsibilities to all the stakeholders. The UP Disaster Management Authority (UPDMA) has been set up. It is headed by the Chief Minister as its Chairperson, with a

14 member Governing Board. The District Disaster Management committee (DDMC) is headed by the respective District Magistrates. Disaster specific Working Groups were also formed for each specific disaster, to dovetail ongoing schemes with specific disaster mitigation efforts and address disaster management concerns in annual and five year plans. The working group is headed by an officer at the level of Principal Secretary. The Remote Sensing Application Center- UP (RSAC-UP) advises each working groups. Natural resources related GIS Mapping of 40 districts has been completed in UP. The Academy of Administration and Management is the nodal institute for all training programmes related to disaster management and disaster management modules have been adopted in all in service training programmes in the state. The subject has been also been introduced in school curricula.

The GOI-UNDP Disaster Risk Management Programme is currently running in 13 pilot districts and 6 cities. Disaster Management Development Plans have been initiated at district block and village levels. The UNDP supported Disaster Risk Programme operative in 17 states, is also being implemented in UP, covering the following 12 districts- Baharaich, Balrampur, Bijnor, Budaun, Deoria, Ghazipur, Gonda, Gorakhpur, Rampur, Saharanpur, Sant Kabir Nagar and Sitapur.

The India Disaster Resource Network (IDRN), under the auspices of MHA-GOI has a district-wise inventory of resources and district magistrates have been directed to update this database on a priority basis. State and District Level Emergency Operation Centers (EOC) have been set up at the office of the Relief Commissioner and 13 pilot districts (Under the UNDP-GOI programme). The district centers are equipped with National Informatics Centre (Nicnet) facilities. SPACENET and POLNET are also being established as per GOI guidelines.

Concluding Remarks

As such, land, water and human resources are the most important endowments in UP. The state has been divided into nine agro-climatic regions, areas with their own agro-climatic specifications and related socio-economic linkages. Small and Marginal holders in agriculture comprise the mainstay of the economy in each of these zones. However, the exact agricultural patterns and livelihood mechanisms vary widely. Essentially, then locale specific and household specific approaches are needed to solve the problems: a "one package fits all" approach does not work. The state faces many problems, including that of structural adjustment and now climate change, all of which impact the sustainability, not only of agriculture and other livelihood mechanisms, but of the very human actors themselves. This is accompanied by oppression at the local level and the distancing of the masses from governance, despite the panchayati raj institutions (PRIs). Forced /distress migration, usually as casual labour to the urban areas of the state and the nation's metros, is the order of the day. The state led solutions, as reflected in major projects like UPDASP and UPLDC, though purportedly eco-friendly and sustainable, end up not being able to achieve their objectives as they are mainly centered around technology transfer and enhancing agri-outputs and do not seek to address the social aspects and long term interests of small, marginal, landless and women farmers, as the approach is not context specific. While components like biodynamic agriculture, organic farming, and indigenous traditional knowledge and so on have been incorporated due to growing pressures, a definite and comprehensive plan is still not in place. Gender equity and other equity issues are largely ignored due to the traditional mindsets of agricultural scientists in northern India.

References

CEAG (2008) Climate change Adaptation Activities in India.

CEAG (2009), Climate Change & Agriculture.

Government of UP, 11th Plan Document, 2007-08.

Kurukshetra (2000), Kurukshetra Magazine, March Issue, Delhi.

Reilly, J. N. Hohmann, and S. Kane, (1994), Climate change and Agricultural Trade: Who benefits who Loses? Global Environmental change 4, No. 1: 24-36.

Yojana (2008), Yojana Magazine, June Issue, Delhi.

Bio-medical Waste Management In Delhi

KIRAN DABAS

Introduction

Every system in nature progresses towards disintegration and it contributes to creating waste (anything which is not intended for further use). Rapid increase in population, industrialization, urbanization, fast changing life style over consumerism result into increase in quantum of urban solid waste. With increasing need of health care in fast changing society, the role of hospitals/nursing homes comes to the forefront. Hospital is a residential establishment which provide short term and long term medical care consisting of observational, diagnostic, therapeutic and rehabilitative services for a person suffering or suspected to be suffering from disease or injury and for parturient. A modern hospital is complex multidisciplinary system which consumes thousands of items for delivery of medical care and is part of physical environment. All products consumed in hospitals have some unusable left overs which is known as biomedical waste. It is ironical that hospitals/nursing homes and clinics, which provide succor to the ailing, also create health hazards through indiscriminate disposal of bio medical waste. Therefore, waste management is important for public health. Today with increased use of disposable material and the presence of dreaded disease like Hepatitis-B and AIDS, it is almost important to take care of the infected and hazardous waste to save the mankind from disaster. Proper management of hospital waste is essential to maintain hygienic, aesthetics, cleanliness and control of environmental pollution.

The Study Area

The national capital Territory of Delhi comprising of nine districts with an area of 1483 sq. km. is situated between the Himalaya and Aravali range in the heart of the Indian subcontinent. It lies between 28° 25' and 28°53' north latitude and 76°50' and 77°22' East longitude. There are three statutory town continuing in Delhi since 1961- (i) New Delhi Municipal Council (ii) Delhi Cantonment Board (iii) Delhi Municipal Corporation. There are nine districts and twenty seven tehsils in Delhi. The total population of Delhi is 137, 82, 976 (Census 2001). 93.01% of total population lives in urban areas in 2001 in Delhi. In Delhi there are about 72 hospitals under government sector, 550 registered nursing homes and 936 dispensaries. In addition to this, there are about 1560 unregistered establishments with different names like nursing homes, Medical centers, Dental Hospitals, MTP centers etc. About 40,000 Hospital beds are available in public and private sector in Delhi. There is big network of health care institutions in Delhi. Although they are not under one banner but these can be utilize by better coordination among different organization. (Table 1). With increasing number of hospitals and nursing homes in Delhi, this number may go up even higher.

Objectives

The objectives of the present study are:

1. To enquire about the nature and volume of bio-medical waste generation in Delhi.

Table 1

Sl. No.	*Organization*	*Hospitals*	*No. of Beds*	*Dispens-aries*	*Allopa-thic*	*Homeo*	*Ayur*	*Unani*	*SHS*	*MHS*	*MCWC*	*Total*
1	Delhi Govt	37	7119	385	187	51	12	0	63	72	0	385
2	MCD	15	3625	274	37	14	99	15	0	0	109	274
3	NDMC	2	200	45	11	12	10				12	45
4	ESIC	4	1000	34	34							34
5	Central Govt.	10	3840	99	84	11	3	1				99
6	Autonomous	6	2994	0								0
7	Defence	3	1850	1	1							1
8	DVB	0		24	24							24
9	DJB	0		15	14		1					15
10	DTC	0		27	27							27
11	SBI	0		9	9							9
12	RBI	0		8	8							8
13	Railways	2	466	12	12							12
14	BHEL	0		3	3							3
15	Indian Airlines	0		3	3							3
16	Regd. NH	606	14950	0								0
17	Un Regd. NH	1550	~ 5000	0								0
	Total	2235	41044	939	454	88	125	16	63	72	121	939

Source: Directorate of Health Services, GNCT of Delhi.

2. To investigate about existing infrastructural and operational machinery of Government for proper disposal & management of bio-medical waste.
3. To give suggestions to combat with the problems relating to bio-medical waste.

Bio-Medical Waste in Delhi

Handling and disposal of hazardous and non hazardous waste has become big task due to growing urban population of Delhi. Improper disposal has become one of the challenging problems of urbanization. The bio-medical waste is of special kind of waste. Bio-medical waste adversely affects the environment and contributes to the overall environmental problem of solid waste disposal. The infectious nature of bio-medical waste itself reveals the danger of mismanagement This waste has capability of being communicated by multiplication of body tissues and capable of causing disease or adverse impact on human body and environment. Moreover new hospitals are coming up in a mushrooming speed to meet the health hazards. Main cause of increase quantity of medical waste is the shift from reusable items to disposal goods. The problem of medical waste has acquired gargantuan proportions and complex dimension. On an average a hospital bed generates 0.5 Kg waste per day, out of which 10-15% is infectious and rest is general waste. Therefore, proper disposal of biomedical waste is needed. The improper handling, treatment and disposal of the biomedical waste leads to spreading of diseases such as HIV and hepatitis and infections like ocular, genital and skin infections.

Sources of Medical Wastes

Hospitals are primary target for regulation, but they are not the only generators of medical wastes. The range of potential generators includes the following:-

1. Hospitals: This category includes facilities providing general medical and surgical care, psychiatrie care, and treatment of tuberculosis, and services in specialties such as obstetrics' and gynecology, eye/ear/nose/throat, and rehabilitation.
2. Intermediate care facilities: This category includes nursing homes and facilities for in-patient care for the developmentally disabled.
3. Clinics and physician offices.
4. Dental offices.
5. Laboratories: This category includes the full spectrum of research, medical, industrial, diagnostic, manufacturing, and pharmaceutical preparation labs.
6. Others: Following are the remaining generators of medical wastes:

Funeral homes	Veterinarian
Agricultural facilities	Blood Bank
Animal care facilities	Ilospics
Emergency medical services	Illicit drug users
Home health care	Medical schools
Public health units	Nursing

Waste Characterization

For waste to fall within the guidelines pertinent to medical waste, they must possess the potential to transmit infection or present a public health or the environment for reasons other than their infectious potential. The following are the seven categories of wastes, by virtue of their characteristics, are universally handled as medical wastes:

1. **Sharp:** This category includes hypodermic needles, syringes, scalpel blades, blood specimen tubes, Pasteur pipettes, and broken glasses.

2. **Cultures and stocks of infectious agents:** This category includes specimen cultures from medical and pathological laboratories; if infectious agents from research and industrial laboratories; culture dishes; devices used to transfer inoculate, or mix cultures; and discarded vaccines.
3. **Bulk human blood and blood products:** Blood and items that are saturated and have the potential for dripping blood as well as blood products such as serum and plasma.
4. **Pathological waste:** This category includes any human tissues, body parts and/or fluids, or organs removed during surgery or postmortem procedures.
5. **Isolation wastes:** This category includes wastes contaminated with blood, excretions, exudates, or secretions from sources infected with highly communicable diseases.
6. **Animal waste:** This category includes body parts, contaminated carcasses, and fluids of animals that have been afflicted or purposely infected with agents during research, production of biological or testing pharmaceuticals.

Generation of Medical Waste in Delhi Government Hospitals

There are 37 Hospitals under government of Delhi. The total no. of beds in different hospitals is different. Total no. of beds in Delhi government hospitals are 7119 (Figure 1). Bio-medical waste generated in Delhi govt. hospitals, bed wise per day in December 2006 is given in figure 2.

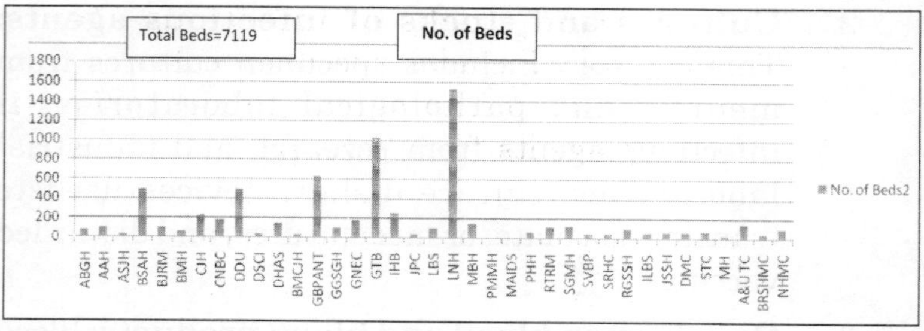

Figure 1: Bed Strength in Delhi Govt. Hospitals

Source: DHS, GNCT of Delhi, 2008.

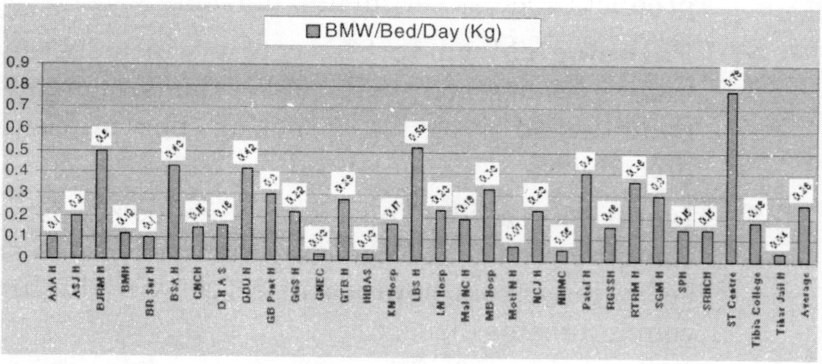

Figure 2: Bio-medical Waste Generated in Delhi Govt. Hospital

Source: Directorate of Health services, GNCT of Delhi, 2006

Dealing The Bio-medical Waste

Under Bio-medical waste (Management & Handling) rules 1998, all health care institutions are required to handle bio-medical waste in specified manner. Delhi is generating approximately 7000 metric tones of waste out of which 70 tones are bio-medical waste. According to the rule, there are a few basic steps that hospital have to follow in order to deal with biomedical waste.

- The first and the most crucial step is segregation. Segregation of Waste refers to the basic separation of the different categories of waste generated at the source of their generation.

- The Bio-Medical Waste (Management and Handling)Rules, 1998 have stipulated the following method of segregation on the basis of a simple colour-coded system (Table 2).

Table 2

Colour Code of the Rules	*Waste Category*	*Treatment as Per Schedule I*
Yellow	· Human Anatomical Waste · Animal Waste. · Microbiology and Biotechnology Waste. · Soiled Waste.	Incineration/Deep Burial. (These containers of bags shall not be made of chlorinated plastic)
Red	· Microbiology and Biotechnology Waste. · Soiled waste. · Solid waste.	Autoclaving/Microwaving/ Chemical Treatment.
Blue White Translucent	· Waste Sharps. · Solid Waste.	Autoclaving/Microwaving/ Chemical treatment. Destruction and shredding.
Black	· Discarded medicines and Cytotoxic drugs. · Incinerator ash · Chemical Waste (Solid)	Disposal in Secured Landfill.

Technology Used for Bio-medical Waste Treatment and Disposal

As bio-medical is a specialized class of waste, that is highly infectious and hazardous, there are specific technologies required to treat and dispose the waste. The standard technologies being utilized in the country for treatment and disposal of biomedical waste are:-

- Microwave
- Autoclave
- Incineration

Microwave

- The microwave is based on the principle of generation of high frequency waves.

- These waves causes the particles within the waste material to vibrate, generating heat.
- This heat generated from within, kills all pathogens.

Autoclave

- The autoclave operates on the principle of the standard pressure cooker.
- The process involves using steam at high temperatures.
- The steam generated at high temperature penetrates waste material and kills all the micro organisms.

Incineration

- A process that works on the simple principle of burning or combustion is technically called incineration. The incinerator, as the machine is referred to, uses either oil or electricity to power itself. Waste material is fed into the incinerator and is burnt in it. Normally incinerators are operated at temperatures between 300 °C to 1100 °C based on the volume of waste, the type of incinerator and the type of fuel used.

There are 37 Hospitals under government of Delhi. 6 Hospitals are having incinerators and 9 hospitals are having autoclaves and shredders and 3 Hospitals are having microwave for scientific management of Bio-medical waste. Bio-medical waste from those hospitals, where such facilities are not available, are segregated and transported in special vans to hospitals where such facilities exist. The government of NCT of Delhi (GNCTD) has been allotted land by Delhi Development Authority for establishment of centralized Bio-medical waste treatment facilities, 1000 sq meter each at Okhla & Gazipur in Delhi.

Occupational Hazard

Every hospital in Delhi generates average of 150 Kg of waste every day, though the figures varies with the capacity of the Hospitals. The major contribution comes from the Govt. Hospitals, at the same time the private Hospital with clean image have been found equally responsible for disusing the medical waste harmfully. The hospital staff manages the waste only, and most of the Hospitals do not segregate their waste in infectious and non infectious waste. Though the waste is treated in the Hospital incinerator but most of the Hospitals have damaged incinerators, they do not work. The issue of incinerators coupled with the pollution that the incinerators, itself creates. Hospital staff and waste handlers in the Hospital run the highest risk f contracting diseases from following improper bio-medical waste management practices as they are exposed to a multitude of diseases and infection.

The bio-medical waste scattered in and around the hospitals invite flies insects, rodents, cats and dogs that are responsible for the spread of communicable diseases like plague and rabies. Rag pickers in the hospital sorting out the garbage are at a risk of getting tetanus and HIV infections. The recycling of disposable syringes, needles, IV sets and other articles like glass bottles without proper sterilization may be responsible for Hepatitis, HIV and other viral diseases. It become primary responsibility of health administrators to manage hospital waste in most safe and eco-friendly manner.

Observations

There are some inadequencies and drawback in the present bio-medical waste management system. They are as follows:

1. The segregation of waste in almost all hospitals is not satisfactory.

2. Colour coding for various categories of waste is not followed.
3. The storage of bio-medical waste is not in isolated area and proper hygiene is not maintained.
4. Personal protective equipment and accessories are not provided.
5. Most of the hospitals do not have proper waste treatment and disposal facilities.
6. Emission monitoring of five incinerators indicated that they do not meet the emission norms.
7. Most of the incinerators are not properly operated and maintained, resulting in poor performance.
8. Sometimes plastics are also incinerated leading to possible emission of harmful gases.
9. Several hospitals have not applied to State Pollution Control Board for authorization under the rules.
10. General awareness among the hospitals staff regarding bio-medical waste is lacking.

Recommandations

1. Hospitals should apply to State Pollution Control Board for authorization to handle and treat the waste.
2. Incinerators, which do not conform the design & emission norms as per rules, must be modified and air pollution control system may be retrofitted to minimize the emission level.
3. The operator should ensure proper O&M of incinerator through attainment of required temperature in both the chambers, regular operation of the incinerator, proper maintenance of the logbook and storage of the waste in isolated area; plastic incineration should not be undertaken.

4. Proper training and personal safety equipment/ accessories should be provided to waste handling staff.
5. Records of waste generation, treatment and disposal should be maintained by the hospital.
6. Various regulatory agencies, hospital, medical associations and municipal corporation should work together for proper management of bio-medical waste in Delhi.

References

Central Pollution Control Board (2006), "Manual on Hospital Waste Management, May 2006.

M. Kela, A. Goel, Ratna Singh, R. Aggarwal (2002), "Implementing Hospital Waste Management. A Guide for Health Care Facilities", Srishti, June 2002.

Planning Commission (2009), Report of High Power Committee on Urban Solid Waste Management, Planning Commission, Govt. of India.

Roos R. L. (2004), Environmental Concerns in Hospital Waste Disposal, Hospitals. 48: 120-122.

Satpathy S., Pandhi R. K. (1998), Manual for Hospital Waste Management at AIIMS, New Delhi.

SECTION - II

DEVELOPMENT

Eco-development: Can Development be a Sustainable Proposition?

VICTOR R SAVAGE

Introduction

Geographers among other academics have been engaged in critical reviews of the recent World Bank's (2009) *World Development Report: Reshaping Economic Geography* (see Buckley and Buckley, 2009; Anant, *et. al*., 2009; Poon and Thai, 2010). With a subtitle like 'reshaping economic geography', economic geographers especially felt compelled to interrogate the World Bank report and most came to the conclusion that despite the geography subtitle the report failed to include any geographical perspective. Yet, while the absence of geography in the World Bank report leaves much to be desired, I will leave the deliberations and discussions to more worthy exponents of geography and economic geography. The World Bank Report has another underlying theme that seems worthy of academic consideration. The question that is begging an answer in developmental dialogues is whether development trajectories can in fact produce greater economic equality among peoples within and between states. Does indeed development create spatial inequalities, and widen income disparities amongst communities. These issues underscore the whole debate and political reasoning for sustainable development – in trying to ensure equity across space and time (over generations). Hence, this paper addresses developmental process and the implications of the environmental variable in relation to sustaining development in developing countries. In doing so, this paper will discuss developmental process in the context of five other variables: lessons, goals, population, capitalism and the ecosystem.

Development: Lessons from the Past?

Since the post World War II, nation-states and academics in the social sciences and humanities have engaged in the quest for development or what is often called the policy and programme of developmentalism for the developing world. Development appeared in economics, political science and sociology text books in varied ways but as Andrew McGregor's (2008: 7) notes development is "elusive, slipping between different meanings for different people and places and changing over time". Rather than providing a single definition, one might define development as comprising a combination of several concepts such as progress, change, modernisation, growth, quality of life, improving social indicators, better standards of living, enhancing equality, releasing the creative potential of people and environmental sustainability. Since 1945, the world became dichotomized into the developed and developing worlds and politically into a tripartite conceptualization of the First World (capitalist democracies), Second World (communist planned economies) and Third World (non-aligned countries), which ended in the late 1980s after the collapse of the Soviet Union. Development programmes have become the cornerstone of many of the United Nation's varied agencies: Habitat, UNESCO with rather dismally critical reviews of their developmental programmes (see Sachs, 2005). The social sciences perpetually are engaged in the developmental challenges in developing countries with seemingly no solutions in sight: poverty, hunger, malnutrition and food security, tropical diseases, slums and squatters, environmental degradation, unemployment and underemployment, corruption and the problems of weak governance.

In the historical context, development was seen in a different light. It was encapsulated in the notions and definitions of civilizations and high cultures. Historians have over the decades been engaged in trying to explain the rise and fall of civilizations through the prism of many single variables: leadership, corrupt governments, population growth, carrying capacities, wars, taxes and more recently environmental issues. Development in all civilizations was rather short lived, ephemeral epochs of great human achievements in technological innovations, enhancement of scientific knowledge, territorial annexations, conquests of peoples, trade, architectural prowess, agricultural production and urban development (Braudel, 1995). Yet, while history concentrates on the successful examples of development little attention has been paid to the other societies that probably survived on marginal resources. Many societies and communities in the past were probably representative of the marginal communities struggling in many developing countries today. The difference in the past 'models' of development is that they were space-bound, civilizations in various places and points in time; today development is global and the integration and relationships of states makes developmental issues in developing countries an integral and relational issue for developed communities. Globalization, the internet, and climate change clearly has given development visibility, transparency and brought humanity closer together despite the political and economic contestations. The study of past human civilizations has given us several issues to reflect upon in trying to achieve development goals in our time.

Firstly, that sustaining development in past civilizations was never an easy task. Over time, most civilizations could not sustain themselves and many fell into decay and destruction. The towering civilizations of Greece, Rome, and those in China and India all underscore the fact that civilizations never found a human political, economic or technological system to ensure sustainable livelihoods for its subjects. Secondly, with climate change looming as the greatest human challenge of the 21st century, a plethora of books seem to point that most civilizations in the past fell due to severe environmental problems. The common environmental denominator in the fall of past civilizations has been prolonged drought, water scarcity, and sporadic rainfall giving rise to the collapse of agricultural production, food insecurity and hunger, thirst and starvation (see Linden, 2007). In separate studies on the lowland Maya cities in Central America and the Khmer Empire in Southeast Asia, civilizations on opposite ends of the world, it was found that their collapse was due to the effects of climate change (Diamond, 2009). Suddenly we see a greater historical interest in the correlation between environmental transformations and the development of civilizations – on a global scale, for example, in Felipe Fernà 3 ndez-Armesto (2001) *Civilizations* and at a regional Southeast Asian level in Lieberman (2003) book. Thirdly, the study of civilizations has become contextualized within the political framework of political states today: Chinese, Indian, Egyptian, Cambodian, Greek, Italian and Mexican. One wonders whether in their own times, the architects of civilizations saw their political entities within the strict confines of ethnic or cultural identities. Indeed as Amy Chua (2008) argues the 'glue' that held civilizations together was their policy and practice of cultural and religious tolerance; if one can generalize, all 'successful' civilizations were exemplifications of cosmopolitan societies. Indeed as Chua

(2008) contends, the fall of many civilizations came when intolerance reigned, when leaders imposed cultural and religious hegemony on people, and when foreign cultural influences were snuffed out in favour of one commanding 'style' of civilization. Ironically, while a 'style of civilization' to use Alfred Kroeber's term, provides the identity of past civilizations, this 'style' also impedes foreign cultural influences.

Fourthly, the discomforting reality for humanists is that no civilization thrived on equity and a classless society. Every civilization thrived on social hierarchies, stratified communities, slavery, and varied classes in varying degrees some institutionally upheld (the Mayans; Greeks) and others culturally endorsed (the caste system in India). This concern with equality and democracy even in the birthplace of democracy in ancient Greece was an ideal rather than a reality. And finally, one might ask which school of civilization we subscribe to? Do we accept Arnold Toynbee's (1976) view that all civilizations are not end goals but journeys in search of human betterment and quality living. Alternatively, do we support Fernand Braudel's (1995) view that civilizations around the world are testimony to a common *Homo sapien* search for human achievement rather than viewing them in competitive terms as exemplified in Samuel Huntington's (1996) 'clash of civilization' thesis. Essentially are all human civilizations following a common global pathway of expressing human creativity, innovations and humanity?

Within the above context, the pursuit of development policies, aspirations and goals today needs to be better framed. My argument is that the whole institutional, political, economic and academic interrogation of 'development' and 'developmentalism' is a Western based project. It stems from the hangover of Western colonialism in the tropical world. The four hundred years of Western hegemonic expansion, territorial annexations, colonization of landscapes and colonial exploitation has indeed from all accounts given the Western world its ballast and economic foundations of its current high levels of development. Without the subjugation of peoples and exploitation of natural resources, Europe never could have achieved its high levels of economic development, its superior economic advantage and the quality of living that it enjoys today. The fundamental issue was that Europeans and later Americans enlarged their ecological footprint globally and ensured an unequal access to natural resources, markets and labour. By the end of World War II, when Third World countries were in the process of decolonization, fighting for independence and becoming autonomous independent states, their former colonial masters found it hard to let go. The independence struggles in Indonesia with the Dutch and Vietnam with the French and then Americans (the Vietnam War) are evidence of a Western mindset that colonization was a god given right to subjugate the tropical 'savages' and civilize them.

Development: Whose Developmental Goals?

The post World War II developmentalism trajectories in the tropical world need to be viewed within a broader global canvas of economic and political changes. The challenge that developing countries face in their goals for development are unprecedented for two reasons. The world never has seen so many states and societies simultaneously seeking to achieve development within their own political boundaries. Secondly, the development of states in the tropical world has had to confront explosive population challenges with a gamut of challenges without hitherto local governmental experience and management expertise. The challenges developing countries faced in the post World War II period need to be understood as part of the colonial histories, the Cold War ideological warfare, and expanding global capitalism. Specifically there are six interrelated issues in the developmentalism trajectories for 'Third World' countries.

Firstly, the Western concerns with 'development' issues in the tropical world underscore a post-colonial guilt expression. After 400 years of exploitation, rapine, bondage and extermination of peoples in the tropical world (see Said, 1994); governments, academics and NGOs in the Western world feel the necessity to rectify the past colonial harm, reduce the excesses of colonial exploitation and neutralize the exploited nations. Nothing can change the historical baggage of the way the Western world subjugated peoples in the tropical world. The scars of the long colonial experiment are not only found on tropical landscapes but more important in the destruction of indigenous cultures, and the transformation of indigenous mindsets into colonized mentalities. When the Western colonialists left, the tropical colonies were exhausted of their natural resources, their environments were degraded, their economies were dependent on the West, their economic system of capitalism had been institutionalized, and inequality became the norm. Andre Gunner Frank (1969), Latin American social scientists and other Marxist intellectuals in Europe were critical of the state of the satellite colonies of the Third World; their dependency theory became the torchlight of the calls for a Marxist revolution to purge the capitalist system embedded in the unequal metropolitan-satellite relationship.

Secondly, the model of the nation-state taken from the European Westphalian model became the cornerstone of the post World War II global political system. New states, former colonies gained independence and hence the proliferation of states in the tropical world. The United Nations now counts 193 states as members a far cry from the 50 states some 70 years ago. The unfortunate tragedy is that none of the new states were established in the European mode of nation-states and hence decolonization created a surfeit of new states with many nations within each state. Governments were hastily put together in many developing countries with inexperienced local leaders and bureaucracies who had no idea of how to govern and manage national issues. Furthermore all the new states inherited political entities with many nations clamoring for independence and autonomy within what Ben Anderson (1991) calls the 'imagined states': the result is never-ending civil unrest. Countries like Myanmar, Indonesia, the Philippines and Sri Lanka have seen the worst of these minority revolutions. The advent of new states with new governments gave the world new challenges and problems. National problems in developing countries like poverty, malnutrition, starvation, population growth, illiteracy and ill-health became global problems for the United Nations, the World Bank and other international agencies to manage, handle and solve. Unlike the Western developed countries who rode on colonialism to expand their footprints and enhance their national economic coffers, the development trajectories of developing countries can count on no such luxury. Furthermore developing countries inherited a fairly institutionalized, well developed global political and economic system which was already uneven and unequal. In short, most developing countries by virtue of their small economies were at the bottom of the totem pole in the hierarchy of states around the world.

Thirdly, the whole developmental dialogue in the post-colonial phase was ironically a battlefield of two ideological systems that became known as the Cold War. The developmental debates and programmes became an economic and political contestation between the *laissez-faire* democratic systems of the Western world versus the planned socialist political economies of the communist regimes. Indeed both systems had their ardent supporters in the Third World (the Philippines and Thailand: US support; Vietnam: Russian and Chinese support) and the rest of the countries pretended to be non-aligned and independent, but this was more in theory than reality. Non-aligned countries were after all recipients of aid from both the West and communist regimes, bowing in different directions depending on the winds of change. India, a founder of the non-aligned movement was squarely in the Russian orbit, while Pakistan, Singapore and Indonesia were clearly in the American sphere of influence. Cambodia was the classic case of receiving aid and military assistance from both the US and China during the Cold War. The theory, policy and practice of development were thus ideological as well as applied. One wonders thus whether the whole trajectory of development over the Cold War years was geared to the specific requirements of Third World stakeholders or whether it was an economic and political war of one-upmanship between the two superpowers: the United States and the Soviet Socialist Republic of Russia. In short, this was another colonial war played out in the environmental turfs of tropical countries. It was not development targeted at national development goals but high stakes ideological warfare aimed at carving out territorial political alignments, propaganda selling economic systems, and underwriting strategic global interest between the superpowers. In the end developing countries suffered once again because national regimes in Third World countries were propped up by their metropolitan patrons not because of their efficient governments or effective powers of governance but by their support of their metropolitan godfathers.

Fourthly, development in the Third World countries became an experiment of many varied external influences rather than the aspirations and visions of national communities and leadership. The major tragedy in developing countries was the way the Western world saw themselves as the models and benchmarks that developing countries should aspire to become. This proved to be the fatal attraction for the developing world. First world models of development were shaped and influenced by usual processes in history which could not be replicated in the last 60 years in the developing world. Yet, for the developed world, the Third World satellites had to be modeled like the West – the political and economic crusade was targeted towards progress and growth based on natural resource exploitation, and industrial outputs. Western social scientists developed a body of knowledge and theories that essentially was fault finding and nit-picking about cultures and societies in the developing tropical world. Societies in the Third World, who were unable to follow a Western developmental prescription, were found to have backward traditional cultures and customs, superstitious religious beliefs, the lack of spirit and creativity, products of environmental determinism, and corrupt regimes. Western social scientists thus viewed Third World communities through Western eyes, edited by Western cultural conceptions and aspirations. The body of Western based social and cultural literature became a damning indictment of the failures of Third World societies. There was only development approach for change and more effort was used to get communities and societies in tune with Western customs, tradition, political and economic behaviour. Hence, the Americans were the arch type crusaders of human rights, democratic government systems and *laissez-faire* economic processes.

Western governments and academics forgot that their political and economic systems took centuries to develop and came at different periods and sequences in history. Now they want both the economic and political systems to be simultaneously implemented in their former satellite colonies. Even a relatively more developed country like Russia, who ventured to change their economic (capitalism) and political (democracy) systems simultaneously after the collapse of communism have never quite recovered for their bold experiment – the system remains chaotic, unmanageable at times, political corruption is rampant, and the institutional embedding is weak. After seeing the near collapse of the Russian political system and its political fragmentation, the Chinese government leaders and other communist regimes are cautious, wary and reticent about such complex instantaneous political and economic dramatic change. Chinese leaders see any Western moves to liberalize their economic and political systems as attempts to destabilize their country as occurred in Russia. Hence, the Chinese stubborn intransigence to Western calls for greater economic transparency and political liberalization; these calls will fall on deaf Chinese political ears.

Fifthly, the whole capitalist economic system has not favoured the developmental experiment for the developing world in terms of creative endeavours, industrial innovation and research and development. The metropolitan powers have controlled the global economy and trade has been lop-sided. The Ricardo's theory of 'comparative advantage' which the colonial masters implemented under colonialism was still being propagated by Western economists like Milton Friedman and Paul Krugman (see Reinert, 2007). Essentially the comparative advantage model of development has instituted a two-tier developmental system: the Western developed countries continued to concentrate on innovative and creative technological changes while the developing world was meant to keep on their subservient role as natural resource suppliers, industrial processing of primary produce, and plantation growers. In short, the comparative advantage system of development imprisoned developing countries to a dependent economic role in the global supply chain and trading system. The developing countries that chose to break from this subservient role and embark on the economic path of economic emulation have flourished and have increased their standards of living. Japan, Taiwan, South Korea, Singapore, and now China are clearly examples of how learning from others has lifted their industrial expertise, enhanced their R & D abilities, and given them competitive products at the global level. The success of the Chinese industrialization is clearly a hallmark of how copying and emulation has taken Chinese manufactured goods to global heights. The irony is that the Chinese economic success model is now seen as a threat to the developed Western world.

Sixthly, the tropical world always fascinated the European mind. This was the region that Greek philosophers portrayed as examples of their environmental deterministic ideas from Herodotus (hard and soft environments and civilizations) to Plato and Aristotle (hot tropics and behavioural responses). Yet, it was only from the late 15th century onwards that Europeans began to experience the tropics firsthand in a sustained manner over the next four centuries. Their historical experiences moved from trade to colonization and colonialism and their views of the tropical world changed from Paradise to the White Man's Burden, from plenitude to tropical environmental challenges, and from environmental determinism to environmental possibilism through colonial administration, and the use of science and technology in agricultural developments. The 'tropical zone' in the era of high colonialism was the 'colonial project' for the western experiment in creating behavioural change through education, political management of societies, and religious Christian indoctrination. In short, Western colonial governments saw themselves as saviors and harbingers of civilization, progress, ethical governance and political stability.

Eco-development: an Experiement in Progress

The last 50 years have been witness to many social, economic and political experiments with the processes of development. Since the post World War II period, as states removed themselves from colonial rule and gained independence, they became new case studies of developmental processes. Some states became xenophobic after their long history with colonialism and tried to follow their own nationalistic paths to development. Myanmar is a good and extreme example of this developmental trajectory. Other states try to embrace capitalism and Western models of development with verve and vigor and Singapore comes to mind. But for most states there is a range of positioning between these extreme cases.

The current issues of development seem increasingly to be undergirded by environmental concerns and appear in academic literature and state policies as conceptual statements under sustainable development, green development and eco-development. Elsewhere I have dealt with the problems of sustainable development (Savage, 2006) and for varied reasons it is difficult to see how developing states can achieve sustainable development goals. This paper turns attention to three variables that need to be considered in realizing any broad agenda of sustainable development amongst developing states: population or demographic issues; capitalism as the economic mode; and the ecosystem's carrying capacities.

The Demographic Variable

Ever since Thomas Malthus wrote about the population challenge to food security and state developmental trajectories, there has been debate over the relationship of demographic issues in the development equation. Up to the end of the nineteenth century, the world still had a manageable population spread somewhat evenly across the five continents. But in the twentieth century, global population escalated to somewhat unmanageable proportions.

In the last 50 years, several Western academics have revisited Malthus' theory of population growth and have given dismal accounts. The most explosive which garnered world attention was the Paul Ehrlich (1968) book, *The Population Bomb,* which sold 3 million copies. The second is the Club of Rome report about dwindling natural resources and the population equation involved in consumption habits of natural resources (Mesarovic & Pestel, 1974). And the third was a chain of articles and books by Garrett Hardin (1980) about the tragedy of the commons, triage, and life boat ethics for the developing world. In Hardin's prescription the developed world needs to exercise triage (selective saving) in rescuing the millions of people in the developing world dying of hunger, poverty, disease and malnutrition. Hardin's hard line, negative views on the population explosion in developing countries in some ways underscores the developed world's ambivalent attitude to the plight of the developing world's poverty, malnutrition, starvation and unemployment. Triage already is practiced by the developed countries, which accept the best knowledge-based, professional and skilled labour into their countries and leave the unskilled, old and uneducated back home. In short, governments in the developing countries must accept responsibility for their unchecked population policies as Western countries undermined by their own economic problems are less willing to offer aid and help to the poverty-stricken millions in the developing world.

There is however no consensus among academics and governments about the demographic issue. The demographic optimists believe that technology is the panacea for demographic challenges and hence governments should not worry. The neo-Malthusian pessimists believe that runaway population growth will undermine economic development, deplete natural resources and degrade ecosystems. A third school of thought embodying Jeffrey Sachs' (2008) thinking is cautiously optimistic based on the belief in the demographic transition in the cycle of population changes. Their prescription is to speed up the voluntary reduction in fertility rates to stabilize population growth (Sachs, 2008:160–162). Countries like China, Thailand, Vietnam and Singapore have successfully slowed fertility rates only to be confronted with new demographic challenges, aging populations and high dependent populations which are unable to support economic growth. The most glaring case of population dependency is Japan.

As population has snow balled over the last century, it is the developing countries that are paying the highest price for their inability to address the multifarious problems confronting their expanding populations. All things being equal, large populations and unchecked demographic growth are the millstones and albatrosses around the governments in the developing countries. These demographic issues are compounded because people are drifting and migrating to already over-taxed cities. We therefore see a shift of poverty and other ills now shifting into urban centres with the influx of rural migrants. In nearly every sector of social, economic, environmental, political and cultural spheres, the challenges facing governments in African, Latin American and Asian states seem at times insurmountable. Coupled with environmental vulnerabilities, dwindling natural resources and degrading ecosystems, feeding, clothing, educating and housing populations seems to be beyond governments and their ability to govern. In an unsettled global economic system where developed countries are themselves reeling from economic problems, the plight of developing countries has become marginalized issues. In short, saving the global commons whether in climate change issues or saving tuna fish is now determined by the adage to each for himself and God for us all.

Three important outcomes are related to population growth at the global scale:

(i) the relationship to technology; clearly hunting and gathering supports lower population densities than sedentary wet rice cultivation. According to Ian Simmons (2008: 97) hunters and gatherers maintain low population densities of 0.01 to 1.0 persons per square kilometre; swiddeners (shifting cultivators) maintain 10–80 people per square kilometre and sedentary farmers with solar power support 100–1000 people per square kilometre. Hence global population increased tremendously with the advent of the agricultural revolution: the world's population expanded from 4 million in 10,000 BC at the onset of the agricultural revolution to 720 million in 1750 just before the industrial revolution (Simmons, 2008: 54).

(ii) the relationship to the quality of life and standards of living. Clearly the countries maintaining high quality of living and lifestyles are also the largest consumers of material goods, natural resources and raw materials. There are numerous comparisons of consumption between the developed and developing worlds, between the US and poor countries, between the richest and poorest peoples in the world. Unfortunately inequality is embedded in the global systems of states and it does not seem to be changing. Large populated countries (China, India, Indonesia and Bangladesh) are going to find greater difficulty lifting the quality of living of their citizens. Furthermore closing the gap between rich and poor is an ideal with little chance of becoming a reality.

(iii) the relationship to environmental, social and economic impacts. Indeed large populations might have ecological footprints but developed countries have far greater ecological footprints. So long as all nature becomes commodified the rich countries will always have command and clout over natural resources, labour and raw materials. This is compounded by the fact that cities, with 50 per cent of the world's population, and their locus of economic and political power command global hinterlands even though they only account for 1 per cent of the world's land area. The demographics of urban agglomerations need urgent attention because an urbanized world will have different movements, consumption and production processes from the former rural populated economic systems.

The Capitalistic Economic System and Consumption

The irony of the debates today is that while capitalism has been the main driver of the global economy for the last 70 years, the series of financial crises has left the world in a sobering mood about the future of capitalism. When the Asian economic or 'tom yam' crisis hit in 1997–98, the Western world and its economic bodies (World Bank and International Monetary Fund) offered a lot of economic advice, prescriptions and methods for salvaging crumbling economies in the developing Asian world especially in Southeast Asia (Thailand, Indonesia, Malaysia, the Philippines). In the 2008–09 sub-prime crisis the Western world got a bigger taste of its own inadequate, corrupt and inefficient economic system and their governments seem to lack the political will and moral responsibility to set things economically right. While the 1997–98 crises was seen as a failure of Asian governments and governance to manage their economic systems, the US sub-prime global meltdown underscored a much larger defect of whether the global capitalistic system was seriously self destructing and defaulting. The moral of the story for Asian and developing economies is to take care of their own national economic systems and be wary of Western pressure, advice and prescriptions for economic development. Indeed America, the engine of economic growth is suddenly becoming more economically vulnerable and politically ambivalent about its future course. The rise of other economic powers such as China, India, Brazil, Russia, South Korea and the Middle East has challenged the Western developed world's position as drivers of the global capitalistic economy. For developing countries it seems like a period of chaos and confusion. No one quite knows whether the clash of the US and China on trade, currency issues, climate change will have economic and political reverberations in the developing world. But developing countries should be mindful that they are in the vortex of a major transition in global political and economic changes, and the world is heading for more multi-polar power relationships. The power brokers are changing, new alignments will form and new satellites and allies will emerge.

Several books over the last few years have made pointed arguments about the un-sustainability of the current capitalistic economic system. The most poignant statements come from books by Joan Martinez-Alier's (2005) *The Environmentalism of the Poor*, Steven Stoll (2008) *The Great Delusion*, John Bellamy Foster's (2002) *Ecology Against Capitalism* and James Gustave Speth (2008) recent book, *Bridge at the Edge of the World the Environment, and Crossing from Crisis to Sustainability*.

Despite the collapse of communism in Russia and Eastern Europe, some scholars are still advocating a socialist path to Green development. The most important statements come from Saral Sarkar's (1999) *Eco-socialism or eco-capitalism* and Leo Panitch and Colin Leys (2006) edited book, *Coming to Terms with Nature*.

The clash between economic proponents and environmentalists is perhaps the most difficult problem in global environmental debates. The economic argument is that in general economic growth is good for the environment. The proponents assert that the empirical evidence of improvements in the quality of life in economically driven capitalistic countries like the US, Canada, Japan and European countries over the last two decades shows that when services overtake manufacturing, environmental improvements take place. To add to this there are the "environmental contrarians" who argue that environmental problems will be solved with more development based on three primary assertions: (i) the environment in Western countries is in better shape than it was decades ago; (ii) environmental regulation costs more than it is worth; (iii) continued growth, free markets and technological progress will solve the remaining problems (see Ayres, 1999:139–141). Robert Ayres (1999) has debunked all three arguments based on evidence in the US and Europe. Furthermore the 2008–09 economic crises which is still brewing in Europe and the US demonstrates that free markets and the "invisible hand" will not cure national economic woes nor change for the better to produce a responsible capitalistic system. As Ayres (1999: 142) notes "I know of very few cases – if any – where totally unregulated competition in the private market place solved an environmental problem for its own sake". This inability to close inequality is what William Robinson sees as one of his four problems in 21st century capitalism – the problem of "global social polarizaton" (Speth, 2008:186).

Given the complexity of the developmental experiment, there is no one size fits all system and solution for development. Every country needs to find its own system for development based on its national histories, its cultural systems, its societal givens and its political and economic abilities. The main challenge seems to have political leadership which is enlightened and nationalistic enough to engage in developmental programmes for their national rather than their personal good and interests.

The Ecosystem Carrying Capacities; Environmental Limitations

The World Wide Fund for Nature (WWF) produces every two years its global living index and each latest issue gives a more dismal picture than the preceding report. The reasons seem simple enough. Global populations across countries are consuming resources at rapid rates in future natural resources are likely to be exhausted or polluted and degraded beyond human use. By 2050 WWF believes that our current consumption habits would require a global population of 9 billion to use resources from two earths. Unfortunately, we do not have another earth to sustain our lifestyle and living standards. So what will the future scenario be like? It does not take much imagination to project what 2050 will be like because all the ingredients of the global environmental, political, economic, social and consumptive challenges are embedded in our society right now.

Ecologists have tried to differentiate global ecosystems according to bioregions and other physical related properties such as bio-capacities in measuring the ecological footprint. Bioregions and ecosystems are unique entities and each supports different types of organisms including human beings. The advent of humans with their arsenal of technologies also complicates the use or misuse of ecosystems and bioregions. Without understanding carrying capacities, heavy over use of ecosystems can indeed lead to permanent damage in which the productive capacities of the ecosystem are severely diminished. Two current trends are likely in future to become more pronounced and their impact on varying ecosystems and bioregions require better comprehension and study.

One is whether the current political-economic system embedded in capitalism, commodification and consumerism is sustainable over the long run. Can increasing materialism in all forms, the opium of our decadent living and lifestyles continue over the decades without economic and environmental fallouts? Clearly, the American consumer and material society has been the subject of the rude shock arising from the 2008–09 sub-prime crisis and the reason for the uneasy relationships between the US and China. Unfortunately it is the American model of lifestyle living, material binges which has become the standard torchbearer of developmental aspirations of poor people around the world. Development unfortunately is all about raising living standards but to what levels, what benchmarks?

The second issue has been the central theme of this paper is whether development trajectories will close inequalities between states. The current ecological footprints between countries and between cities and their rural hinterlands demonstrate that we live in an unequal world. Clearly this is the case and the only reason why we do not need two worlds by 2050 to support American standards of living is because global inequality will support 20 per cent of the world's population on high living standards while 80 per cent of the population will live in poverty. The development paradigms which aim to close the gap between the have and have-nots, the rich and poor will remain ideals but in reality disparities of wealth are likely to persist in the future. This does not mean that the overall quality of human living will stagnate. Indeed as both Bjørn Lomborg (2001) and Julian Simon (1995) have demonstrated the quality of human life has indeed improved over the centuries despite disparities of wealth, increasing poverty and problems of malnutrition. One positive indicator is the widening human life span: in 1900 the average life span was 30 years, whereas in 2000 it was 67 years (Lomborg, 2001:4). Using an array of indices to measure "the real state of the world", the Danish political scientist was emphatic in declaring that despite the dismal news reports, "things are getting better" (Lomborg, 2001:3–33).

Where Do We Go From Here?

The environmental and climate change challenge has given the global community four areas to think about.

Firstly, environmental and climate change issues is a platform for greater communication between the sciences and social sciences/humanities. Unlike the dismal recounting of C.P. Snow's "culture wars" between the sciences and humanities, we need to endorse a cultural dialogue between the hard and soft sciences. Essentially we need to place greater emphasis on what Bryan Norton (2003:11–12) calls "pragmatism" in environmental philosophy, an "adaptive approach" in environmental management, and greater "convergence" in dealing with environmental ideas, issues and philosophy. These convergences and pragmatic approaches in the humanities will help to consolidate greater communication and dialogue between scientists and the social scientists. One link between the sciences and arts was in Barry Commoner's view to solving the environmental crisis. He argued that the scientific community's primary duty was to disseminate scientific knowledge which will then enable and empower the public to make valued judgments and tackle problems (Egan, 2007:131). After all, the environmental and climate change mandate is too large an agenda for any one discipline to tackle. Without good science we can never know the speed, extent and magnitude of environmental and climate change challenges. And without the social science and humanities inputs, we will not be able to devise and implement adaptive and mitigation mechanisms to save human beings and other organisms from the ravages of environmental hazards. Despite all the mass media noise about the reliability of climate change science, the problem lies more with the fact politicians and bureaucrats tend to view scientific findings as predictable statements when in fact science is neutral. It is for the humanities (moral issues) and social sciences (adaptive mechanisms) to determine the way societies and governments respond to scientific findings. Hence both the scientists and social scientists need to work and collaborate closely to ensure that environmental solutions and policies are focused and targeted at the right challenges confronting human communities.

Secondly, the environmental and climate issues have become representations of the global political-economy because the unit of decision-making now rests with the governments of the 193 member states of the United Nations. What unfortunately we are witnessing at global environmental and climate change forums and conferences are the spatial division of the earth into national sovereign territories. Thus, we have a jigsaw of ecosystems which are defined by the political divisions and jurisdictions of state boundaries and borders and have nothing to do with the natural flows of ecosystem processes. Countries thus negotiate from national territorial positions with little concern about the holistic workings of Gaian Earth. Both the Kyoto (1997) and Copenhagen (2009) climate change meetings underscore how national interests, state priorities and territorial sovereignty issues command the global stage and the holistic, universal and humanitarian issues become marginalized. Yet, the tragedy is that if states and communities around the world do not see environmental and climate change issues within global perspectives and as the global commons, there will be no effective and holistic solutions for the global community. Countries cannot believe that the safeguarding of their national spaces and ecosystems will help improve the global commons and solve global environmental challenges. In short, while climate change and economic globalization threaten the world's community, the international system still runs on an outmoded system based on the nation-states which is concerned with national sovereignty issues rather than the global community.

Thirdly, the climate change and environmental issues in developing countries throw up major collisions between the economy and environmental issues. It would seem that both camps have widening differences and the question is how governments in developing countries are going to swallow the bitter pill of taming economic growth and moderate capitalism for their long term development. Given their fledgling and fragile economies, many governments in the developing world find it difficult to adjust economic growth to fall in line with ecosystem limitations. The environmental prescription for development requires a reorientation of development goals, a change in economic trajectories and decoupling of the logic that a free-market capitalistic system is good for development, environment and reducing inequalities. Can countries in the developing world follow the cycle of development in the developed countries of moving from agriculture to industrialization and finally services in the belief that the final stage will reduce environmental problems? Can in fact the global community absorb the massive negative environmental fallout of China, India, Indonesia, Vietnam and Brazil's industrial developments before they become service economies? What will the situation become in Africa as several states there now seem to be on an economic rebound with the credo, Africa Rising?

Fourthly, the environmental and climate change problems of the 21st century will require working, re-aligning and readjusting the relationships between cities and their hinterlands. As cities become the norm for living, the question is how do we create sustainable cities? While we currently pay lip service to the ideas of sustainable cities and eco-cities, in reality none of our cities currently are sustainable. Indeed in the history of human societies, it will be rare to find a city that has been sustainable over hundreds of years. Indeed swidden communities have led sustainable lifestyles over thousands of years but rarely have city based civilizations lasted hundreds of years. Thus, the challenge for humankind this century is to find the system of sustaining cities over time. Unfortunately with globalization and better transport technologies, urban populations have been able to tap into global hinterlands for their sustenance. But once again urban residents have been faced with inequalities: the rich urban residents live sustainable affluent livelihoods while the poor scrimp and live in poverty on the margins of urban society. Hence, development while increasing urban living has done so albeit unequally. The urban question once again poses major developmental challenges. What it seems to have done is shifted the issues of poverty, disease and malnutrition from the rural to the urban areas in many developing countries. In many cities in developing countries, some 40 per cent of the population lives in slums and squatter settlements and these settlements in many ways represent the levels of poverty in cities. If the rural areas have failed to provide sustainable livelihoods for peasants and the underbelly of society, the cities likewise are sites of temporary refuge for rural environmental refugees. In the long term, if cities cannot find employment, economic stability and sustainable livelihoods for its migrants and domestic residents, the cities will become the tinderbox for riots, rebellions and civil strife in future. Compounding this precarious situation is the destabilizing impact of climate change which is creating more prolonged droughts, flash floods, sea level rises, and heat waves which are impacting directly and indirectly (food insecurity) on urban populations.

Concluding Thoughts

What after all today's ecological movement is advocating is a return to isolation, and the abandonment of treasure and knowledge to tribes and nations in foreign lands who pose no threat to us. Consciously of otherwise, this is a death-wish. We are not talking here about eschewing food additives and colouring matter, whole food in a whole land, as were the earlier ecologists, but something different and deathly. For today's ecologists, their hope of regeneration presupposes a return to primitivism, and thus, whether they wish to enunciate it or not, concomitant anarchy, the burning before the replanting, the cutting down of the dead tree. The father of the movement is an utter rejection of all that is, and for at least three millennia all that was. (Bramwell: 1990:248)

Anna Bramwell's dismal concluding statement in her history of ecology in the 20th century would come as a surprise to most environmentalists. It seems difficult to imagine that deep ecologists are actually advocating a return to primitivism or a return to a state of indigenous living by adopting an ecological approach to our way of living. On the other hand even if the key to sustainable living lies with indigenous peoples should we write them off as unrealistic examples? What is needed is not to write off the value of indigenous livelihoods, folk culture and folk science as an "oddity" (Orr, 2004: 10) but to embrace such indigenous knowledge as part of the wider fund of knowledge that can undergird the building of a sustainable society. And within the kernel of indigenous folklores and culture lies important knowledge about human-nature relationships and ecological ideas. It seems a myth to believe that education is about giving students the means for upward mobility and success in life framed within a unilineal view of development and state progress. The tragedy as David Orr (2004:12) notes is that our planet does not need more "successful people", rather it needs more "peacemakers, healers, restorers, storytellers and lovers of every kind".

In his interesting and sobering treatise on evolution, *A Japanese View of Nature*, the Japanese primatologist, Kinji Imanishi (2002:71) concluded that the history of the evolution of living things is the "history of the rise and fall of the ruling class". At different times in geological history, different species became the ruling class as human beings are today in our world: "Mankind arose within the community of mammals, and has replaced other mammals temporarily as the ruling class of the society of living things" (Imanishi, 2002:73). In his view, the discontinuity of a species, including the ruling species could come about due to an "irregular environment" and "discontinuity in the environment" (Imanishi, 2002:63). The only difference is that as human beings, we changed the environment we evolved in and have now created an environment which as a species we possibly cannot adapt rapidly enough to survive in it. Hence, as a species, humankind also will become discontinued in the process of evolutionary changes of organisms: "Those who replace us, though they perhaps should no longer be called mankind, will originate and be created from the human race. This is what evolutionary history teaches" (Imanishi, 2002:72). I do not think Imanishi expected that his pessimistic prophesy of the demise of the human race was going to take place so soon.

Indeed, two books seem to place the 21st century as the definitive critical century for human survival. In Lord Martin Rees (2003) book, *Our Final Century*, the question is whether we will survive this century as a species and in James Martin's (2007:26) *The Meaning of the 21st Century* there is a sense of foreboding that this century could see the termination of *Homo sapiens* or "if *Homo sapiens* survives, civilization may not". Suddenly global warming seems to loom larger than it has ever been before in our print and mass media. For *Time* magazine's 28 April 2008 front page heading equated global warming as a "war" and the issue at stake was "How to win the war on global warming" (Walsh, 2008). Such 'war' prescriptions are *deja vu* in the way socialist regimes such as the former Soviet Union, China, Vietnam and Cambodia in the past viewed nature and the environment as 'enemies' and human beings as victors in their conquest of nature. The rhetoric of war on nature is once again sadly misplaced for two reasons. Firstly, the enemy is not climate change and global warming, the enemy lies within each of us human beings. We are the creators of the global warming monster. Secondly, throughout history, communities around the world lived sustainably within their given ecosystems because they never fought wars with nature; they learnt to live and adapt with nature in a harmonious and *relational* manner as peaceful participants (Peterson, 2001) rather than as conquerors of nature and their hinterlands. Hopefully, the human race in our comfortable urban abodes will be able to find the peaceful answers to our predicament, the political will to make the difficult decisions and the bold initiatives to make changes to our economic system.

Yet the urban basis of civilization, its ability to sustain human high culture and the survival of the human species is now in question. With growing global environmental problems and the unpredictable climate change outcomes point us in another direction of caution, circumspection and reflection of the trajectory of global civilization. Apart from Fernàndez-Armesto's (2002), Eugene Linden's (2007) Brian Fagan's (2008) and Ian Whyte's (2008) books, which discuss the impacts of climate change and the environmental basis, circumstances and limitations of varied civilizations, few historical studies have considered the environmental dimensions, the abuse of nature in the equation of what makes civilizations sustainable. If one accepts Braudel's (1995:7) view that civilization is "the common heritage of humanity" then all cities and civilizations around the world should share a responsibility in sustaining not just their cities but also the global human civilisation which I call the Gaian civilization. Hence, the software of Gaian civilization should include the way society relates to Nature in an ecologically responsible manner as well as the use and conservation of environmental resources.

If by the end of this century, dramatic environmental and climatic changes and transformations take place as forecasted scientifically by Tim Flannery (2006) and economically by Nicholas Stern (2007), it will not be our cities or civilizations that will be in question, it will be the *Homo sapiens* at stake. The scientific and technological arrogance of civilised *Homo sapiens,* as the dominant and intellectual species, who have become the new deity of all living matter, might be our own undoing. Will such a situation arise in the future? Maybe if one takes a long look at evolutionary history.

Acknowledgement

I thank to Prof Kenneth Corey for his editorial comments of my earlier draft. The ideas and expressions here however remain solely mine.

References

Anant, Maringanti, Sheppard, Eric and Zhang, Jun (2009), 'Where is the geography? World Bank's WDR 2009', *Economic & Political Weekly*, XLIV(29): 45–51.

Anderson, Benedict (1991), *The Imagined Communities,* London: Verso.

Ayres, Robert U. (1999), *Turning Point: The End of the Growth Paradise,* London: Earthscan Publications Ltd.

Bramwell, Anna (1990), *Ecology in the 20th Century: A History*, New Haven and London: Yale University Press.

Braudel, Fernand (1995), *A History of Civilizations*, New York, New York: Penguin Books.

Buckley, Robert M. and Buckley, Thomas D. (2009), 'Discontent with the World Bank's Excursion into Economic Geography: Lions and Butterflies Once More?' *Environment and Planning A,* 41(12), pages 2806-2815.

Chua, Amy (2008), *Day of Empire*, New York: Doubleday.

Diamond, Jared (2009), 'Maya, Khmer and Inca', *Nature,* 461(24):479–480.

Egan, Michael (2007), *Barry Commoner and the Science of Survival*, Cambridge, Massachusetts: The MIT Press.

Ehrlich, Paul R. (1968), *The Population Bomb*, New York: Ballantine Books.

Fagan, Brian (2008), *The Great Warming: Climate Change and the Rise and Fall of Civilizations*, New York: Bloomsbury Press.

Fernández-Armesto, Felipe (2001), *Civilizations: Culture, Ambition, and the Transformation of Nature*, New York: Touchstone.

Flannery, Tim (2006), *The Weather Makers: How Man is Changing the Climate and What it Means for Life on Earth*, New York: Grove Press.

Forbes, Steve (2009), 'Capitalism: A True Love Story', *Forbes Asia*, 5(16):8-12.

Foster, John Bellamy (2002), *Ecology Against Capitalism*, New York: Monthly Review Press.

Frank, Andre Gunner (1969), *Latin America: Underdevelopment or Revolution*, New York: Monthly Review Press.

Hardin, Garrett (1980), *Promethean Ethics: Living with Death, Competition, and Triage* Seattle: University of Washington Press.

Huntington, Samuel P. (1996), *The Clash of Civilizations and the Remaking of World Order*, New York: Simon & Schuster.

Imanishi, Kinji (2002), *A Japanese View of Nature: The World of Living Things,* New York: RoutledgeCurzon.

Leiberman, Victor (2003), *Strange Parallels Southeast Asia in Global Context, c. 800–1830, Volume 1, Integration on the Mainland*, Cambridge: Cambridge University Press.

Linden, Eugene (2007), *The Winds of Change: Climate, Weather, and the Destruction of Civilization*, New York: Simon & Schuster.

Lomborg, Bjørn (2001), *The Skeptical Environmentalist: Measuring the Real State of the World,* Cambridge: Cambridge University Press.

Martin, James (2007), *The Meaning of the 21st Century: A Vital Blueprint for Ensuring our Future,* New York: Riverhead Books.

Martinez-Alier Joan (2005), *The Environmentalism of the Poor: A Study of Ecological Conflicts and Valuation*, New Delhi: Oxford University Press.

McGregor, Andrew (2008), *Southeast Asian Development*, London: Routledge.

Mesarovic, Mihajlo and Pestel, Eduard (1974), *Mankind at the Turning Point*, New York: E. P. Dutton & Co., Inc.

Norton, Bryan G. (2003), *Searching for Sustainability: Interdisciplinary Essays in the Philosophy of Conservation Biology*, Cambridge: Cambridge University Press.

Orr, David W. (2004), Earth in Mind: On Education, Environment, and the Human Prospect, Washington: Island Press.

Panitch, Leo and Leys, Colin (2006), *Socialist Register 2007: Coming to Terms with Nature*, New Delhi: Leftword Books.

Peterson, Anna L. (2001), *Being Human: Ethics, Environment, and Our Place in the World*, Berkeley: University of California Press.

Poon, Jessie P.H. and Thai, Diep T. (2010), 'Micro-credit Development in Northern Vietnam', *Geografisca Annaler*, 92(1): 65–79.

Rees, Martin (2003), *Our Final Century: Will the Human Race Survive the 21st Century?* London: Heineman.

Reinert, Erik S. (2007), *How Rich Countries Got Rich... and Why Poor Countries Stay Poor*, New York: Carroll and Graf Publishers.

Sachs, Jeffrey (2005), *The End of Poverty: How We Can Make It Happen in Our Lifetime*, London: Penguin Books.

Sachs, Jeffrey (2008), *Common Wealth: Economics for a Crowded Planet*, New York: The Penguin Press.

Said, Edward W. (1994), *Culture and Imperialism* London: Vintage.

Sarkar, Saral (1999), *Eco-socialism or Eco-capitalism? A Critical Analysis of Humanity's Fundamental Choices*, London: Zed Books.

Savage, Victor R. (2006), 'Ecology Matters: Sustainable Development in Southeast Asia', *Sustainability Science*, (1):37–63.

Simmons, Ian (2008), *Global Environmental History 10,000 BC to AD 2000*, Edinburgh: Edinburgh University Press.

Simon, Julian L. (ed.) (1995), *The State of Humanity*, Oxford, UK and Cambridge, USA: Blackwell.

Snow, C. P. (1961), *The Two Cultures and the Scientific Revolution*, Cambridge: Cambridge University Press.

Speth, James Gustave (2008), *The Bridge at the Edge of the World: Capitalism, the Environment, and Crossing from Crisis to Sustainability*, New Haven: Yale University Press.

Stern, Nicholas (2007), *The Economics of Climate Change*, Cambridge: Cambridge University Press.

Stoll, Steven (2008), *The Great Delusion: A Mad Inventor, Death in the Tropics, and the Utopian Origins of Economic Growth*, New York: Hill and Wang.

The World Bank (2009), *2009 World Development Report: Reshaping Economic Geography*, Washington, D.C.

Toynbee, Arnold (1976), *Mankind and Mother Earth*, New York: Oxford University Press.

Walsh, Bryan (2008), 'Why Green is the New Red, White and Blue', *Time*, 171(16): 31–42, 28 April.

Whyte, Ian (2008), *Worth Without End? Environmental Disaster and the Collapse of Empires*, London: I. B. Tauris.

Agricultural Development and its Impact on Environment and Farming Population in Haryana

KULDEEP SINGH

Prologue

In the Copenhagen Summit (December 7-18, 2009), the differences between the North and the South sharpened and no consensus could be reached. Dr. Manmohan Singh, the Prime Minister of India said that future negotiations must continue on the basis of Kyoto Protocol, 1997 and Bali Action Plan, 2007. However, it is beyond controversy that environment has become a matter of great concern in the 21st century. The relationships between the richness of the natural environment and farming practices, *ipso facto*, are complex. Many valuable habitats in the world are maintained by extensive farming, and a wide range of wild species rely on environment for their survival. But inappropriate agricultural practices and land use can also have an adverse impact on natural resources, like-pollution of soil, water and air; fragmentation of habitats and loss of wildlife. The Food and Agriculture Organisation (FAO) of the United Nations has stressed that agriculture is an essential component of societal well-being. It occupies 40 per cent of land surface, consumes 70 per cent of global water resources and manages biodiversity at genetic, species and ecosystem levels. The Common Agricultural Policy (CAP) in Europe has identified three priority areas for action to protect and enhance the European Union's rural heritage, e.g., biodiversity, preservation and development of 'natural' farming and forestry system along with traditional agricultural landscapes; water management; and finally dealing with climate change. Thus producing enough food for a rapidly growing population and taking care of our planet are two of the world's biggest challenges in the 21st century.

The Green Revolution in India undoubtedly provided food security for billions of people by substantial increase in foodgrains. It could be possible by improved crop varieties and excessive use of inputs like chemical fertilizers, pesticides and new farm machinery. The recent explosion of knowledge in the fields of information technology and biotechnology particularly in the last decade of the 20th century and the first decade of the 21st century has offered scope for brighter prospects of Indian agriculture. However, The agriculture and allied sectors, which contributes about 17.2 per cent to the country's GDP, grew by just -0.2 and 1.6 per cent in 2009-10 and 2008-09 respectively as against an average growth rate of about 4.9 per cent over three years, i.e., from 2005-06 to 2007-08. Such a tendency is not a welcome trend in a country like India where more than 60 per cent of its total work force is employed in the agricultural sector.

Need of the Study

Like Indian economy, agriculture is also the mainstay of Haryana's economy. Agriculture has been accorded high priority since the inception of the State. Manifold expansion of infrastructure took place and support services related to agriculture in Haryana were created. About 21.89 per cent of the total income of the State comes from agriculture and allied activities. According to Population Census of India, 2001, the total population of Haryana is 211.45 lakhs people out of which 150.29 lakhs people reside in the rural areas who are directly or indirectly engaged in agricultural pursuits. In order to meet the ever increasing demand for food, feed and fibre for growing population of the country, which has already crossed a billion mark, the production of foodgrains will have to keep pace with population. It appears that agricultural chemicals will continue to remain as a key input in the modern agricultural strategy to fulfil the country's needs. Pesticides are heterogeneous group of chemicals whose desired activity is the killing of unwanted living organisms in a specific manner. The increased use of pesticides, insecticides, chemical fertilisers and farm machinery since the end of Second World War has introduced a new hazard to human beings and animals arising out of the misuse or excessive use of such poisonous compounds. Inspite of their highly important use in agriculture, these chemicals produce a number of hazardous residues in agriculture and dairy products, food, water and in meat causing a great public health concern.

Review of Literature

Numerous studies have been done to assess the impact of agricultural development on environment from local to global level. Tiwana, N.S. *et al.* (2009) analysed that Punjab has only 1.5 per cent landmass of the country. However, it consumes about 17 per cent of total pesticides used in India. The per hectare pesticides use is the highest in Punjab, i.e., 923g./ha as compared to other agriculturally advanced states like Haryana, Andhra Pradesh, Tamil Nadu, Karnataka and Gujarat. Although the use of pesticides has helped to enhance economic gains through crop protection yet they have had serious implications to human health and non-target plants and animals by accumulating in food and water.

In a recent study done by Bai, Z.G., Dent, D.L., Olsson, L. and Schaepman, M.E. (2008); land degradation has been described as the long-term decline in ecosystem function and productivity. They estimated that land degradation occurred at 24 per cent of land worldwide, with cropland over represented. Steinfeld, Henning (2006) also analysed that livestock are one of the most significant contributor to the environmental problems in the modern times. Livestock production occupies 70 per cent of all land used for agriculture and 30 per cent of the land surface of the planet. It is one of the largest sources of greenhouse gases, responsible for 18 per cent of the world's greenhouse gas emissions as measured in CO_2 equivalents. Comparatively, all transportation emits 13.5 per cent of the CO_2. Livestock produced 65 per cent of human-related nitrous oxide which is 296 times the global warming potential of CO_2 and 37 per cent of all human-induced methane which is 23 times as warming as CO_2. It also generates 64 per cent of the ammonia, which contributes to acid rain and acidification of ecosystems.

In another study by Steinfeld, H., Gerber, P., Wassenaar, T., Castel, V., Rosales, M. and de Haan, C. (2006); livestock expansion is a key factor driving deforestation in the basin of Amazon which is equal to 70 per cent of the earlier forested area and is now occupied by pastures and the remainder used for feedcrops. Tegtmeier, E.M. and Duffy, M. (2005) concluded that cropland in the U.S. imposed \$5 to 16 billion external costs in 2005 which accounted for \$30 to 96 per hectare. The study also calculated external costs of livestock production that imposed \$714 million in the U.S. in 2005. Horrigan, Leo *et al.* (2002) analysed that the modern industrial agriculture system consumes fossil fuel, water and topsoil at unsustainable rates. It contributes to numerous forms of environmental degradation including air and water pollution, soil depletion, diminishing biodiversity and fish die-offs. The pesticides used heavily in industrial agriculture are associated with elevated cancer risks for workers and consumers. Such cases are coming under greater scrutiny for their links to endocrine disruption and reproductive dysfunction. Thus environmental and human health problems associated with current food practices are discussed at length in the study.

Pretty, J. *et al.* (2000) assessed that agriculture imposed external costs upon society through pesticides, nutrient runoff, excessive water usage and assorted other problems. The study calculated total such costs of £2,343 million in the U.K. in 1996. It accounts for £208 per hectare. Both of the studies of the U.S. and the U.K. did not include agricultural subsidies in their studies. However, it seems true that the inclusion of agricultural subsidies will also increase the external costs of agriculture to society.

Objectives of the Study

The specific objectives of the present study are the following:

1. To identify the nature and extent of the path ways of modern agriculture and its relationships with environmental concerns along with the prevailing notion in the regard world over.
2. To examine the growth of agricultural activities and the consequent environmental problems in Haryana.
3. To provide overall linkages between agricultural sector and environment with empirical findings related with farming population of Haryana.

4. To seek the possible ways out to overcome the problem of agriculture sector and environmental dilemma.

Methodology

The primary as well as secondary data has been collected. The primary data has been collected from two villages, i.e., one named Darbi randomly selected from the district of Sirsa where the crop rotation of wheat and cotton is most popular. Another village named Jyotisar has been selected at random from the district of Kurukshetra where the most popular crop rotation is that of paddy and wheat. The fifty farmers and fifty farm labourers from each district engaged in different farm activities such as sowing, thrashing, spraying of pesticides and insecticides including sowing of chemical fertilizers randomly selected from each village were contacted and data was collected for various aspects with the help of a questionnaire specifically designed to meet the need of the study. Thus, in all a sample of 100 farmers and 100 farm labourers of the respective farmers was taken. The secondary data was collected from the different issues of the Statistical Abstracts of Haryana published by the Economic and Statistical Advisor, Planning Department, Government of Haryana, Chandigarh.

In Haryana, the most popular crop rotation is either wheat and paddy or wheat and cotton. Sirsa is the district of Haryana where the area and production under the crops of wheat and cotton is the highest throughout the state. Similarly, Kurukshetra is a district where the area and production of paddy and wheat is the highest in the state. This is the reason that two districts Sirsa and Kurukshetra from Haryana have been selected. To achieve the desired results, crop-wise compound annual growth rates in relation to area, production and yield for the period under study are computed by the use of least square technique fitting an exponential function :

$$Y = ab^t$$

where Y = area/production/yield of each crop

a = constant

b = 1+r

r = Compound Annual Growth Rate (CAGR)

t = time variable in years

Empirical Results

With the adoption of new agricultural technology, agriculture in Haryana has gone under a sea change since 1966-67 till date. No doubt, the area, production and productivity of both food and non-food crops has gone up many times. However, the present study has taken only three crops, i.e., wheat, paddy and cotton into account, which are the chief food and fibre crops of the state. The area, production and productivity trends of wheat, paddy and cotton; have been presented in the Table-1. It is found that the growth rates for these variables for the said crops were quite significant which are statistically confirmed by the t-values and the value of R^2.

Table 1
Crop-wise Growth Rates of Area, Production and Yield in Haryana (1966-67 to 2008-09)

Crop	*Area*			*Production*			*Yield*		
	CAGR	*t-value*	R^2	*CAGR*	*t-value*	R^2	*CAGR*	*t-value*	R^2
Wheat	19.35*	9.1529	.92	12.83*	10.0136	.93	8.53*	9.7888	.93
Paddy	9.65*	14.1008	.93	13.89*	7.6954	.89	6.91*	2.9504	.65
Cotton	11.49*	10.3211	.94	14.81*	4.1672	.71	-10.89	0.6892	.60

Note : 1. Computed from the data taken from various issues of Statistical Abstract of Haryana from 1966-67 to 2008-09.

2. Data pertaining to Cotton has been taken from 1966-67 to 2007-08. Moreover data for the year 2007-08 are provisional estimates.

* Denote significant at one per cent level of significance.

Table 2
Crop-wise Growth Rates of Area, Production and Yield in Sirsa and Kurukshetra Districts

Sirsa
Table 2(A)

Crop	*Area*			*Production*			*Yield*		
	CAGR	*t-value*	R^2	*CAGR*	*t-value*	R^2	*CAGR*	*t-value*	R^2
Wheat	10.30*	15.6302	.84	19.84*	7.1132	.66	7.96*	27.70	.69
Paddy	7.99*	2.0932	.96	-9.84*	2.5542	.96	1.96*	5.4823	.90
Cotton	11.91*	8.3262	.65	25.19*	8.3921	.83	2.63*	13.04	.88

Kurukshetra
Table 2(B)

Crop	*Area*			*Production*			*Yield*		
	CAGR	*t-value*	R^2	*CAGR*	*t-value*	R^2	*CAGR*	*t-value*	R^2
Wheat	8.76*	4.1432	.59	6.7*	4.5231	.76	9.67*	15.4302	.91
Paddy	9.16*	13.3342	.68	12.39*	13.5012	.82	8.78*	30.6331	.68
Cotton	-2.23*	68.8362	.74	-5.23*	32.8734	.75	-1.94	1.2415	.71

Note : The results presented in Table-2 (A) & (B) are computed from the data taken from various issues of Statistical Abstract of Haryana from 1966-67 to 2008-09.

* Denote significant at one per cent level of significance.

The behaviour of the area, production and yield for the crops has been analyzed separately for the districts of Sirsa and Kurukshetra. The results have been presented in Table 2. It is revealed from the table that the growth rates of area, production and yield of these crops increased significantly in Sirsa except the production of paddy. The CAGR of production for paddy declined drastically by –9.84 per cent. In comparison to Sirsa district, the district of Kurukshetra has been bestowed by the bounty of nature an edge in terms of favourable agro-climatic conditions for wheat and paddy crops. As a result, the CAGR of the area, production and yield for the crops of wheat and paddy has increased substantially. As far as the CAGR with respect to cotton crop is concerned, it has revealed the negative trend. These results are confirmed by the corresponding t-values and the value of R^2.

Among the various factors contributing to such an increase in the production and productivity of the crops under the study are the excessive use of fertilizers, chemicals and pesticides. Presently, more than one thousand agro-chemicals are being used all over the world. Developed countries consume 80 per cent of the total produced agro-chemicals whereas remaining 20 per cent are consumed in developing countries. India consumes about 3.7 per cent of total world consumption (Kathpal and Kumari, 1998). Several global reports have ranked India among leading pesticide consuming countries. However, currently pesticide consumption is the lowest in India. A report from the Assocham says pesticide consumption in India is the lowest at 0.5 kg per hectare as against 17 kg per hectare of Taiwan, 12 in Japan, 6.6 Korea, 7 in the US and 2.5 in Europe. The decline was primarily because of ban or restriction on use of organochlorine pesticides mainly, HCH and DDT. Before the imposition of complete ban on the manufacture and use of HCH w.e.f. April 1,1997, annual production of HCH alone was about 25,000 MT (Agnihotri, 2000). The position of the use of the pesticides and chemical fertilizers in Haryana has been presented in Table 3 and 4.

Table 3
CAGR of Pesticides and Fertilizers Consumption in Haryana (1966-67 to 2008-09)

	CAGR	*t-value*	R^2
Pesticides	12.80*	5.0909	.76
Fertilizers	12.42*	6.4648	.82

Note : 1. Computed from the data taken from various issues of Statistical Abstract of Haryana for the period 1966-67 to 2008-09.

2. Data for Pesticides for the year 2008-09 are provisional estimates.

Table 4
CAGR of Pesticides and Fertilizers in Sirsa and Kurukshetra Districts

	Sirsa			*Kurukshetra*		
	CAGR	*t-value*	R^2	*CAGR*	*t-value*	R^2
Pesticides	7.15*	4.5226	.63	8.79*	2.6905	.91
Fertilizers	18.15*	4.8739	.93	12.20*	5.5347	.68

Note: 1. Computed from the data taken from various issues of Statistical Abstract of Haryana.
2. Data for Pesticides for the year 2008-09 are provisional estimates.

The Tables 3 and 4 show that the CAGR of both, i.e., pesticides and fertilizers consumption in Haryana as a whole and both the districts of Sirsa and Kurukshetra has increased significantly during the study period. Similarly, the CAGR of farm machinery during the period under study also showed an impressive growth that further worsened its environment.

Table 5
CAGR of The Farm Machinery In Haryana

	Tractors			*Diesel Pumps*		
	CAGR	*t-value*	R^2	*CAGR*	*t-value*	R^2
Haryana	11.33*	3.1996	.84	11.86*	4.7741	.77
Sirsa	14.81*	7.6301	.86	9.64*	5.5501	.73
Kurukshetra	12.20*	4.4628	.67	2.32*	3.9827	.86

Note : Computed on the basis of the data taken from various issues of Statistical Abstract of Haryana from 1966-67 to 2008-09.

The farm machinery has also contributed as significant input for the agricultural development in Haryana. The same is true for both the districts under the study, which is confirmed by the CAGR of the use of tractors and diesel pumps as shown in the Table 5.

It is revealed from the Table 6 that wheat, cotton (American and Desi), grams and paddy are the dominant

Table 6
Percentage Distribution of total Cropped Area of Selected Farmers in Sirsa and Kurukshetra Districts

Sirsa		*Kurukshetra*	
Cropping Pattern	*%age*	*Cropping Pattern*	*%age*
1. Paddy	7.60	1. Paddy	48.63
2. Wheat	42.76	2. Wheat	45.22
3. Jawar	0.05	3. Jawar	0.17
4. Groundnut	0.15	4. Groundnut	0.04
5. Cotton (American)	25.81	5. Cotton (American)	0.55
6. Cotton (Desi)	13.56	6. Cotton (Desi)	0.09
7. Sugarcane	.43	7. Sugarcane	5.12
8. Grams	8.72	8. Grams	0.09
9. Bajra	0.92	9. Bajra	0.09

Note : Computed from the primary data

Kharif and Rabi crops in Sirsa district which accounts for 42.76, 39.37, 8.72 and 7.60 per cent of the total cropped area, respectively, whereas paddy and wheat followed by sugarcane are the dominant Kharif and Rabi crops of the district Kurukshetra which claim 48.63, 45.22 and 5.12 per cent of the total cropped area of the selected farmers, respectively. The crops of paddy, wheat and cotton are highly sensitive to many pests and diseases. To safeguard the production of these crops, all farmers selected in the study in both the districts used a heavy amount of the pesticides, weedicides and insecticides.

No doubt, the chemicals, fertilizers, pesticides alongwith the use of tractors and diesel pumps contributed a lot in increasing the productivity of the crops under the study. However, it cannot be denied that the use of these inputs also adversely affected the public health and environment by spreading a number of diseases like heart and lungs cancer, allergy, bronchitis, asthma, tuberculosis, hyper-acidity, fatal accidents and certain neurological problems. To know the testimony of the fact, the data was collected from 100 farm labourers who were engaged in different farm operations.

To know the actual position, a question was addressed to the farm labourers – 'Are you suffering from any disease – Yes or No?' It is found that out of 100 farm workers, 76 per cent were suffering from one or another type of disease. A detailed analysis is presented in the Table 7.

Table 7
Distribution of Farm Labourers By Type of Disease in Sirsa and Kurukshetra Districts (Combined)

	Type of Diseases	*No. of Farm Workers*	*Percentage*
1.	Heart Attack	4	05.26
2.	Lung Cancer	6	07.89
3.	Allergy	26	34.24
4.	Bronchitis	2	02.63
5.	Asthma	15	19.73
6.	Tuberculosis	9	11.84
7.	Hyper-acidity	11	14.47
8.	Neurological Problems	3	03.94
	Total	76	100

The most common disease which are caused by pesticides and insecticides are allergy, asthma, hyper-acidity and tuberculosis, which account for 34.24, 19.73, 14.47 and 11.84 per cent, respectively.

Conclusion

The study revealed that Haryana state has played a pivotal role in ushering in an era of green revolution in India. The agricultural development plans of the state laid a greater stress on increasing agricultural production and achieving high growth rate. Due consideration has not been given to maintain the ecological balance. The increase in the productivity of the foodgrains particularly wheat, paddy, and cotton have been found due to the significant increase in the pesticides and fertilizers causing a great concern for the public health. The greater use of insecticides has contaminated the environment and these chemicals have entered the food chains including the ground water. Likewise, a significant increase in the number of tractors and diesel pumps also resulted in the environmental pollution emitting carbon dioxide (CO_2), carbon monoxide (CO), Nitrogen monoxide (NO) and Nitrogen dioxide (NO_2), Hydrocarbons (HCs), smoke and smog. These emissions polluted the environment in such a manner that the diseases of heart and lungs cancer, allergy, bronchitis, asthma, tuberculosis, etc. have become quite common in the society.

In essence, there is already an alarming bell. There is a dire need to get up and act at the global level. Hence modern agriculture may prove harmful if not planned with foresight and adequate regard for environmental pollution.

Limitations

The present study is highly relevant in the modern day context of sustained development. However, the study is a modest attempt in this regard. A detailed, comprehensive, inter-disciplinary study in this direction for Indian economy as a whole is the need of the hour on the pattern of the studies in various developed countries if the policy-makers are really interested to translate their dreams of prosperous society in its true perspectives.

References

Acquaah, G. (2002), "Pesticide Use in U.S. Crop Production", in *Principles of Crop Production, Theories, Techniques and Technology*, Prentice Hall, Upper Saddle River, NJ.

Agnihotri, N. P. (2000), "Pesticides Consumption in Agriculture in India – An Update," *Pesticide Research Journal*, Vol. 12 (1): 150-155.

Bai, Z. G., Dent, D. L., Olsson, L. and Schaepman, M. E. (2008), "Global Assessment of Land Degradation and Improvement; Identification by Remote Sensing", *UN Food and Agriculture Organisation, Report 2008/01*, FAO/ISRIC-Rome/Wageningen.

Borger, Julian (2008), "Feed the World? We are fighting a losing battle, UN admits", *The Guardian*, February 26, London.

Horrigan Leo, *et al.* (2002), "How Sustainable Agriculture can Address the Environmental and Human Health: Harms of Industrial Agriculture", *Environmental Health Prospectives*, Vol. 110, No. 5.

Kathpal, T. S. and Kumari, Beena (1998), "Pesticidal Contamination of Environment, its Implications and Management." *In* Dhaliwal, G.S., *et. al.* (Eds), *Ecological Agriculture and Sustainable Development*, Vol. 2: 535-551.

Malla, G. (2008), "Climate Change and Its Impact on Nepalese Agriculture", *Journal of Agriculture and Environment*, June-July, Vol. 9: 62-71.

Mc Kie, Robin and Rice, Xan (2007), "Millions Face Famine as Crop Disease Rages", *The Observer*, April 22, London.

Mortished, Carl (2008), "Already we have Riots, Hoarding, Panic: The Sign of Things to Come?", *The Times*, March 7, London.

Pretty, J. *et al.* (2000), "An Assessment of the Total External Costs of UK Agriculture", *Agricultural Systems*, 65(2): 113-136, doi:10.1016/S0308-521×(00)00031-7.

http://www.esse.ac.uk/bs/staff/pretty/AgSyst%20pdf.

Qaim, Matin and Janvry, Alain De (2005), "Bt Cotton and Pesticide Use in Argentina: Economic and Environmental Effect", *Environment and Development Economics*, Vol. 10, May, No. 2: 179-200.

Rao, E.V.S. Prakasa and Puttanna, K. (2000), "Nitrates, Agriculture and Environment", *Current Science,* Vol. 79, No. 9: 1163-1168.

Sample, Ian (2007), "Global Food Crisis looms as Climate change and Growth Strip Fertile Land", *The Guardian*, August 31, London.

Shah, Binod P. and Devkota Bhupendra (2009), "Obsolete Pesticides: Their Environmental and Human Health Hazards", *Journal of Agriculture and Environment*, June-July, Vol. 10: 60-66.

Simoens, Steven (2009), "Health Economic Assessments: A Methodological Primer", *International Journal of Environmental Research and Public Health*, Vol. 6, No. 12: 2950-2966.

Singh, B. and Dhaliwal, G. S. (2000), "Pesticide Contamination of Fatty Food Commodities." *In* Dhaliwal, G.S. and Singh, B. (Eds.) *Pesticides and Environment*: 155-190.

Smith, Kate and Edwards, Rob (2008), "2008: The Year of Global Food Crisis", *The Herald*, March 8, Glasgow.

Steinfeld, Gerber, P., Wassenaar, T, Castel, V., Rosales, M., de Haan C. (2006), "Livestock's Long Shadow-Environmental Issues and Options", *UN Food and Agriculture Organisation Report*, Rome.

Steinfeld, Henning (2006), "Livestock a Major Threat to Environment", *UN Food and Agriculture Organisation Report,* November 29, Rome. *http://www.fao.org/newsroom/en/news/2006/1000448/index.html.*

Tegtmeier, E.M. and Duffy, M. (2005), "External Costs of Agricultural Production in the United States", *The Earthscan Reader in Sustainable Agriculture. http://www.organicvalley.coop/fileadmin/pdf/ag_costs_IJAS2004.pdf.*

Thakur, J. S., *et. al.* (2008), "Epidemiological Study of High Cancer Among Rural Agricultural Community of Punjab in Northern India", *International Journal of Environmental Research and Public Health*, Vol. 5, No. 5: 399-407.

Tiwana, N. S. *et al.* (2009), Pesticide Pollution in Punjab: A Review, *Asian Journal of Water Environment and Pollution,* Vol. 6, No. 2, pp 89-96.

UN Food and Agriculture Organisation (2007), *Environment and Agriculture*, 20th Session, Item 6 of the Provisional Agenda before Committee on Agriculture, Rome, April 25-28, COAG/2007/6.

Watts, Jonathan (2007), "Riots and Hunger Feared as Demand for Grain Sends Food Costs Soaring", *The Guardian*, December 4, London.

Yang, Xiaohuan and Ma, Hanqing (2009), "Natural Environment Suitability of China and its Relationship with Population Distributions", *International Journal of Environmental Research and Public Health*, Vol. 6, No. 12: 3025-3039.

Zargar, M. Y. and Khan, M. A. (2004), *Agriculture and Environment*, Aph Publishing Corporation, New Delhi, *http://www.aphbooksindia.com*

Population, Poverty and Environment in Khyber Pakhtoonkhwa Pakistan

IFFAT TABASSUM, M.MUMTAZ AND FAZLUREHMAN

Introduction

The problem of environmental degradation, which is not simply the wastage of resources rather it is also a threat to the complex structures that support human development, is often aggravated by poverty (UNFPA, 2001). Population dynamics usually accelerate the rate of poverty-oriented environmental degradation. As the resource base of the poor is subdivided and fragmented, intensive and often-inappropriate land use becomes imperative for immediate survival. Beside availability, the proper utilization of natural resources is also critical in determining the extent of poverty because poor management of resources can severely damage its base. In areas where natural resources are scarce and other economic opportunities like off-farm activities lack, poverty is closely related with natural resources provision and management; particularly land and its biological products. Population pressure and the rash desire to use resources extensively can have disastrous consequences leading to environmental degradation (Wolfgang Meckelein 1986).

The impact of population pressure in any region cannot be assessed by considering population growth and density. The major concern is with the population resource ratio, resource use system (Israr ud din, 2001) and ways in which population, resource and environment are linked together. This requires detailed consideration of affluence, consumption, technology and population growth, but also previously ignored underrated social concerns such as gender roles and relations, political structure, institutions, and governance at all levels (UNFPA, 2001).

To feed the rapidly growing population of the world more and more land is brought under cultivation and agriculture has encroached over the fragile land. The utilization of natural resources particularly land, beyond its capacity is considered as basic cause of environmental degradation. The world is faced with numerous environmental threats; including loss of biodiversity, soil erosion, water scarcity and contamination of ground and surface water resources due to explosive rate of population growth, rapid urbanization and high per capita consumption of resources etc. These are further accelerated by concentration of population in few large cities and unequal distribution of wealth, facilities and services.

Poor have always been blamed for environmental deterioration as the short-term interests put great pressure on environmental carrying capacity which could be harmful over longer periods. Dasgupta (1993) challenges the statement that the poor degrade their environmental resource base. He argued that poverty forces them to discount future incomes at unusually high rates because difficulty in meeting today's household needs means that short-term exploitation of the environment must take priority over long-term protection. In turn decline in environment and destruction of resources base accelerates poverty through decreasing yield. Hence it is clear that poverty, environment and population are closely linked. The positive view of the paradigm is that progress in one of the interlinked sectors is likely to generate positive effects on the others, for instance: efforts to slow down population growth, to reduce poverty, to achieve economic progress, to improve environmental protection, and to reduce unsustainable consumption patterns are mutually reinforcing (ICPD 2001). A basic policy proposal stemming from this hypothesis is to limit population growth. Mehra, *et al.* (1992), for example, argues that the deteriorating living conditions of the poor lead to lower productivity, which in turn provides an incentive for raising large families. Economic, environmental and governance circumstances in which population growth takes place are usually the driving forces behind poverty-environment interactions; these, in turn, can be strongly influenced by external policy and institutional factors.

Rethinking about Poverty

Sen (1981) has identified two essential questions regarding poverty: Who are the poor? And at what level is poverty defined? Conventional definitions of poverty have focused on income and consumption. However, both income and consumption-based definitions of poverty have been widely criticized for not capturing all the elements of poverty. For example, there is a growing debate about whether income should be assessed in terms of flows (e.g., the sale of agricultural products) or assets (e.g., land that may be rented or used as collateral on loans) (Riordan, 1992). Secondly, neither measurement includes the consumption of state-provided commodities (such as health and education) nor common property resources (e.g., forests and rangelands). Thirdly, the measurements do not contain information about the distribution of income within the family (e.g., the proportion which is available to women). Fourthly, access to income does not necessarily indicate whether people have access to common public services, such as clean water, sanitation, education and health. These concerns have led to the development of a different set of indicators, based on the fulfillment of basic needs and the quality of life. Although these indicators are more robust than the income-based definitions, they too have come under criticism. Among other weaknesses, they do not capture all the manifestations of poverty. They also tend to focus on inputs rather than processes and outcomes.

In an effort to address these various shortcomings, another assessment methodology known as the sustainable livelihoods guidance sheet has been developed (DFID, 1999). This framework has been prepared particularly for the understanding of vulnerability context and sustainability of livelihood among the poor. Especial attention has been paid for the understanding of different capitals including Human, Social, Physical, Financial and Natural (BOHLE, 2001 and TRÖGER, 2002). Most of the scientists are of the view that social capital plays a major role in alleviating poverty (Coleman, 1988).

It is these broader concepts of livelihoods and wellbeing that tend to emerge from self-assessments, such as the participatory poverty assessment recently undertaken in Pakistan by the Asian Development Bank (ADB). The degree to which people draw on criteria other than income during these self-assessments of poverty can be striking (Judah, 1991; ADB, 2002). Of particular significance in the ADB assessment is the inclusion of environmental goods and services, which conventional definitions of poverty tend to overlook. In nutshell, it is perhaps meaningful and appropriate to define poverty in terms of vulnerability; and to look at multiple dimensions of deprivation, including: lack of income and other material means; lack of access to basic social services such as education, health and safe water; lack of personal security; and lack of empowerment to participate in the political processes and decisions that influence one's life (Bohle, 2001).

Problem

In few geographical regions, it seems that Pakistan's population has exceeded the existing carrying capacity of land, while inappropriate land use has resulted in environmental degradation and increased poverty through reduced land productivity (Figure 3). Evidence shows that over past years our forests in Northern region shrunk to half of original extent (IUCN, 1996) and women average journey to collect fuel wood have increased many folds in rural areas (Iffat, 2005) in some parts of the country. This can also be attributed to the anthropogenic factors and non-equilibrium dynamic patterns (Nüsser 2002).

Khyber Pakhtoonkhwa *(KPK), one of the most fragile provinces of the country is faced with numerous problems. It occupies 9.36% of the country land area and has 13.65% share in the total population of the country. Compared to166 persons per sq.kms in Pakistan, the average population density of the province is higher than the national average, i.e. 238 persons per sq.kms (Table 8). However, crude population density varies greatly from place to place within the province ranging from 1606 in Peshawar to 21 in Chitral (GoNWFP, 1998). Uneven population distribution is the result of highly diverse physiography and resource availability (Table 6).

The province is divided in three distinct geo-ecological regions: the central plains, the north and northeastern mountainous and southeastern semi arid to arid plains. Natural resources in these areas are underutilized due to various natural constrains for example large southern plain (districts of D.I. Khan, Lakki and Kark) and northern mountains cannot be used for extensive cultivation due to rough topography, unfavorable climatic conditions and water shortage. The central part is well irrigated and has fertile plain which is extensively cultivated. This region is thickly populated and major urban centers including the provincial capital Peshawar and second largest city Mardan are located here (see map 1).

Rural to urban migration rate in the region is greatest which has increased population pressure on services and land. Fertile agricultural land is rapidly converted into built up area. Additionally the province is also highly prone to different types of disasters like, floods, earthquake, drought, land sliding and desertification on account of physical and social factors previously discussed. The combined impacts of environmental problems, population pressure and re-current disasters have resulted into increased poverty in the last few decades. Poverty depletes natural resources, which in turns increase sufferings of people. Majority of low-income people are found in slums, shantytowns and rural areas; where basic needs are never provided to the communities with high density leading towards intensity (Gans, 1973).

Most cities and towns in the province are growing with leaps, bounds, and not fit for healthy living environment. For instance, in urban areas; a high proportion of the population have been living in shelters, slums and shanty towns with little provision of facilities like clean drinking water, safe disposal of solid waste, liquid waste and deficient fulfillment of basic needs. Houses in these areas are normally of poor quality and overcrowded, built in low lying and marginal lands subjected to periodic floods and other natural hazards. People are forced by the economic condition to live in these service–deficient areas as most of them have un-secured jobs within marginal incomes and cannot afford to live in better place. The urban economy in most of these areas has not diversified

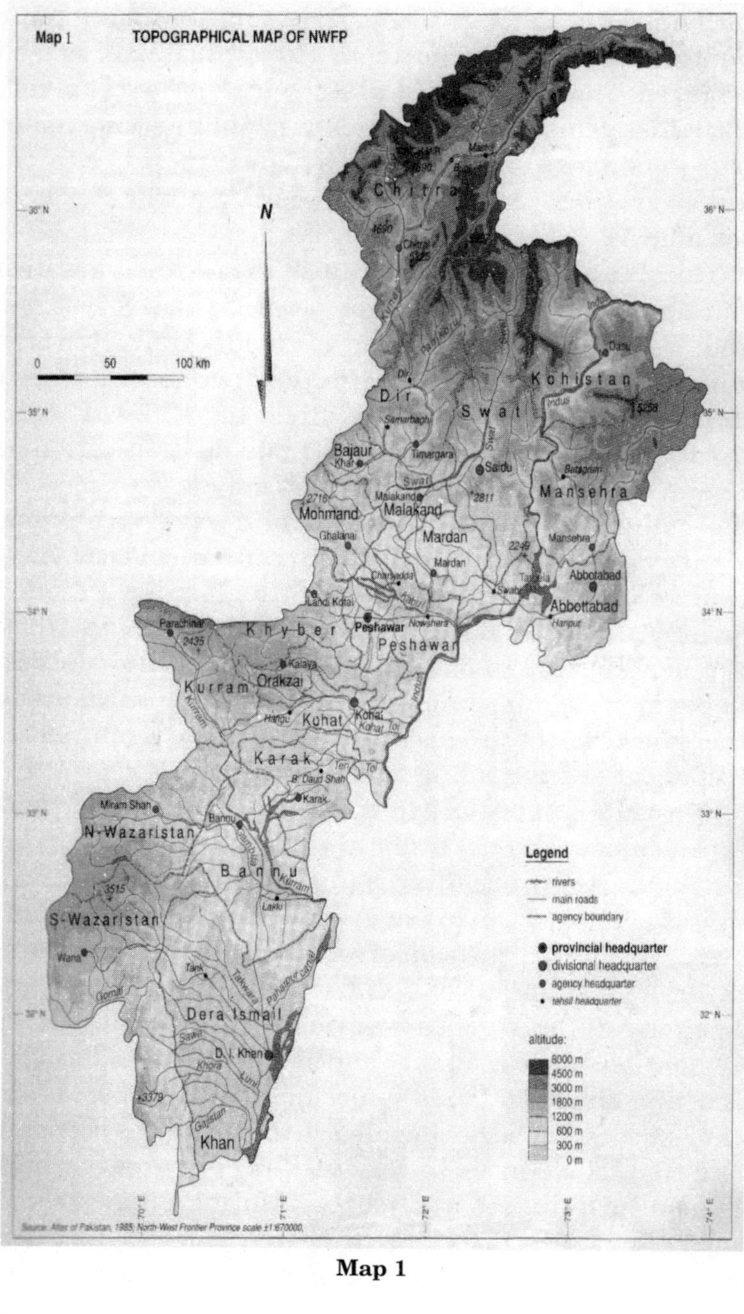
Map 1
TOPOGRAPHICAL MAP OF NWFP
N
0 50 100 km
Chitral
Dir
Swat
Kohistan
Bajaur
Mansehra
Mohmand
Malakand
Mardan
Abbottabad
Khyber
Peshawar
Kurram
Orakzai
Kohat
Karak
N-Wazaristan
Bannu
S-Wazaristan
Dera Ismail
Khan
Legend
rivers
main roads
agency boundary
provincial headquarter
divisional headquarter
agency headquarter
tehsil headquarter
altitude:
8000 m
4500 m
3000 m
1800 m
1200 m
600 m
300 m
0 m
Source: Atlas of Pakistan, 1985; North-West Frontier Province scale ±1:670000.

Map 1

resulting into environmental degradation. For the last few years the worsening energy crisis and political instability has also accelerated poverty in the province through closure of industries and deterring tourists.

Results and Discussion

While the increasing environmental threat, land degradation, loss of biodiversity and water scarcity and contamination is largely the by-product of mis-management of resources, given in conceptual framework in figure 1, poverty can drive overexploitation of limited resource base. Incidence of poverty indicates that the Khyber Pakhtoonkhwa has high rate of poverty index (46.5%) compared to national figure of 38% (GoNWFP, 2008: 301). Most districts of the province are in the lower three quartiles of the social and economic development. Only 25 per cent of the total districts are in first quartile (Table 1).

Table 1
KPK: Ranking of Districts based on Social and Economic Indicators (1999)

Rank of the districts	*Number of districts*	
	Social	*Economic*
1st Quartile	5	5
2nd Quartile	6	8
3rd Quartile	3	5
4th Quartile	6	2

Source: UNWFP, (1999)

The population in KPK is growing at annual growth rate of 2.82% higher than national population growth rate, (2.69%). Average household size of the province has increased from 6.8 persons in 1981 to 8.0 in 1998 (last Census), which is highest among all the provinces and national figure (Table 8), widening the gap between consumption and the resource availability. Short terms imperatives to produce more food and feed for the growing population both human and livestock is leading to unsustainable development pattern within the province. Disparities in development, resource use allocation is found in different regions, societies and different groups developing a sense of inequality and causing poverty among different sections and groups of societies, particularly southern region of the province (IUCN and Go NWFP, 1996).

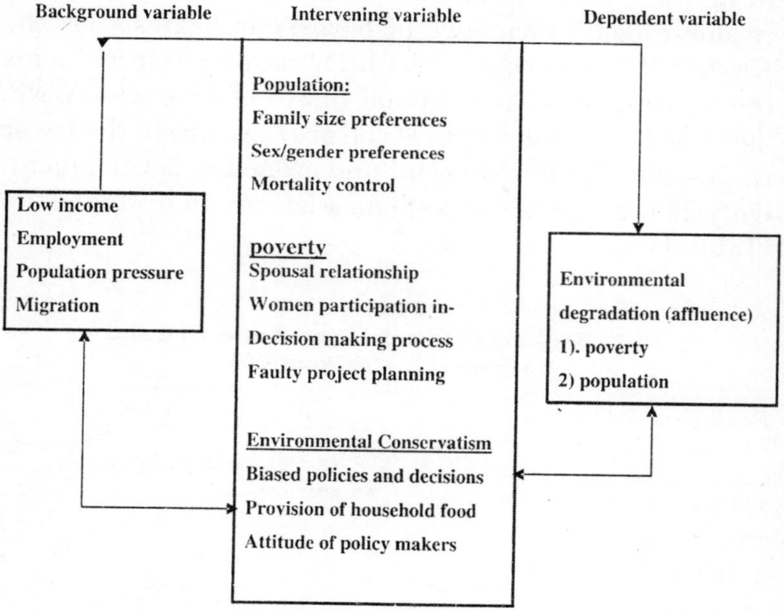

Figure 1: Showing the Relationship between Predictor and Response Variables

Population problems, poverty and environmental change are linked in many ways and through multiple social and economic mechanisms, at various social, economic and geographic levels. It is shown in Figure 1 and Figure 2.

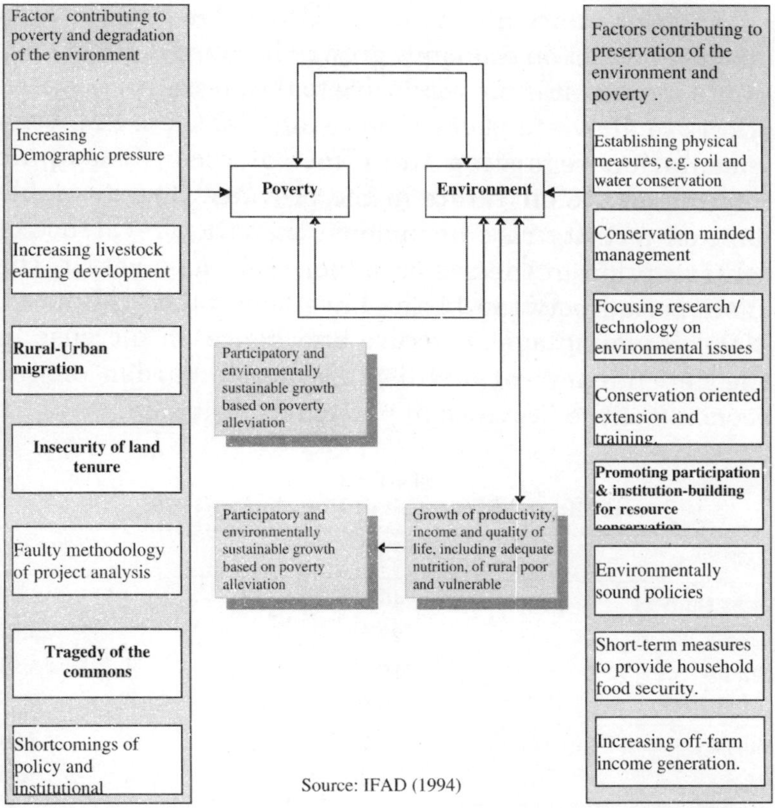

Figure 2: Showing Link among Population, Poverty, Environment and Development

Demographic processes are linked with socioeconomic development and poverty in a complex manner. Rapid population growth is the consequence of high fertility and lower mortality. According the population reference bureau data sheet 2009 the crude fertility rate is 30per thousand and death rate is only 7 per thousand (Population Reference Bureau 2009). In addition trend of large family, gender roles in deciding the size of family and preference to male children are also very important factors of population growth in KPK (Mumtaz, 2005).

High population density in KPK is leading towards adverse impact on economic growth because it affects per capita income. It is not conducive to the success of targeted efforts to alleviate poverty, for example, there had been speculation regarding the effect of recent economic reforms on the birth rate in the province. The available data on fertility rate has shown that the prevalence of contraception in the region is not high, as a result, the short period between births have been noted (Mumtaz, 2005). A young age structure and hence an increase in the dependency ratio will place heavy burden on the economy of the province in particular (Table 2).

Table 2
KPK: Age Structure of Population(1998)

Age	*1981(%)*	*1998 (%)*
Less than 14	46.82	47.2
15-29	22.54	25.47
30-44	15.17	13.91
45-59	9.04	8.38
60-74	5.02	3.91
75+	1.40	.98

Source: GoP, 1981 & 1998 Provincial Census Reports of NWFP by Population Census Organization Islamabad

An increasing young population has serious implication for the budget allocations in education and heath sectors. Moreover, provision of employment opportunities for growing adult population is also a serious problem in the province that has been further aggravated by political instability. As for situation of rapid population growth in KPK is concerned there are few major contributing factors (Mumtaz, 2005) e.g.

- Prevalence of traditional society where preference of male sex results in increased number of children.

- Lack of access to basic needs including water supply, fuel and energy increases the need for children to help in fields and homes in poor families. It also includes lack of security in illness and old age as sixty per cent of people are living below the poverty line.
- Another important factor is female education, her subordinate social and economic status and awareness regarding women reproductive health. This means less awareness of family planning methods, contraceptive measures and use of facilities and availing benefits. Low status of women is often associated with poverty; they have less or no power to control fertility.

High population results over burden and overstretching of social services as well, like schools, health centers, reproductive health centers, family planning clinics, water and sanitation services. School participation rate of boys and girls is one of the important indicators in this regards. The general situation of the province is not promising even if compared with national figures. The rate is further decreasing with the passage of time particularly in higher classes (Table 3). The situation of health facilities in the province is also below satisfactory level. According to the official figure (GoNWFP 2008:146) person per hospital bed is 1654 in 2007 compared to 1530 in the country (GoP, 2007). The number of doctors per person is seven times higher in the province than the national average of 1370.

Factors Contributing to Link between Poverty and Degradation of the Environment in KPK

Physical factors like physiography and climate are the major natural determinants of environmental and economic marginality in the province. Mountain specificities (Jodha, 1992) also play a crucial role in

Table 3
School Participation Rate of Primary, Middle and High School Age Group Population of Pakistan and KPK, for the year 2006-2007

	Pakistan in percentage			*KPK in percentage*		
	Primary	*Middle*	*High*	*Primary*	*Middle*	*High*
Both sex	91	51	48	82 (70)*	53 (29)*	45 (26)*
Male	99	57	57	96 (78)*	68 (37)*	64 (35)*
Female	81	44	35	67 (61)*	35 (20)*	24 (17)*

Source: GoNWFP 2008: 118 and 120

* Indicate school participation for the year 2007-2008.

determining the socio-economic situation of mountain inhabitants. The mountainous and plain areas of the province are prone to different types of natural hazards and calamities, such as landslides and earthquakes are more common in mountains (the 2005 earth quake and artificial lake of Ataabad due to landslide 2010 are witness to it) while floods and droughts are frequent in plains of the province. In addition various social problems have increased the magnitude of natural hazards many folds. These factors have been summarized in figure 3.

Uneconomic Landholdings

Due to its geo-ecological conditions very small portion of the land in the province is cultivated. Landholdings in the province are relatively small both because of land scarcity and because of the prevailing laws of inheritance which results into division of land among the male children. Fragmentation of land has serious implications over productivity and mechanization. More than ¾ of the total households (79%) own less than one to five acre of agricultural land while 99% own less than 50 acres (Table 4). In such a situation, the costs of inputs are usually higher than the outputs and thus farming become an uneconomical activity. Therefore, the farmers are also looking for other employment to supplement their livelihood. Consequently, many of these farmers are migrating outside their respective villages to participate in off farm economic activities.

Table 4
Number of Farms by Size of Farms in KPK for the Year 2000

Total farms (area in Acres)	*Number*	*In per cent (%)*		
	1086044	100	100	100
Under 1.0	273377	25.17	79.26	99.33
1.0 -2.5	378152	34.81		
2.5-5.0	209417	19.28		
5.0-7.5	96898	08.92	19.97	
7.5-12.5	69323	06.38		
12.5-25.0	33929	03.14		
25.0-50.0	16598	01.53		
50.0-100	6058	00.56	00.77	00.77
100-150	1275	00.12		
150 and a above	1017	00.09		

Source: GoP, Agriculture Census 2000 (NWFP) by Agricultural Census Organization.

Poor Accessibility

Inaccessibility affects the communities in many districts of the province. In many areas, access to biological resources (such as forests) has become increasingly restricted, as state bodies have taken over the role of natural resource management from local institutions. Secondly, local communities lack access to markets, technologies and credit; this makes it difficult for them to sell their products in the market of the lowland, and to obtain inputs and services that could enhance their productivity. Thirdly, because of the province's physical isolation and political factors, it has not been able to influence national socio-economic or political processes in its favor; this has been further exacerbated by a misperception within central government that the KPK lack development potential or resources (Table 5 to 7).

Sex Selective Rural Urban Migration

Weak resource base and lack of employment opportunities at their origin has resulted into extensive out migration from rural to urban area as well to abroad like Middle Eastern countries. This is revealed in sex ratio which is lowest in the province when compared with other regions of the country (Table 8). Out migration from rural areas has negative implication for the indigenous resource management and utilization. It is both symptoms of, and cause for the degradation of environment and natural resources. As the migrants are usually young male in search of improved economic opportunities for themselves and their families, their absence often results to deepen rural poverty. In their absence the management of agriculture production and other associated activities becomes the responsibility of aged and women. The number of female household head is increasing. Meanwhile remittance sending into these households is normally consumed in the daily needs and in raising the life standard squeezing the investment in agriculture and education sectors due to unavailability of capable community members. This has adverse effects on agricultural productivity as well as children education. These out migrants also face multiple problems at the destination and are often forced to live in areas lacking in minimum basic facilities. They become the agent of environmental degradation by exerting pressure on already scares services and facilities in these areas (Khan Amir, 1994).

Pattern of Land Use

The carrying capacity of the land is one of the important factors contributing towards the way the land is used. The inappropriate land use pattern normally results into over cultivation of marginal lands, deep tilling, application of chemical fertilizers and pesticides which increase the environmental marginality. Intensive annual cropping in central districts like Peshawar, Mardan and Charsadda and cultivation of steep slopes of mountains in northern parts of the province contribute towards land degradation. Moreover, rapid deforestation and over grazing on the marginal land further deteriorate the land resources of the province.

Unsustainable Population Growth

Increasing population besides squeezing the availability of space for natural resource generation also exerts high pressure on services and facilities. All these services and utilities require heavy capital investments, the resources for which must be raised through taxes. This presumption that the region has a sound economy with a well-developed and growing social, economic, financial and industrial base, and the citizens are equitable beneficiaries and have the capacity to be taxed. If these prerequisites are absent – as they are in the KPK – then

Table 5
KPK: Population Growth 1951-2011

Census year	*Population (million)*	*Inter-censal Increase*	*Annual growth rate*
1951	4.556	-	-
1961	5.731	25.77	2.34
1972	8.388	46.37	3.32
1981	11.061	31.86	3.32
1991*	15.507	40.19	3.2
2001*	20.084	43.39	3.0
2011*	26.934	29.26	2.6

Source: IUCN and GoNWFP, 1998: 71

*Estimated Population

the fast growth in population is a luxury the province cannot afford.

Comparing the increase in population with the available land in past decades shows a high population increases while the productive land resource increase is much lower, further indicating that land resources are under tremendous pressure. Overexploitation of the resource is currently increasing, which is likely to continue as the province attempts to meet the demand for food. The growing pressure and high rate of exploitation is having many adverse environmental impacts and is increasing the pressure on other natural resources.

The capacity of cultivable land to support larger numbers of people can be enhanced through changes in cropping patterns and intensity of cropping, which could be achieved by investments in human resource development and technology. But converting presently unused lands to productive use appears to be prohibitively capital-intensive, involving heavy investments in complex civil works. Therefore, the potential of the total land resources of the KPK to support larger population is not encouraging in future.

Trends and Traditions of Large Family Size

As discussed in the beginning that most people in the KPK live in small villages. Household size in the province is large enough (8.1 persons per household). The KPK society is predominantly Muslim and is dominated by strong conservative values exhorted in many cases by local religious leaders. Men have been brought up in a male-dominated patriarchal society, where it is men who deal with the outside world and women are supposed to limit their lives to childbearing and housekeeping. Subjects other than this about which women are believed to know little are taboo, and women therefore have a limited role in decision-making.

Conservative and traditional norms dictate that women follow a socially acceptable behavior. In few areas of the province women have generally not been educated and are permitted only limited mobility, restricted to their family houses. Women accept and some even believe that

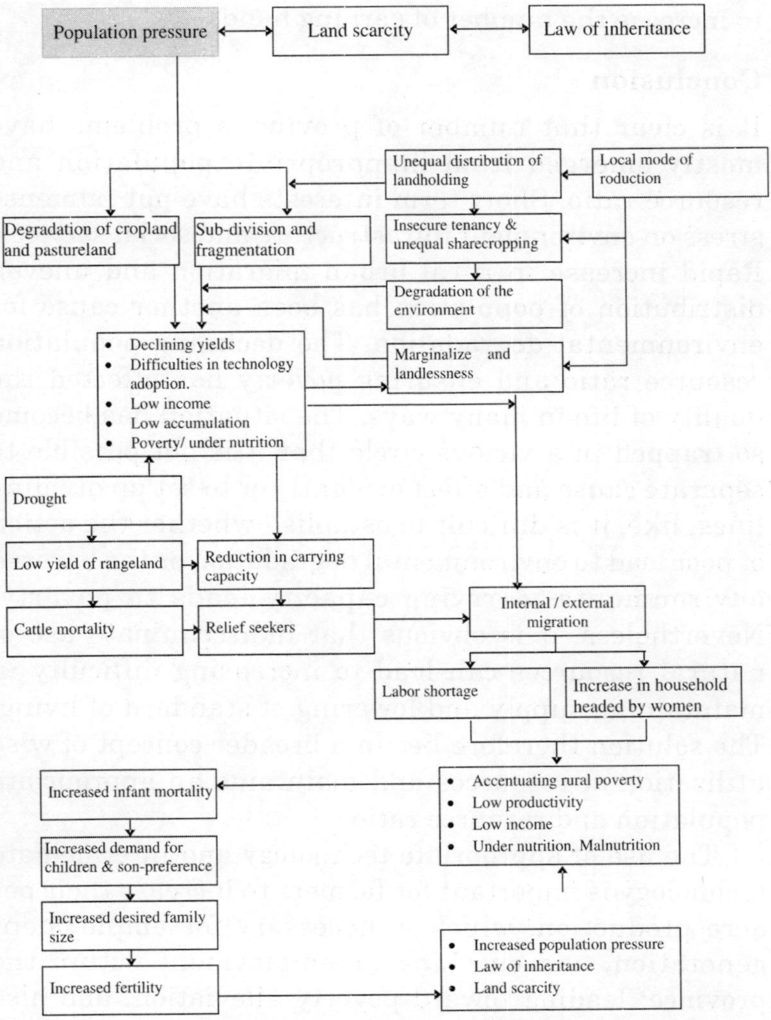

Figure 3: The Population, Poverty and its Impact on Environment

their only purpose in life is to produce children. Therefore, producing many children is the norm and not the exception. Rural society is still tribal. Sons are a matter of great pride and give strength to the father in the village. Attempts to have many sons generally lead to large families. Poverty also dictates the need for more children to increase the number of earning hands.

Conclusion

It is clear that number of province's problems have mostly emerged from inappropriate population and resource ratio. Short term interests have put immense stress on environment infrastructure and social services. Rapid increase in rural urban migration and uneven distribution of population has been another cause for environmental degradation. The declining population resource ratio and ensuring poverty has affected the quality of life in many ways. The situation has become so trapped in a vicious circle that it is not possible to separate cause and affect evidently or to set up dividing lines, like, it is difficult to establish whether the action of poor lead to environmental degradation or the lowered environmental carrying capacity leads to poverty. Nevertheless, it is obvious that indiscriminate use of natural resources can lead to increasing difficulty in maintaining supply and lowering of standard of living. The solution therefore lies in a broader concept of wise utilization of resources and maintains an appropriate population and resource ratio.

The use of appropriate technology and intermediate technology is important for farmers to increase their per acre production, which is necessary for employment generation, and checking un-employment within the province; leading toward poverty alleviation, and also sustainable development.

References

ADB, (2002), Language, Ideology and Power: Language-learning Among the Muslims of Pakistan and North India Karachi: OUP.

Bohle, H. G. (2001), Neue Ansätze der Geographischen Risikoforschung: Ein Analyserahmen zur Bestimmung nachhaltiger Lebenssicherung von Armutsgruppen. In: Die Erde 132(2), 119-140.

Casterline, J. B., Z. Sathar and M. Ul Haq (1999), "Preferences and Behaviors: A Study of Contraceptive Intentions in Pakistan", Paper Presented at the Annual Meeting of the Population Association of America, New York, 25-27 March.

Coleman, J. S. (1988), Social Capital in the Creation of Human Capital. In: *American Journal of Sociology* 94 Supplement, S95-S120.

Dasgupta, (1993), An Inquiry into Well being and Destitution. Oxford: Calderon Press.

DFID (Department for International Development) (1999), Sustainable Livelihoods Guidance Sheets. London.

GoNWFP (2008), NWFP Development Statistics 2008, by Bureau of Statistics Planning & Development Department Government of NWFP.

GoNWFP (2002), A District Based Multiple Indicator Cluster Survey, Govt. of N.W.F.P., Planning and Dev. Department: in Collaboration with Federal Bureau of Statistics and UNCF.

GoPakistan (1990-95), "District wise Socio-economic Database on Pakistan". By Federal Bureau of Statistics (FBS).

GoPakistan, Population Census Reports 1998.

GoPakistan, Economic Survey 2007-08.

Herbert Gans (1973), The Urban Villagers, Group and Class in Life of Italian –Americans Glencoe: the Free Press, p. 11.

ICPD. (2001), International Conference for Population and Development Governments, UN System Organizations and NGOs are Urged to Disseminate the Programme of Action.

IFAD. (1994), The Diagram is Borrowed from IFAD: State of the World's Poverty 1994.

Iffat, T., (2005), Women's Role in Household Income Generation by using Natural Resources of Rangeland in Semi Arid Mountainous Region of Pakistan: A Case Study of Karak District. Depart-ment of Geography, University of Peshawar, Pakistan.

Israr ud din, (2001), "Role of Community in Management of Natural Resources in Chitral District", Paper Presented in, "Workshop on NRM" at Sustainable Development Planning Institute Islamabad Pakistan.

IUCN Pakistan (1996), Sarhad Provincial Conservation Strategy.

Jodha, N. S., M. Banskota & T. Partab (1992), (Eds.): Sustainable Mountain Agriculture: Perspectives and Issues. 2Vols. New Delhi.

Joseph, A., J. Prasad and S. Abraham (2002), "The Risk of Infection: Gynaecological Program among Young Married Women in Tamilnadu", in, S. Bott and others (eds.), Adolescent Sexual and Reproductive Health: Evidence and Facts.

Khan, Mumtaz (2005), "Examination of Women Reproductive Health within Socioeconomic and Cultural Context", A PhD Research Dissertation, U A F , Pakistan.

Khan. Amir, (1994), "Inter Relationship between Demographic Factors Development and the Environment in the ESCAP Region". In *Asia Pacific Population Journal*. Vol. 9, 3, pp. 37-54.

Mehra, Rekha *et al.* (1992), Engendering Development in Asia and the Near East: A Sourcebook. Washington, DC: ICRW.

Nüsser, M. (2002), Pastoral Utilization and Land Cover Change: A Case Study from the Sanqebethu Valley, Eastern Lesotho. In: Erdkunde 56(2), 207-221.

PRB (2009) URL *http://www.prb.org/pdf09/09wpds_eng.pdf*

Riordan, T. (1992), Environmentalism Pion Limitted London NW2-5JN 207 Brondsbury Park London. Pp. 315.

Sen, Amartya K. (1981), "Poverty..AN Ordinal Approach to Measurement Conometrica". 44(2) 219-231, 1981.

Tröger, S. (2002), Sustainability in the Context of Natural and Social Environments: An Actor-oriented Interpretation from South-west Tanzania. In: Erdkunde 56(2), 170-183.

UNFPA. (2001), Rapid Population Growth and Poverty in Pakistan. Population Headliner.

Wolfgang Meckelein (1986), Geography –Planning for the Future. In: *Applied Geography and Development*. Vol. 28, pp. 7-18.

Annexure

Table 6

District-Wise Area and Population (1998) of the Province & District Wise some Land Use Indicators in KPK for the Year 2004-05

District	*Area*	*Population*	*Crude density*	*Land use Intensity %*	*Sex Ratio Males / Hundred Females*	*Forest Area as % of Reported Area*	*Cropping Intensity %*
KPK				**61.63**			
Abbottabad	1967	880666	448	72.89	**105**	**22.80**	**99.30**
Batgram	1301	307278	552	82.13	100	46.64	102.06
Bannu	1227	675668	236	47.47	107	0.13	91.93
Buner (Daggar)	1865	506048	271	90.86	107	40.86	121.59
Charsadda	996	1022364	1026	88.98	100	23.78	181.35
Chitral	14850	318689	21	85.94	108	0.06	50.28
D I Khan	7326	852995	116	39.23	103	42.57	99.35
Hangu	1097	314529	287	69.57	111	0.55	47.88
Haripur	1725	692228	401	72.84	96	7.42	170.32
Karak	3372	430796	128	84.33	100	30.84	109.39
Kohat	2545	562644	221	68.33	96	3.15	46.92
Kohistan	7492	472570	63	23.86	101	2.41	82.62
Lakki	3164	490025	155	57.64	124	28.58	82.78
Lower Dir	1582	717649	453	97.71	104	-	52.07
Malakand	952	452291	475	97.81	98	64.36	120.22
Mansehra	4579	1152839	252	82.02	107	8.46	123.31
Mardan	1632	1460100	895	97.24	98	75.61	134.52
Nowshera	1748	874373	500	53.50	107	4.91	112.16
Peshawar	1257	2026851	1606	80.03	109	6.24	107.01
Shangla	1586	434563	274	45.50	111	0.09	118.77
Swabi	1543	1026804	554	76.57	106	28.99	191.61
Swat	5337	157602	125	54.44	101	17.83	107.69
Tank	1679	238216	142	43.93	106	26.74	193.02
Upper Dir	3699	575858	156	98.47	109	27.42	35.49
	27220	3176331			103	62.57	118.37

Source: Go NWFP, 2007. Bureau of statistics, P&D and GoNWFP Agriculture Statistics Peshawar

Table 7
District Wise Population of N. W. F. P., 1972, 1981 & 1998 Censuses

District	*Population*			*%age variation*		
	1972 (Census)	*1981 (Census)*	*1998 (Census)*	*1981 over 1972*	*1998 over 1981*	*1998 over 1972*
NWFP	8388551	11061328	880666	31.86	60.34	111.43
Abbottabad	551144	679328	675667	23.26	29.64	59.79
Bannu	350629	434668	307278	23.97	55.44	92.70
Battagram	174009	339119	506048	94.89	-9.39	76.59
Buner	187984	265517	1022364	41.24	90.59	169.20
Charsadda	513193	630811	318689	22.92	62.07	99.22
Chitral	159000	208560	852995	31.17	52.80	100.43
D.I. Khan	380229	494432	314529	30.04	72.52	124.34
Hangu	125721	182474	692228	45.14	72.37	150.18
Haripur	426068	490104	430796	15.03	41.24	62.47
Karak	191204	249681	562644	30.58	72.54	125.30
Kohat	264465	326617	472570	23.50	72.26	112.75
Kohistan	204521	465237	472570	127.48	1.58	131.06
Lakki	216590	276118	490025	27.48	77.47	126.25
Lower dir	277481	404844	717649	45.90	77.27	158.63
Malakand	185872	257797	452291	38.70	75.44	143.33
Mansehra	473597	727469	1152839	1.94	58.47	61.55
Mardan	696622	881465	1460100	26.53	65.64	109.60
Nowshera	433287	566594	874373	30.77	54.32	101.80
Peshawar	784443	1084347	2019118	38.23	86.21	157.40
Shangla	179813	251546	434563	39.89	72.76	141.67
Swabi	507631	625035	1026804	23.13	64.28	102.27
Swat	520614	715938	1257602	37.52	75.66	141.56
Tank	93389	141062	238216	51.05	68.87	155.08
Upper Dir	251045	362565	575858	44.42	58.83	129.38

GoPakistan, (2008).

Table 8
Province Wise Important Characteristics of Pakistan

Item	*Pakis-tan*	*Islama-bad*	*Punjab*	*Sindh*	*KPK*	*Baluch-istan*	*FATA*
Area (Sq.Kms)	796095	906	205344	140914	74521	347190	27220
Population(000)	132352	805	73621	30440	17736	6566	3176
Density (Persons per sq. kms)	166	889	359	216	238	19	117
Sex ratio (Males/ hundred females)	108	117	107	112	105	115	108
Average HH size (Persons per HH)	6.8	6.2	6.9	6.0	8.0	6.7	9.3
Urban population (%)	32.5	65.7	31.3	48.8	16.9	23.9	2.7
Annual growth rate (%)	2.69	5.19	2.64	2.80	2.82	2.47	2.19

Source: Go Pakistan, (2002) Hand Book of Population and Housing Census Pakistan by Population Census Organization Statistics Division Islamabad.

Development Gap and Spatial Patterns of Environment and Development in Himachal Pradesh

KAUSHAL KUMAR SHARMA AND
S. K. BANDOONI

Introduction

Immediately after independence, India was forced with, among so many problems viz. an acute of mass poverty and economic stagnation due to a century of colonial exploitation, World War II and partition of the country. Therefore, since independence, India has made concerted efforts for her economic development through various five years plans. Moreover, concentrated development efforts were made in those sectors of the economy where returns were likely to be the greatest. Although it did bring some progress in aggregative terms at national level and state level, but failed even initiating a trend towards achieving a balanced regional development. In the second decade of planning, therefore the idea of balanced regional development of less developed areas and weaker section of the society got emphasis. As a result various development programmes, directed towards the development of the backward and the weaker sections of the society were launched. In the following decade of planning, a new dimension of development with social justice was added with heavy reliance on 'Area

Development' and "Target Group' approach for enhancing the benefits of backward areas and exploited section of the society. Many programmes conceived of progressive reduction in the regional inequalities, in the pace of development and in the diffusion of technological benefits, uniformly to the rural masses by introducing the concept of micro-level planning. In spite of all these applicable changes, the benefit of planning has not reached equally or equitably to all the people and in all the regions of the country. There is growing evidence to show that regional disparities have further increased during these periods both in time and space, which opens up a new vista of research and investigation. It is in this spirit that the present study, pertaining to the level of development in Himachal Pradesh has been taken up. It is hoped that the finding of the study would be of direct relevance and use to those entrusted with the responsibility of planning social and economic development.

As a concept it controls overall development with a view to improve the quality of life of people. In this sense it is comprehensive and multidimensional concept and encompasses the development various economic activities. Development must recognize three points viz. rate of transfer of people out of low productivity agriculture and related activities into more rewarding pursuits, lowering the degree of poverty and improving the quality of life.

Problems and Issues

Most of the people are not happy with the current social and economic situation and it firmly believe that poverty can not be removed form India. Therefore, present state of inequality in the living standard will continue to prevail indefinitely. The level of consumption of material goods such as food or steel can be measured numerically with reasonable accuracy. The level of provision of services such as health and education is more difficult to quantify. The development gap can be viewed at the level either of individuals or group of people. All nations work through regional and local agencies to look at the problems in regional perspective. The gap between the economically better off and worse off individuals and regions has tended to widen in the last two hundred years or so. In India in general and Himachal Pradesh in particular it has not been able to understand the root cause to this problem. Therefore, funds allocated for development are under-utilized or sometimes over-utilized and development becomes hazardous and multi-directional. For balanced development, all efforts should be made to reduce the gap for meaningful planning. We want to explore from the above evidence, the relationship between inequality, rural poverty and agricultural output growth in general and food in particular. The presupposition is that by reliance on agricultural growth alone without reducing land concentration, it would require, a long time for most of the present generation of rural poor to rise above the poverty line. Therefore, it follows that output growth is the prime aim of rural development where land distribution is highly unequal. In other words, where economic activities in agriculture and distribution of income are governed by indifferent market behaviour, no one can be blamed for inequalities and the condition of rural poor. In such situation the problem of poor and poverty will become common and a way of life.

Therefore, development should sensitise and allow people to articulate their needs and feelings, make people capable of transforming their environment themselves into more congenial ones of their choice, help people to meet their basic needs, provide reasonable security, and ensure to encourage cooperation, collective problem solving, create self respect and self confidence and finally create, nurture and rehabilitate their internal capacity to interest and innovate. Hence, for development to be successful it should not only be environmentally and economically sound but also socio-culturally acceptable and within the capability of management and administration as well as existing infrastructure. The problem of development is increasingly of interest to regional planners, agricultural scientists, sociologists, geographers and policy planners.

The emphasis of development policies and programmes have been to counter this by increasing the assets of rural people and to overcome traditional values which act as constraints in promoting better health, sanitation, education and family planning. Thus any attempt at integrated rural development must focus on increasing the productive capacity of rural poor and make provision for social services. In order to take the full advantage of the emerging technological environment, the problems and potential of the people of each and every part of the state has to be seen from a micro rather than macro scale. Regionalisation, no matter how it is done, serves only academic purpose if it is not linked to a strategy of development that helps in bridging the development gap between the regions: big and small; rural and urban; industrial and agricultural; mountain and plains etc.

Study Area

The state of Himachal Pradesh is situated in the northern part of India between 75°47‘ 55" East to 79°04‘ 22" East longitude and 30° 22’ 40" North to 33°12’ 40" North latitude and it covers an area of 55,673 square kilometers (Fig.1). It was formed as a separate state of Indian Union in 1971. Since 1st September, 1972, there have been no changes in the administrative structure of Himachal Pradesh except carving out of new sub-divisions, sub-tehsils, raising of sub-tehsils to the level of tehsils within the district boundaries. Presently, there are 12 districts, 52 sub-divisions, 109 tehsils and sub-tehsils in Himachal

Pradesh. From development point of view, the Pradesh is divided into 75 development blocks. The smallest unit for development- cum-administration is Panchayat and their number is 3243. The State has a three tier Panchayati Raj structure comprising of 12 Zila Parishads, 75 Panchayat Samitis and 3243 Gram Panchayats on the rural side; and 1 Municipal Corporation, 20 Municipal Councils and 28 Nagar Panchayats on the Urban side besides 7 Cantonment Boards.

Lying in the Himalayan region, the entire state with few exceptions, is mountainous. The state is mainly drained by five rivers viz. Yamuna, Sutlej, Beas, Ravi and Chandrabhaga or Chenab. They all are snow-fed rivers. During monsoon they become raging torrents, carrying enormous quantity of water while in winter, they shrink in volume. The climatic conditions vary from hot and sub-humid tropical in the southern low tracts to temperate, cold alpine and glacial in the northern and eastern high mountains. Himachal Pradesh is predominantly rural in character as 91.30 per cent of the population lives in rural areas and the remaining 8.70 per cent, is located in 58 towns.

Agriculture being the single largest industry and main occupation of the people of Himachal Pradesh has an important place in the economy of the state. It gives direct employment to 75 per cent of the total working population

Figure 1: Location of Himachal Pradesh in India & Administrative Divisions

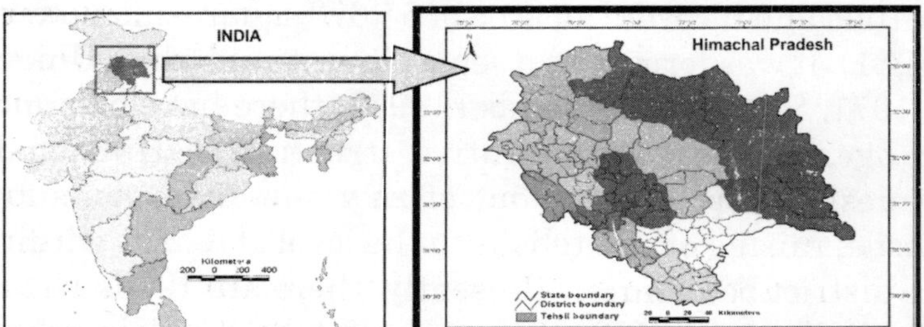

of the state. Income from the agriculture and allied sector accounts for 37.31 per cent of the State Domestic Product. Out of the total 55.67 lakh hectares of area, a total number of 7.53 lakh holding command an area of 9.80 lakh hectares. Net cultivated area forms 17.2 per cent to total area with cropping intensity of 169 per cent. The total food production is 13.97 million metric tonnes (1993). The horticulture promotion in the state lies in the national priority because the state is most suitable for growing variety of fruits and vegetables. Horticulture is not only the supplement of food but it plays vital role, in the environmental conservation in Himachal Pradesh. Nearly 44.75 per cent villages have been connected with motor roads with density of 32.62 kilometers per 100 square kilometers of area. The state does not have favorable minerals and industrial base and it is comparatively poor. The state has vast hydro-electric potential of about 20,000 MW. The state has achieved 100 per cent electrification. Drinking water is available to nearly 98 per cent of villages. There has been large-scale improvement in health, medical services, education etc. Briefly the chief characteristics of under development in Himachal Pradesh can be described as under;

- Unfavorable land–labour nexus is found in the state.
- The continued threat of unemployment looming large on agricultural sector.
- Phenomenally high rents reducing the share of the operators of hired land.
- Continued inflow of labour in the occupation due to lack of alternatives, increases the consumption burden, shrinking the level below poverty line.
- Daily-wage workers are on the increase.
- The combination of land scarcity, economic distress and demographic pressure together enhance problems.

- Investments in IRDP try to divert the investible funds back to the rural areas so as to protect rural proletariat against their looming unemployment.
- Generation of alternative employment avenues by rural industrialsation and rural construction.

The experiences of development in Himachal Pradesh are most interesting. Before its establishment as a separate and full hill state, various development programmes had limited impact. The highest priority was given to road expansion. Development of hydroelectric power and irrigation potential were the other components of infrastructural development. These efforts helped the state to harness its vast potential of mountain resources. Thus infrastructural development contributed immensely to the horticulture development programmes undertaken by the state. However, there are certain limitations in achievement of the targets. Traditional subsistence farming continues as major occupation and does not supplement meager income of the farmers. In Himachal Pradesh with regard to non-viable production processes the situation of poverty is further aggravated due to uneven topography, lack of irrigation facilities, poor quality of soils and fragmented tiny holdings. Over the years there have been qualitative and quantitative changes in the rural areas of Himachal Pradesh. The improvement has been visualised in rural education, health, transport and communication, water supply, sanitation, rural housing, industries, energy, employment, labour and social welfare. The present study is focused on the efforts of the government to improve the living conditions of the people within the existing social structure. Government has so far not been able to integrate the methodologies and approaches of development in Himachal Pradesh. The state has vast diversities in physical, social and economic spheres. Due to this, people's participation in all development activities at all levels becomes necessary. Hence, democratic decentralised planning has assumed great significance these days. The basic idea behind decentralised planning is to minimize the burden of decision making at the apex and also to ensure greater involvement of the people in planning process at the grass root level. In developed plan period the emphasis has been laid on people's participation. The community development programme failed to develop planning capability at local level in Himachal Pradesh. The district level planning is consistent with the specific needs of the people for gainful employment and alleviation of poverty.

Objectives of the Study

- To identify the Land Resource Regions.
- To identify the role of water resource particularly irrigation for development.
- To examine the variation in the human resource in context of density of population, per capita income, educational and medical institutions.
- To adjudge the variations in the transport and power.
- To overview the poverty.
- Finally, to find out the levels of over-all development.

Research Design and Methodology

The concept of development is multi-dimensional one. But in this study a holistic view of development has been taken in a more comprehensive way so that the development in different sectors is measured. For this purpose large numbers of indicators have been selected. Every care has been taken to select the indicators so that they reflect the level of development of the sector to which it belongs. Few sectoral developments have been measured and then the level of development has been calculated. The indicators of study are as follows (Table 1):

Table 1
Order of Development

Order of Development-1	*Order of Development-2*	*Order of Development-3*
Agriculture	Agricultural Condition Agricultural Infrastructure Agricultural Production	Per Cultivator NSAIntensity of Cropping, Irrigation Consumption of Fertilizer, Use of Tractors, Land Holding Production of Food grains Per Person
Socio-Institutional	Educational Institutions Health Centers	Type of Schools Number of Teachers Primary Health Centers Dispensaries, Hospital Beds Doctors
Infrastructure	Transport Telecommunication Electrification	Metalled RoadsDensity of Roads, Post-offices Electricity, Telephone Facilities
Human	Population Characteristics	Density of Population SC and ST Population, Urban Population, Literacy

Source: Compilied by the authors

The present study is an enquiry into the spatial variations in the level of development in Himachal Pradesh. In the conformity with the focus of the study of the economic and social dimensions of development, spatial variations have been examined in terms of:

(i) GIS and its Application: Suitable GIS Modules has been applied to overlay the maps to show the integration of various variables of development. This technique has scope to update the data from time to time for planning. Wherever, need raised cartographic techniques has been used to show the variations.

(ii) Uses of the outcome in overall development strategy: The outcome of the work can be considered as a decision support system for development planning. It can provide an information base for the inequalities, which will help in the identification of problem and provide the base for future policy to be implemented in the development procedure.

Unit of Study

Block as a unit of study has been taken for study because the data is easily available for comparison. In total, 72 blocks are there which lie in the different physical, climatic, social and economic zones of Himachal Pradesh, from different development zones, some village have been identified for cross examination of development levels.

Sources of Data Collection

The present study is based on secondary data, which has been collected from government and quasi government departments. The major sources of data are:

- Statistical Abstract of Himachal Pradesh, Economic and Statistical Organization, Government of Himachal Pradesh.
- District Census Handbook of all the Districts.
- Census of India, 2001.

Geographical Conditions for Agriculture

Relief

Relief of the land exercises a direct influence on land use, particularly through elevation, ruggedness and slope. Relief also affects the ease of cultivation and degree of accessibility. Moreover, prediction of land capability has generally been used upon land types, combing common features of landforms, soil and vegetation. It is essential to know the landform type of agricultural relevance for understanding the suitability of our area for irrigated farming, commercialization and farm mechanization. On the basis of heights and slopes, we can recognize (i) Hills - high and low, (ii) Mountains - simple and complex, (iii) Valleys at different levels in the foothills and outer and inner zone.

Rainfall

Rainfall is the most important climatic factor as it determines the potential of any region in terms of crops to be produced, farming systems to be adopted, the nature and sequence of farming operation to be followed and the targets to be achieved in agricultural productivity. Most of the rainfall received in the state is due to the south-western monsoon which starts from June and lasts up to September. The rainfall in the months of winter season is surely the result of western disturbance, which yields good amount of rainfall and snowfall throughout the state. The highest rainfall is recorded in parts of Kangra, Chamba and Mandi districts, which embrace the Dhauladhar range. Another segment viz. Nahan and Ponta which receives high rainfall is located in Sirmaur district. On the side of lowest recording of average rainfall, Lahaul and Spiti, Kinnaur and Chamba are first, second and third respectively. The value of mean annual rainfall declines towards east and north gradually. In the central part of the state the annual rainfall ranges between 800 to 1200 mm. About 70 per cent of the annual rainfall is received during July and September; about 20 per cent from October to March and 10 per cent from April to June. Taking both temperature and rainfall distribution, the climatic types in Himachal Pradesh range from the sub-tropical typified by Kangra, Mandi, Bilaspur and lower Sirmaur to cool humid temperate in the interior and cool dry at higher elevation of the outer Himalaya. The third type is that of cold moist temperate marked in the inner parts of the lesser Himalaya. To the north of the Pir Panjal and the central axis of the Great Himalaya, a cold dry temperate verging on Tibetan type is found in Lahaul Spiti and Kinnaur.

Winter Season

The season has a spell of three months i.e. December, January and February. The rabi crop, which is usually a post-monsoon crop, start growing after germination during this season. This crop depends to a large extent on local condition and timely rain. During this season the average rainfall is 180 mm. The higher reaches in the state receives moderate to heavy snowfall.

Pre-Monsoon Season

This season lasts from March to May. During the pre-monsoon season timely and proper distribution of rain is essential to the rabi crops grown during the previous season. The stage of fruit bearing to fruit maturity lasts up to the end of this season. The average rainfall during this season is 160 mm. Thus fluctuation from the normal rainfall pattern can ruin the rabi crop. After completing the harvest operation of rabi crops, the farmers start preparing their fields for transplantation of paddy and sowing of kharif crops before the commencement of monsoon.

Monsoon Season

This season has a long span of four months - June to September and the amount of rains recorded during this season is also higher than rainfall recorded in any other season of the year. The crops grown during this season require regular rainfall. In most of the districts, the rainfall is above normal which is 840 mm The districts of Kangra, Sirmaur, Hamirpur, Bilaspur and Shiwalik belt of Chamba receive the highest rainfall as compared to other districts in the state. The lowest rainfall is recorded in Lahaul and Spiti and Kinnaur districts. The highest number of rainy days is also recorded during this season. With the close of the season, the kharif crops in most parts of the Pradesh are ready for harvesting and sowing of minor crops like pulses and oilseeds begins.

Post Monsoon Season

Post-monsoon season is from October to November. After harvesting the kharif crops the sowing of rabi crops is almost complete during this season. The average rainfall during this season is 100 mm. The rainfall and spells of snow is a result of the western disturbances. The high reaches in Chamba, Kangra, Lahaul and Spiti, Kinaur, Shimla receive snowfall and the district of Mandi, Bilaspur, Una, Sirmaur and lower parts of Kangra and Chamba receive rainfall.

Land Resource Regions

Land is the prime resource for man. Since the beginning of human existence man has directed his activities towards this earth resource. Thus land must be carefully utilized so that it may fulfill our multifarious requirements. The best use of each parcel of land requires adaptation of scientific and modern technology. By investigating the land use problems we can work out appropriate techniques for the optimum utilization of land according to various categories of land. Gupta and others have classified the land resources regions into various categories. According to them the land resource region of Himachal Pradesh is divided into three categories viz. snow clad mountains semi-arid region, alpine grass and meadow region and forested region.

Snow clad Mountain and Semi Arid Land Resource Region

Around half of the north-eastern parts of the district of Lahaul and Spiti and 10 per cent of the northern tip of the Kinnaur district fall into this category. The soils found in this area are glacial in origin and limited extent of udolfs is found in different locations. The severe land and climatic limitation do not allow agriculture to be practiced. Most of the land available can be utilized for grazing and wild life. The land supports goat and sheep rearing.

Alpine and Meadow Land Resource Region

This region encompasses the snow clad areas of Himachal Pradesh. It consists of approximately 20 per cent of the total land area of the state. The region covers 2 per cent area of the eastern part of Chamba district, 25 per cent of Lahaul and Spiti district, 30 per cent of Kinnaur, 20 per cent of north-eastern parts of Kullu and 5 per cent of Kangra district. The soils found in this region are mostly udolfs, orthent and orchrepths. The land is mostly suitable for forestry and grazing. The area supports the growing of dry fruits like walnut, almond, grapes and cash crops like zira, saffron, kuth and millets can also be grown. The region does not get much rainfall and is not very productive in terms of agriculture.

Forest Land Resource Region

It comprises of 50 per cent of the total land resources of the state. It covers the whole districts of Sirmaur, Solan, Shimla, Bilaspur and more than 80 per cent of the district of Mandi, Chamba, and Kullu district. This is the most important resource region of the State. The importance can be seen by the fact that more than 50 per cent of the total revenue is received from the forest. The soil of this region is rich in organic matter and nitrogen, but deficient in other plant nutrients such as phosphorus, calcium etc.

This region is the main agriculture zone of the state. According to the altitude various crops are grown. The foot hill zones are utilized for growing of grains like rice, wheat and citrus fruits. In the mid-hill areas, covering districts of Shimla, Kullu and parts of Sirmaur, temperate fruits and cash crops like ginger, potatoes, vegetables are grown. This resource region is, therefore, the major land resource region of the state. The soil and climate is most suitable for agriculture and horticulture. In the absence of a strong industrial base the state's economic development depends to a very large extent upon sound agriculture base. Therefore, careful and scientific evaluation of land utilization is important for agricultural planning.

Water Resources

Water is a priceless natural gift but is not an ultimate constant. Human history has revolved around conflicts over water and many civilizations were ruined because of mismanaging this most prime natural resource. Rivers, lakes, canals, swamps constitute the inland water resources. Rivers have a definite relation to man, though it varies from place to place and from time to time. Among the primary land resources of Himachal Pradesh, its water resources occupy a vital position. The rich perennial water resources in the state is due to heavy snowfall in the higher Himalayas, good amount of rainfall and high amount of underground seepage of water. The state has the head-waters and mountain catchments of four of the five rivers of the Indus System, three of them having their sources in its territory. The catchments area of river Yamuna's major tributaries, the Ton, the Giri and the Pabbar are in the Sirmaur-Shimla region of the state. The Sutlej, which is the largest among the five rivers of Himachal Pradesh drains into the state from Shipki (altitude 6608 Mts.). At the Zanskar and the Great

Himalayan crossing, the Sutlej receives the Spiti from the north-west. In Kinnaur, it is joined by Baspa river and just below Rampur Bashahr, it is joined by Nogli stream. Finally, it enters into Punjab at Bhakra. The Beas, which forms the world famous valley of Kullu and Kangra, originates from 3978 meters high Rohtang Pass. The main course of this river is southward to Larji and then to the west. Its main tributaries in the east are Parbati, Spin and Malava Nala in the west, Solang Manalsu, Sujoin, Phojal and Sarvari streams. Its northern feeders are Uhal, Luni, Ran and Bina and the southern feeders are Hansa, Tirthan Bukhli, Jinni, Suketi, Panddi, Son and Bather. It enters Kangra district at Sandhol. In Kangra district it is joined by Binua, Neogal, Banganga. The rivers like Gaj, Dehar and Chakki join from the north and Kunah and Man from south. The northern and eastern tributaries are snowfed and perennial, while the southern affluents are seasonal. The river Ravi has its source in the Dhauladhar range. It flows westward through a trough separating the Panjal from Dhauladhar range and then turns southward, cutting a deep gorge through the Dhauladhar range. On its way from Bara Bagahal to Punjab, it receives several streams from the adjoining areas. The left bank tributaries are not important except the Chirchind Nala. The right-bank tributaries are of considerable importance viz. Budhil, Tundahan, Baljedi, Saho and Sial. Chenab is another important river, runs parallel to the mid Himalayan range and at Bhujind, it enters Pangi Valley of Chamba district. It enters into Kashmir at Sansari Nala. Despite vast resources of water, the pace of hydrology has been very slow. With the increasing pressure of population, improved agriculture, industrial development, a need is felt for storage reservoirs for multiple purposes. Since the fifties, systematic hydrological data have been gathered by the Central Water and Power Commission for designing dam's structures, diversions, tunnels and water heads on all the important rivers of the Himalayan region. Beside geological structures, the nature of terrain, volume of water and behavior of river flow, the availability of technical equipment and finance are the essential factors for exploiting the water resources.

Irrigation

The total geographical area of Himachal Pradesh is 55.67 lakh hectares. Out of this a high per centage of the area is under perpetual snow or under forests and steep barren slopes. As per the latest available figures only 5.83 lakh hectare is the net area sown in the Pradesh. It is estimated that the ultimate irrigation potential of the state is approximately 3.35 lakh hectares out of which 50,000 hectares can be brought under irrigation through major and medium irrigation projects and remaining 2.85 lakh hectares can be provided irrigation through minor irrigation schemes. Irrigation projects are classified into three categories, namely, major, medium and minor project. Project, which has culturable command area (CCA) of more than 10,000 hect. are classified as Major Irrigation Project. Project, which has a CCA of more than 2,000 hect. but less than 10,000 hect. is classified as Medium Irrigation Project and projects with CCA of 2,000 hect. or less is classified as Minor Irrigation Project. At present, there is only one major and one medium irrigation project in the state viz. Shah Nehar and Giri Irrigation Project respectively. Rest 18 irrigation schemes have a command area of less than 2000 hectares each, viz. Shah Nehar Project, Giri Irrigation Project, Balh Valley Project, Bhabar Sahib Project - Phase I and II, Dhina Singh Project, Churu Project, Beet Illaqua Project, Anandpur Hydel Changer Area Project, Kirpal Chand Project, Sarwari Project, Baldwara Valley Project, Kala Kund Project, Tikkar Dam, Dhaneta Barsar Project, Sakhral

Project, Jougle Bari Barsar Gadiglore Shah Talai Project, Bara Sold Nagrota Suriyan Project. The assessed and created irrigational potential of Himachal Pradesh is shown in Table 2.

Human Resources

The study of population aims at an assessment of the quantity and quality of human resources. This resource base is of great importance for the exploitation of natural resources in the whole of Himachal Pradesh and more particularly in the higher Himalayas. In the absence of many other wherewithals, man and his labour inputs are the key factors in the development of this region. Man plays a unique role in the overall scheme of resource development. He is both the producer as well as the consumer of resources. As a resource producer man contributes his labour, mental and physical. He creates culture to make his labour more productive and to lessen the impact of resistances he discovers new ways of living.

Density of Population

According to Surveyor General of India, the total area of Himachal Pradesh is 55,673 square kilometers. Out of this total area, 45,318 square kilometers is the measured area according to the revenue papers. Area-wise, Hamirpur is the smallest district of the Pradesh which covers an area of 1,118 sq. kilometers (2.01%) and Lahaul & Spiti has the largest area of 13,835 sq. kilometers (24.85%). The total population of Himachal Pradesh, according to 2001 Census was 60, 77,900, which gives a density of population of 109 persons. There are wide variations in area and population of the districts and the district-wise density varies from 2 persons per sq. kilometer in Lahaul and Spiti to 369 persons in Hamirpur district. Out of the total population, the number of males and females is 30, 87,940 and 29, 89,960, respectively, which means that the number

Table 2
Assessed and Created Irrigation Potential

Sr. No.	*Item*	*Unit*	*Area*
1.	Total Geographical Area	Lakh Hect.	55.67
2.	Net Area Sown	Lakh Hect.	5.83
3.	Ultimate Irrigation Potential Available		
(i)	Major & Medium Irrigation	Lakh Hect.	0.50
(ii)	Minor Irrigation	Lakh Hect.	2.85
	Total (3)	Lakh Hect.	3.35
4.	C. C. A. created upto the end of 31.3.2005		
(i)	By Rural Development and Agriculture Department	Lakh Hect.	0.93
(ii)	I & PH Department Schemes	Lakh Hect.	1.11
	Total (4)	Lakh Hect.	2.04

Source: Department of Irrigation and Health, Himachal Pradesh.

In the First Five-Year Plan (1951-56), irrigation was given second priority and investment of Rs. 55.18 lakh was envisaged against which the actual expenditure was of the order of Rs. 33.70 lakh. By the end of the year 1999-2000 which was the third year of 9th plan (1997-02), the State achieved addition to CCA potential of an area of about 1.00 lakh hectares under major/minor irrigation schemes implemented through plans.

of females per 1000 males is 968. The sex ratio of females per thousand males has been rising continuously in Himachal Pradesh since 1951 Census but declined from 976 in 1991 Census to 968 in 2001 Census. The decline is pronounced in the 0-6 year's age group and is a matter of concern for policy planners. The total per centage of rural population is 90.20% of the total population residing in 17,495 inhabited villages. Himachal Pradesh has the highest per centage of rural population among all the States of the country. The scheduled caste population in the State is 15, 02,170 persons (2001 Census) which is 24.72% of the total population as per 2001 Census. The scheduled tribe population of the Pradesh, which has its concentration in districts of Kinnaur and Lahaul Spiti and parts of Chamba, district and scattered in other districts is 2, 44,587, which is 4.02% of the total population. About 60 per cent of the State's tribal population falls under the tribal sub-plan areas. There are 56 Urban Local Bodies viz. one Nagar Nigam, 20 Municipal Councils and 28 Nagar Pachayats and 7 Cantonment Boards in Himachal Pradesh. In addition to it, there are seven cantonment areas and one Census town. The population of these urban settlements is 5,95,581 as per 2001 Census. The largest one is the Shimla Town with a population of 1,42,161 and the smallest one, is Narkanda with a population of 712.There are 19,63,882 main workers and 10, 28,579 marginal workers in the State as per 2001 Census. This means that there is one main worker for every 3.09 persons and one marginal worker for 5.91 persons. Combining the number of main and marginal workers, it works out that there is one worker for every 2.03 persons.

The density of population is high in Shiwalik and mid-hill districts of the state. It is low in the inner parts of the state viz. Chamba, Kinnaur, Kullu, Lahaul Spiti and Shimla. The most populous areas are in the southern half, a large portion of the middle zone makes multi-sized empty quarters set apart from the populated tracts. A thin line of fringe runs through a wider vacant portion forming the northern part of the state. We can identify the following arrangements of concentration and scattering on the states population map;

(i) Southern Himalaya of dense population

(ii) Hillside moderate to thin distribution

(iii) Northern empty quarter of sparse population

(iv) Relatively vacant zones of little or no population

The densely populated are found in the valleys of southern low-lying areas, flat strips of land, riverside terraces and broad inter-fluvial tracts. A large portion of the area is under agriculture and has greater scope for irrigation. With the density of streams becoming less, relief less favorable, and the cultivated area becoming scarer towards valley margins, the distribution of population becomes thinner. Everywhere to the south of the Pir Panjal range, the rest of the population is dispersed along hillsides and mountain slopes marginal to countless stream valleys. The top of spurs and the hillsides are sprinkled have a sprinkling of population. To the north of the Pir Panjal range, a relatively empty quarter and a thin line of riverside distribution mark the outposts of the regions population in Pangi, Lahul Spiti and Kinnaur.

Per Capita Income

The era of planning started in Himachal Pradesh in 1948 along with the rest of India. The first five year plan allocated Rs. 5.27 crore to Himachal. More than 50% of this expenditure was incurred on road construction since it was felt that without proper transport facilities, the process of planning and development could not be carried to the people, who mostly lived an isolated existence in far away areas. Efforts were also made to involve people more in the process of economic and social advancement.

The community development programme launched in 1952, in certain selected areas of Himachal was later extended to the entire rural area. New ideas and forming techniques suitable to different areas and climatic zones were introduced bringing a new awareness amongst the people about the production possibilities of there respective lands which for centuries were using the ancient time worn methods. In Mandi and Kangra districts, package programmes were undertaken in collaboration with the West German government for popularizing modern techniques of cultivation among the farmers. Suitable agricultural machinery and good breeds of cattle, sheep and poultry were introduced in these areas. Well equipped soil testing laboratories, dairy farms and agricultural workshops were established at various centers, besides an Agriculture University at Palampur (Table 3).

Educational Institutions

According to 2001 Census, the overall literacy per centage of Himachal Pradesh was 76.5% (85.30% for males and 67.40% for females). Comparatively, it is much higher than the all-India literacy rate, which is 65.38%. The literacy rate in Himachal Pradesh has been improving faster than the all-India figures. Himachal Pradesh is characterized by a very strong correlation between sex ratio (females per thousand and males) and literacy. Districts with higher density of female population vis-à-vis male population have high literacy rates. As against nearly 200 Educational Institutions, mostly Primary Schools in 1948, the state had 14,531 educational institutions including 41 colleges, 1,645 higher secondary and high schools, 2,194 middle schools and 10,651 primary schools in 2004-05 (Table 4 and 5).

Table 3
Per capita income at 1990-91 Prices (in rupees)

District / State	*1990-91*	*1999-2000*	*Annual Compound Growth Rate (1991-2000)*
Bilaspur	4, 515	7,547	5.27
Chamba	4,822	6,058	2.31
Hamirpur	2,753	4,243	4.42
Kangra	4,128	5,736	3.34
Kinnaur	8,746	7,930	–0.97
Kullu	6,039	6,098	0.10
Lahaul & Spiti	11,417	12,559	0.96
Mandi	3,394	5,313	4.58
Shimla	7,525	8,304	0.99
Sirmaur	3,934	5,650	3.69
Solan	6,052	11,231	6.38
Una	2,394	4,480	6.47
Himachal Pradesh	4,618	6,507	3.49

Source: Computed by the Department of Planning, Himachal Pradesh based on the information supplied by the DESHP (Table 3.3, Pg. 81 in HPHDR)

Table 4
Educational Institutions

Sr. No.	*Institutions*	*As on 31st March* 1970-71	1980-81	1990-91	1991-92	1992-93	1996-97	1997-98	1998-99	1999-2k	2000-01	2001-02	2002-03	2003-04	2004-05
1.	Primary Schools	3768	6093	7548	7548	7611	9138	10484	10553	10633	10633	10633	10634	10651	10651
2.	Middle Schools	742	1032	995	994	994	1030	1214	1474	1672	1674	1674	1814	2194	2199
3.	High Schools	435	582	860	862	882	949	980	980	981	977	978	962	952	949
4.	Senior Secondary Schools	-	3	150	150	178	282	424	464	525	536	536	572	687	696
5.	Colleges	15	25	25	27	27	32	37	37	37	37	37	40	40	41

Source: Compiled by Department of Planning Himachal Pradesh.

The literacy per centage which was only 6.7% in 1951 and 31.32% in 1971 rose to 42.48% in 1989 and in 1991 it was 63.54%. The state got its first university in 1971.Then two more universities, one for Agriculture and the other for Horticulture and Forestry have come up at Palampur and Solan respectively.

Table 5
Movement of Net State Domestic Product and Per Capita Income

Year	*State Income*		*Per Capita Income*		*Per centage Change Over the Previous Years at 1993-94 Prices*	
	At Constant Prices (Rs. in crore)	*At Current Prices (Rs. in crore)*	*At Constant Prices (In Rs.)*	*At Current Prices (In Rs.)*	*Net State Domestic Product*	*Per Capita Income*
1993-94	4250.03	4250.03	7870	7870	-	-
1994-95	4663.85	5192.46	8489	9451	9.7	7.9
1995-96	4920.52	5930.24	8801	10607	5.5	3.7
1996-97	5198.86	6802.87	9140	11960	5.7	3.9
1997-98	5571.01	7806.98	9625	13488	7.2	5.3
1998-99	5966.28	9507.46	10131	16144	7.1	5.3
1999-2000	6621.85	10881.50	11051	18160	11.0	9.1
2000-2001	6736.56	12108.31	11085	19925	1.7	0.3
2001-2002	7002.66	13336.65	11326	21570	4.0	2.2
2002-2003	7332.34	14262.44	11655	22671	4.7	2.9
2003-2004	7870.81	15932.65	12302	24903	7.3	5.6
2004-05 (Tentative)	8460.12	17127.59	13005	27000	7.5	5.7

Health Institutions

During the last 40 years, there has been tremendous increase in the public health facilities, both preventive and curative. In 1989, the public health institutions numbered 899, including a state hospital, 12 district hospitals,189 primary health centers, besides a number of allopathic and Ayurvedic dispensaries and specialized medical institutions. To meet the shortage of doctors, a medical college also started functioning in 1967, which now has post-graduate teaching facilities in some branches. Death rate has come down by 70% due to various public health measures. The incidence of venereal disease, which was about 17% in 1951, came down to 2% in 1989. Malaria and small pox have been eradicated. The T.B. control programme has been a great success. People have taken enthusiastically to the family planning programme and its interesting feature is that women have out-numbered men in its acceptance (Table 6).

Transport and Power

Development of tribal areas in the state is another step taken by the government for its economic development. The two border districts of Kinnaur and Lahaul and Spiti and the Pangi and Brahmour tehsils of chamber districts are the main tribal areas of the state. They are located on the borders of Tibet. Cut off from the country and even from the rest of Himachal Pradesh for lack of motorable roads, life in these areas had grown in isolation. Kinnaur opened up in early nineteen sixties with the completion of the Hindustan-Tibet National highway and the construction of link roads to it. Land tenure system has improved, and tenants have become land owners. They are using improved seeds, fertilizers and pesticides and most of the households have fruit trees, mostly apple, they are engaged in cottage industries, the most important being the spinning and weaving of wool. Lahaul was opened up in the late sixties with the construction of motorable road over Rohtang pass. High passes like Rohtang (130,50 ft), Kunzum (14,913 ft) and Baralacha (14,000 ft) provide the only contact to Lahaul. Lahaul has

Table 6
Health Institutions in H.P. as on 31st March of Each Year

Sr.	*Items*	*1971*	*1980*	*1995*	*1990*	*1996*	*1997*	*2002*	*2003*	*2004*	*2005*
1.	Hospitals*	39	58	73	73	72	74	103	106	106	116
2.	PHC/CHC/RH/SHCs	72	77	145	225 $	275	310	369	507 ∈	504	505
3.	Allopathic Dispensaries**	119	186	214	197	165	167	155	21	22	22
4.	Ayurvedic Dispensaries***	363	404	430	592	674	826	1128	1135	1135	1122
5.	HSCs	256	856	1299	1851	1907	1980	2068	2067	2067	2068
6.	Dental College	0	0	0	0	1	1	1	1	1	1
	Total	849	1581	2161	2938	3091	3358	3824	3837	3835	3793

Source: Directorate of Health and Family Welfare & Directorate of Ayurveda, Himachal Pradesh.

Note: PHC = Primary Health Centre, CHC = Community Health Centre, RH = Rural Hospital, SHC = Subsidiary Health Centres & HSC = Health Sub-Sentre.

* Including Govt., Private, Voluntary Organisations, Cantonment Board and Trust Hospitals + Ayurvedic Hospitals

$ Subsidiary Health Centres were converted into PHCs during the year 1986-87.

^ Rural Hospitals were converted into community Health Centres during the year: 1993-94 to 1997-98

∈ Allopathic dispensaries functioning in Rural Areas were classified as Primary Health Centres during the year 2002-03

** includes Unani and Homeopathic Dispensaries.

fertile valleys. Spiti is almost a hill desert and has seen much less development. A hill desert development plan under which, it is proposed to link irrigation channel to major glaciers and streams which can provide adequate amount of water to large number of fields. Lift irrigation schemes are being undertaken, in the areas of Spiti, Chandra and Bhaga rivers. Building up of an efficient transport system was the top priority in the 1st Five year plan. The state has 2 airports at Bhintar in Kulu district, and at Jubbarhatti near Shimla. Another airport at Gaggal in Kangra district and sites at Mandi, Hamerpur and Chamba have been identified. It was only by Jan 1991 Himachal was linked with the broad gauge system in UNA by extending the Delhi-Nangal rail line. But railways and airways in Himachal can serve very limited transport needs of the people of the state. Road transport constitutes the very plasma of economic and social infra structure in Himachal.

Roads

During the First Five-Year Plan (1951-56), "Roads" were given highest priority. During this period Rs. 225.41 lakh were spent on the development of roads, which amounted to 42.75% of the total plan investment of Rs. 527.25 lakh. With this investment, the state achieved an additional 2413 K.M. length of roads. The progress made in the development of roads by 1971 at the time of formation of a full-fledged state and level reached by the end of Sixth, Seventh, Eighth and Ninth Five Year Plans and achievements made by the end of Annual Plan (2004-05) is given in the following table:

Besides increase in the motorable road density from 13.66 Kms. per 100 sq. kms. of area in 1971 to 40.89 kms. between 1971-2002, the quality of this crucial infrastructure has also been very significantly improved. Whereas only 29.1% of the motorable roads were metalled and tarred in 1971, the per centage at the end of March, 2005 stands at 57.05%. Similarly, the per centage of roads with cross drainage increased from 36.2% in 1971 to 58.03 % in March, 2004.There is vast difference in the road density of non-tribal and tribal areas. The road density per 100 sq. km. of area according to 2004 data for these areas works out to 68.80 Kms and 8.30 Kms respectively.

Power

Himachal is extremely rich in hydel resources. The state has about 25% of the national potential in this respect. It has been estimated that about 20,300MW of hydel power can be generated in the State by constructing various major, medium, small and mini/micro hydel projects on the five river basins. The state government has been giving the highest priority for its development, as hydel generation can not only meet the growing need of power for industry, agriculture and rural electrification, but can also be the biggest source of income to the state by way of sale of electricity to the neighbouring states.

The most prestigious and major project on river Sutlej in the state is the Nathpa Jhakri (1500MW). It is constructed in collaboration with the central government. The project is funded by the World Bank. The foundation stones were laid of 300MW Chamera II Hydel Project in June 1999, of 2051 MW Parbati Hydel Power Project in December 1999, and of 800 MW Kol Dam project in June 2000. Other major ongoing projects are: Bhaba Augmentation Scheme, Ghanvi Hydro Electric Project (22.5 MW), Larji Hydel Project (126MW) and Khauli Hydro Electric Project (12MW). The state government has given eight hydel projects for private sector participation e.g. Baspa Hydro Electric Project (300MW), Holi Hydro Electric Project (231 MW), Dhamwari Sunda Hydro Electric Project (70MW), Allian Duhangan Hydro Electric Project (192 MW), Swara-Kuddu (162MW) and Budhil

(70MW) etc. The state has electrified each one of its 16,807 inhabited villages. It is very impressive, looking at the location of its villages in far off areas and their isolation. The water resources also provide valuable food in the shape of fish. A national fish seed farm and fish farmers development agencies have been set up. A pilot project is being set up with Norwegian assistance for commercial trout farming in Kulu valley.

Rural Electrification in Himachal Pradesh

According to 1991 Census, the number of Census villages was 19,388. Of these, 2391 villages were un- inhabited and the rest 16,997 villages were inhabited. The state achieved 100 per cent rural electrification target during 1988-89 in the then 16,807 inhabited Census villages. After 1991 Census, 255 more villages have been declared as un-electrified. However, 157 of these were electrified by the end of November 2005, 20 still remain to be electrified whereas 78 are de-populated/ nonexistent with migrated population. In addition to above, 4025 hamlets (out of total of 4182 as per 1988 survey) also stand electrified up to November 2005. Besides above 438 un-identified hamlets have also been electrified in the Pradesh.

Poverty

Poverty in its most general sense is the lack of necessities. Basic food, shelter, medical care, and safety are generally thought necessary based on shared values of human dignity. However, what is a necessity to one person is not uniformly a necessity to others. Needs may be relative to what is possible and are based on social definition and past experience (Sen, 1999). Valentine (1968) says that "the essence of poverty is inequality. In slightly different words, the basic meaning of poverty is relative deprivation." A social (relative) definition of poverty allows community flexibility in addressing pressing local concerns, while objective definitions allow tracking progress and comparing one area to another.

The most common "objective" definition of poverty is the statistical measure established by the federal government as the annual income needed for a family to survive. The "poverty line" was initially created in 1963 by Mollie Orshansky at the U.S. Department of Agriculture based on three times her estimate of what a family would have to spend for an adequate but far from lavish diet.

Poverty Caused by Geographical Disparities

Rural poverty, ghetto poverty, urban disinvestment, Southern poverty, third-world poverty, and other framings of the problem represent a spatial characterization of poverty that exists separate from other theories. While these geographically based theories of poverty build on the other theories, this theory calls attention to the fact that people, institutions, and cultures in certain areas lack the objective resources needed to generate well being and income, and that they lack the power to claim redistribution. That poverty is most intense in certain areas is an old observation, and explanations abound in the development literature about why regions lack the economic base to compete. Recent explanations include disinvestment, proximity to natural resources, density, diffusion of innovation, and other factors (see Morrill and Wohlenberg, 1971:57-64). In a thorough review of the literature on rural poverty, Weber and Jensen (2004) note that most literature finds a "rural differential" in poverty, but that the spatial effect is not as clearly isolated from individual effects as needed for confidence. Goldsmith and Blakely offer a comprehensive perspective on the link between development and poverty in urban contexts. In their book, Separate Societies they argue that the joint processes of movement of households and jobs away from poor areas in central cities and rural regions creates a "separation of work, residence, and economic, social and political life" (1992: 125). These processes which we already discussed are multiplied by racism and political indifference of the localities in which they flourish.

A geographical theory of poverty implies that responses need to be directed to solving the key dynamics that lead to decline in depressed areas while other areas are growing. Instead of focusing on individuals, businesses, governments, welfare systems, or cultural processes, the geographical theory directs community developers to look at places and the processes by which they can become self-sustaining. Interestingly, a few disadvantaged communities around the world are finding their way out of poverty and as such show that it can be done. Some who view regional poverty analyses made proposals in the 1970s to encourage out migration under the premises that it would reduce poverty to have people in a place where there was a growing economy. Instead, the rural poor people moving to the city became urban poor, with much the same hopeless situation. It has been said that much of urban poverty is actually displaced rural poverty.

Several approaches have been taken to build stronger geographical areas like watershed management and community development approach because community developers understand community; their efforts often leverage community assets, integrate economic development in an area with housing and other spatially allocated factors, and hope that the changes will increase opportunities for residents.

Physical geography of a region can influence its economic performance. The economies valley regions, with their easy access to transport, usually outperform the economies of highland areas. Although most economists today follow Sometime linking prosperity with free markets, tend to neglect the role of geography. They implicitly assume that all parts of the world have the same prospects for economic growth and long-term development and that difference in performance are the result of differences in institutions. Our findings, based on newly available GIS data and research methods, suggest otherwise. We have found strong evidence that geography plays an important role in shaping the distribution of world income and economic growth. Moreover, climate can also influence economic development. Districts in tropical climate zones generally face higher rates of infectious disease and lower agricultural productivity (especially for staple foods) than do districts in temperate zones. The very poorest regions in the Himachal Pradesh are those saddled with both handicaps: distance from plains and harsh topography.

Geography affects agricultural productivity. Of the major food grains – wheat, maize and rice – wheat grows only in temperate climates and maize and rice crops are generally more productive in temperate and subtropical climates than in tropical zones. Farming in tropical rain-forest environments is hampered by the fragility of the soil: high temperatures mineralize the organic materials, and the intense rainfall leaches them out of the soil. In tropical environments that have wet and dry seasons farmers must contend with the rapid loss of soil moisture resulting from high temperatures, the great variability of precipitation, and the ever present risk of drought. Moreover, tropical environments are plagued with diverse infestations of pests and parasites that can devastate both crops and livestock.

People's Perceptions of Poverty in Himachal Pradesh

Perceptions of poverty and the factors that contribute to upward or downward mobility are; landlessness or possession of only poor quality, non-irrigated lands, dependency on intermittent casual wage labor, especially in the agriculture sector, dependency on foraging or begging, lack of basic literacy, job skills, united access to social networks, particularly networks that extend outside the village and help residents find better jobs, particularly jobs in the non-farm sector, chronic indebtedness, desertion, widows, or women living alone without an adult male, a high dependency ratio, many daughters, also lack of sons, health or disability, particularly of the primary breadwinner, poor quality mud and thatch housing, insecurity of tenure, social or caste identity, in several villages, low-caste informants assumed all high-caste households to be rich, while they carefully ranked lower caste households by various economic criteria. Among the higher castes, low caste status was assumed to be a strong indicator of poverty. Less tangible factors such as dignity and autonomy were often mentioned. Urban poverty has a distinctly different face than rural poverty. Many of the urban poor included in the consultations were first generation migrants, and the majority felt that migration had improved their economic position. In urban areas, unlike rural villages, employment opportunities - if only scavenging, petty hawking, and begging - are available year round, typically at higher wages than paid in rural areas. Caste was far less frequently mentioned as a barrier to economic mobility. Many also complained about lack of job security and most aspired to more permanent salaried jobs.

Additional factors cited include, very poor housing, lack of secure tenure (extremely important in the eyes of the urban poor), polluted and inadequate water and sanitation services, lack of appropriate job skills and connections-surprisingly, few of the poor in illegal settlements cited formal education as an important factor in determining access to employment, although virtually all respondents felt that basic literacy, including for girls, was important.

The poorest in Himachal Pradesh may be described as follows:

- *Resource poor:* those who do not have access to adequate resources to sustain themselves.
- *Access poor:* Himachal Pradesh's predominantly rural population includes some communities which are as much as 30 km away from the nearest road and in high altitude areas, may be cut off for several months of the year; government services are unlikely to reach them at any time of year. These are also the people who are most dependent on forests for everyday needs like fuel wood, fodder and water.
- *Opportunity poor (primarily herb collectors):* those who do not have opportunities for livelihoods other than their traditional ones, which are primarily agriculture based and supplemented by labour employment. Educational facilities are basic and vocational education is unheard of, they rely on whatever government schemes have to offer and are absolute passive beneficiaries. Girls are encouraged to marry very young and boys to take to traditional occupations. Their exposure to, and interaction with the outside world is very limited.
- *Poor by virtue of gender:* women in Himachal Pradesh are particularly disadvantaged, and in some part of the state traditions dictate that the husband is literally the lord. There is a need to target livelihood strategies at women, especially with SC and the poor, because they are available at home, are more dependent on natural resources and quality of life indices show improvement with women due to their better spending priorities.

- *Poor based on equity:* there is much caste-based discrimination in Himachal Pradesh, as well as social discrimination based on assets and income levels. The equity parameter affects access to basic sustenance needs, social discrimination and sometimes ostracism.

All development strategies are designed to address the problem of poverty. On the basis of available data, the State believes that the incidence of income poverty is much lower than the national average; in fact, for the year 1993-94 the poverty ratio of the State was the seventh lowest amongst 16 states analysed. Poverty levels are more pronounced in rural areas as also in some districts compared to others. One of the possible reasons for an increase in the poverty level could be the failure of development programmes to generate employment corresponding to the rate of growth of population. A sustained effort to address this failure and to more energetically pursue poverty alleviation programmes needs to be accorded high priority. However, the shortage of relevant data as well as the lack of sufficient data-gathering measures points to the need for further examination to make a clearer assessment of the extent of poverty in the State. The main Rural Development Programmes / Schemes for Poverty Eradication and Income equality in Himachal Pradesh are Community Development, Swarnjayanti Gram Swarozgar Yojana, SGSY Special Projects, Installation of Hydrams under Development of Wastelands and Marginal Lands, Sampoorna Grameen Rozgar Yojana, Indira Awas Yojana, National Family Benefit Scheme, Watershed Development Projects, Innovative Rural Housing Projects, Total Sanitation Campaign Project, Mahila Mandal Protsahan Yojana, Environmental Conservation through Mahila Mandals, Rajiv Gandhi Awas Yojana and Mahatma

Gandhi National Rural Employment Guarantee Programme (MNREGA).

Levels of Development in Himachal Pradesh Conceptual Framework

The word 'development' usually implies of 'growth' and 'change', especially for the betterment of a place, a district, region, or a country. Usually when development of a state or country is discussed improvement in its economic states is understood which implies that more and better-paid jobs will be available and that other benefits, such as better hospitals facilities, sport facilities and school, will then follow. Very often, 'development is considered to be the equivalent of economic growth (Courtenay, 1985). According to distinguished economist Seers 'development may be seen as meaning impfovement in the standard of living includes social, cultural and political welfare as economic opportunities'. Development, therefore implies improvement in opportunities to obtain education, participation in cultural and sporting activities receive, necessary health care, line in pleasant surrounding, and to be free from servitude to other people or institution, as well as the chance to be properly fed, clothed and sheltered.

All development is relative. Most of the element of development is interrelated and their analysis become complex. To use such element or indicators to measure such a concept as development is called surrogate measurement. Development is a significant process that has potential for improving the poor and untroden people in the world. However, with time it has been recognized more widely that it consist not just of economic or even of social changes alone but a basic shift in the whole economy and society qualitative as well as quantitative. Development is thus not merely economic growth or more of the same, but rather it is multi-dimensional change.

The concept of development represents the application of the general idea of progress in socio-political sphere. Progress and development are often described as synonyms. Both designate change in a forward direction and hence a change that is not the expression of a simple modification in time, but rather of a modification in the positive sense, of an improvement (UNESCO-1988). In the recent years, and especially since the end of second world war in 1945, the development of the poorer countries has been an important international issue. The study of development has involved economists in particular, and in political area the specialised agencies of the United Nation Organisation - FAO, WHO and IMF in general have been active in assisting developing countries with their main problem.

During the last 50 years some success have been achieved e.g. increase in life expectancy, sharp decline in infant mortality rates, control on communicable disease, eradication of small-pox and plague, improvement in education (all level - both technical and non-technical) and more availability of food supplies.

The interest of geographers in the subject of development is especially in seeking to understand the reasons for difference in the level of development between places and in the changes that occur on the earth surface as development proceed. The geographical study of development is concerned with more than discovering patterns formed by levels of development. It seeks to understand the reasons for creation of such pattern. If the factors that have contributed to the particular pattern can be recognized, it may then be possible to understand more about the processes involved in making some parts more developed than others. If this can be done, then a useful step can be taken towards the possibility of raising the living standards of people in those less developed regions.

The Centre for Monitoring Indian Economy in 1987 had developed an aggregate index to show economic development using district data on a variety of aspects of development. The indices, a very high proxy of indicators for the Gross District Product of district have been computed from a selection of nine base-component indicators: two from agriculture, three from mining and manufacturing and four from service sectors. The conclusion was that all urban district demonstrate very high level of development. Districts which are predominately agricultural (deltaic as well as interior districts) also show greater levels of development from the index value, eight states are above all-India average while thirteen states and five union territories are lower. Level of development reflects a very high correlation with physio-climatic condition in India.

It is evident that no attempt has perhaps, been made to assess the level of development in Himachal Pradesh. Therefore, the levels of development of the state has been analysed with the help of few indicators and the results are as follows-

Economic Infrastructure Development

The economic development of an area is best reflected in infrastructure facilities. The full development of an area not only achieved through productive investment but by investment in basic economic infrastructure - roads, electricity, water communication etc. which facilitate and integrate economic activities. Hence economic infrastructure development has been one of the important components of development planning. Moreover regional evaluation will set trend and guideline for future progress. The present study examines the level of economic infrastructure in terms of villages served by post office and telephone only.

Telecommunication

Telecommunication infrastructure by means of post office, telegraphs offices, telephone etc., have came to play an important role in the lift of the rural people. In the present context of fast expanding world, the absence to telecommunication facility is an indicator of impediment to economic growth of an area. Telecommunication facilities are a pre requisite for effective interaction between rural and urban settlement for domestic, economic, commercial and administrative purpose.

Post office and Telegraph Services

The post and telegraph service in Himachal Pradesh is very well developed. All those areas which had low level of institutional facilities, agriculture and horticulture etc., they show high to very high index value for the people who live in the interior location with post and telegraph service. For them this is only way form link with outside areas. The highly developed conditions of this service lie in the temperate wet and dry region. In the sub-tropical and sub-temperate region the numbers of villages are more as compared to other region. Due to this per centage of village served by post and telegraph service range from medium to high. The least service exist in Chamba and Kinnaur districts.

Socio-institutional Development

Social development is concerned with people and how best to provide them with the material means to help them to realize their full human potential. Social development of an area is best reflected in quality of life of its people. The level of socio-institutional analysis focus on the sequential process through which the economic, industrial and institutional structure of an economy is transformed over time to permit new changes as an engine of economic growth. Hence, socio-institutional development has been one of the important components of development planning. The present study is concerned with the selection of most appropriate indicators of socio-institutional development and an analysis of the relationship between these indicators of different levels of development. The result is contributed on the composite development index. The educational institutions and services and medical/health institutions and services were selected on the basis of high inter correlation to from a development index.

Educational Institutions

The principal institutional mechanism for developing human skill and knowledge is the formal education system. The rapid quantitative expansion of educational opportunities is the basic key to material development. For each variable of educational institution the index value were calculated. Finally the socio-institutional development was classified in the categories viz. least, low, medium, high and very high and the blocks were put under these categories for analysis. Education has been long recognized as vital to development not only fostering attitudes and aptitude conductive to economic and social change but also in meeting a basic need for all individual. The educational institutions deserve first priority as education is one of those instruments that illuminate the social fabric, speed up modernization process and development. Educational institutions are broadly classified as primary, middle, high school, higher secondary collage and university. All of them vary in time and space. Himachal Pradesh is mostly rural in nature and has great physical diversity. The dual nature of the state has extensive variation in educational service from one region to another region.

Primary School

The analysis has been done on the basis of per centage of villages having primary school. The primary schools lie at the base of whole educational infrastructure. Their basic aim is to provide education to all eligible children. The situation of primary school education in villages is very high in the state. The blocks of Chamba district, Spiti, Jogindernagar , Kasuli, Arki, Nalagrah, Kandaghat constitute low level of primary education. Rest of state has medium to very high facilities in primary education.

Middle and High Schools

The middle schools act as a bridge between primary education and secondary education. Due to sparsely distributed villages in the state the contribution of middle school is limited to few central locations. They are nearly one third as compared to primary school. The spatial variation of middle school is more or less similar to primary schools. The importance of high school education is very high because it is almost a prerequisite to have a high school certificate to get a government job. Where as the presence of a high school in a village enhances its importance and makes it a central place for its surrounding area.

Senior Secondary Schools

The grant importance is attached to senior secondary school as it qualifies students to enter into higher education or institutions. Senior secondary schools are very less in the state. They are mostly located in the urban centers and the villages are deprived of their benefits. For greater detail, three more variables viz. per centage of villages with educational facilities, per centage of population served by educational contributions and per centage of villages with the educational facilities have been taken into consideration.

The per centage of population served by academic institution is an important indicator to know the literacy level of an area. In Himachal Pradesh at present a serious mismatch between the supply of educational resources and institution and the demand particularly in rural areas. It is glaring to note that blocks with high level of educational institutions at the same time have highest number of villages with no educational facilities. It is mainly because the numbers of villages are many and it is difficult to provide educational facilities to all. On the other hand, blocks with less educational facilities have high per centage of villages served with such facilities. These blocks have fewer villages and scattered all over the state. In such condition, due to the directive of the government to provide at least primary school at a radius of one to three kilometers, primary schools are found more. To take the facility of education nearer to the masses the, state government is steadily expanding the educational infrastructure and the number of education institutions.

Medical Health Institution and Services

Health services are only one among many forum of provisions needed for improved health as is well known, so that the variety of other indicators are atleast equally important for necessary development in relation to health. Health care of people is a pre-requisite to become employable productively. Most of the health services are available in the urban centers. Health is also a very important component for the success of family planning programmers and therefore, deserves special attention. Medical facilities are commercial to check on mortality rates of population. The human misery related to diseases is reduced and longevity of life enhanced in consonance with availability of medical facilities. Hence, a higher frequency of occurrence of medical institutions is indicative of a higher level of development.

Dispensary

It is well known fact that the nature of spatial distribution of health care facilities has a far reaching impact on the utilization of health care by the cross section of the society having an uneven distribution on space. In such cases, dispensary is the first unit of planned medical facilities. Looking at the cross section of physical diversity in Himachal Pradesh these facilities show variation form one tehsil to another.

Primary Health Centers

Primary health centers are of great importance, it provides most of the basic infrastructure for the welfare of women, child, family planning programmes. These centers form a good link with the villages and educate them about the problems and consequences of population growth. These centers are located by keeping in mind the concept of time and space. Even then, the distribution of Primary Health Centers is uneven in the whole rate. The primary health centers are fairly well distributed in sub- tropical and temperate zone. In the high hill and dry zone health centers facilities have low index value.

Hospitals

This is the highest order of medical institutions which provide treatment to all diseases a mankind suffer. In Himachal Pradesh, there are sixty eight hospitals (Allopathic, Ayurvedic, Government Board, Missionaries, Voluntary Organizations, Projects and Police etc.). Most of these hospitals are located at the district headquarter. The accessibility and population served by there is very low. People from the rural area have to come to these centers for treatment faces many problems.

Other Medical Health Institutions

They include maternity centers, family planning centers, primary health sub-centers, and community health centers. All these facilities correspond to the region of high population infrastructure facilities. The sub-tropical and temperate regions have more medical facilities for the village then that of mid hill and dry region of the state. The mid hill and dry region constitute sparse population and poor infrastructure facilities.

The nature of distribution of health care, the locational dynamics of the facility points in relation to the users, the nature of physiography and climate and their effect on transport network and spatial accessibility, economic status and affordability of users, their socio-cultural level and ability to perceive the need of health care and use it all go to form the basis of comprehension (Dev, etc; 1994.). The present analysis of spatial variation in medical/health services is based on, per centage of villages having medical facilities per centage of population served by the medical facilities and per centage of villages without medical facilities. Index has been calculated and five categories were carved out to show the difference. The medical facilities for villages are high to very high in sub-tropical zone. It constitutes mostly the plain area and low hills of Shiwalik. In sub-temperate region mostly covered by mid-hills the medical facilities are low to medium, very low facilities exists in temperate and dry region of the state. The pattern is almost identical when the analysis is based as population served by medical facilities. The condition is little better in the temperate region where these facilities range from high to medium. Inspite of the radical changes which have taken place in the state in term of spatio-temporal extension of the medical/health institution there are many areas in the state which are poorly served by these facilities. Although sub-tropical and sub-humid region medical institution are more but per centage of population served by them is less. Where as, in temperate wet and dry region facilities are less but per centage of population served by there is more. The main reason of this trend is that is former two region the density of population is very high and widely scattered beet the later two regions the population density is low and concentrated in few place.

Conclusion

Before the formation of Himachal Pradesh, no attention was paid to the economic development and social welfare of the people of this region. None of the erstwhile hill states was an economically viable unit and utilization of vast natural resources of the area was not even thought of. Small princely states had fewer resources and they were not interested to develop their areas. The princes were indifferent, and the people were helpless. They were extremely poor and conservative in outlook. Steeped in superstition and looking mobility due to the absence of proper mean of transport, they stock to their ancestral occupation. The British cared merely for the cantonment towns and for those places which they chose as their summer resorts. For the rest people continued to depend on land. Land was scare due to hilly terrain and then the best of it was with the rulers or their relatives (Singh, 1992). It was only after the formation of Himachal Pradesh that the people and the government of this hilly region started making concerted efforts to improve their own economic condition and thus that of the Pradesh. Today the economy of Himachal Pradesh is based in number of factors like agriculture, horticulture, forest, hydel power and road transport. It is one of the most prosperous and progressive states of India. In short span of time Himachal Pradesh has come to occupy a place of pride in the field of horticulture. Even in the field of food crops it has made fast progress and some region have become food surplus. In spite of multifarious development in the state spatial variation do occur in the level of development mainly due to geographical and other factors. At last it can be concluded that the condition of Himachal Pradesh is much better as compare to many states of the India.

Acknowledgement

We are deeply indebted to University Grants Commission, New Delhi, for financially supporting the work. We also thank Kirori Mal College and Shaheed Bhagat Singh (Eve.) College for providing infrastructure facilities for the smooth functioning of the work. Survey of India, Geological Survey of India, NRSA Hyderabad, ONGC, Dehradun, Wadia Institute of Himalayan Geology, National Bureau of Soil Survey and Land Use Planning Nagpur need special mention for data inputs for the work. Dr. Surinder Singh, Deputy Director, UGC, New Delhi, Dr.(Mrs.) Renu Batra, Joint Secretary, UGC, Dr Bhim Sen Singh, Principal, K M College, Dr P K Khurana Principal, SBS(E) College, Dr. Bhoop Singh, Director (NRDMS), Dr. G. S. Chauhan, Education officer(UGC), have played a special role in identification of problems and outcome. The work would not have been complete without the support from the Government of Himachal Pradesh. Valuable contribution for the project have come from, Nitin Punit, Ms Ranjeet Kaur, Vijender Kumar Pandey, Ritesh Raj, Ms Manorma, Dr. N.P.Singh, Dr. Ravi Shekhar, Dr V S Negi, Ms Poonam Kumria, Surinder Kanojia and Dr. Dilip Sharma. They require special thanks for their sincere efforts. Our families had to bear the stress of our working on odd days when they needed us at home. Their sacrifice is duly acknowledged and due to that the project is a reality.

References

Alier J. M. (2004), The Environmentalism of the Poor: A Study of Ecology Conflicts and Valuation, Oxford, Delhi.

Bandooni, S. K. (2004), Land Resource Management and Development in Hill Areas, Research India press, New Delhi.

Census of India (1991), District Census Handbook(all Districts), Series – 9, Himachal Pradesh, Village and Town Directory, Part – xii – A & B, New Delhi.

Goldsmith, W. W. and Blakely, E. J. (1992), Seperate Societies: Poverty & Inequality in American Cities: Philadelphia, Temple University Press.

Kapadia, H. (1999), Across Peaks and Passes in Himachal Pradesh, Indus Publishing Company, New Delhi.

Kim S. U. *et al.* (Eds.) (2002), Population and Development, Sunrise Publication, Delhi.

Lister, R. (2004), Poverty, Atlantic, New Delhi.

Mishra S. N. (Ed.) (2005), Poverty and its Alleviation, Deep and Deep, New Delhi.

Morril R. L. & Wowenberg E. H. (1971), The Geography of Poverty, New York: McGraw Hill.

Rahman K. and Rahman M. M. (1998), Participation for Development: A Resource Book on Strategy, Programmes and Procedures, Manak Publication, Delhi.

Rahman, S. A. (2006), The Beautiful Himalaya: Himachal Pradesh, Reference Press, New Delhi.

Sen, A. (1999), Development as freedom, Oxford: Oxford University Press.

Sharma K. K., *et al.* (Eds.) (2006), Environmental Geo-Hazards: Science & Society, Research India Press, New Delhi.

Sharma K. K. *et al.* (Eds.) (2009), Himalaya Parvat Mei Sansadhan Prabhandan, Research India Press, New Delhi.

Valentine, C. (1968), Culture and Poverty, Chicago University of Chicago Press.

Weber B. Jensen (2004), Poverty and Place: A Critical Review of Rural Poverty Literature. Oregan State University. Rural Poverty Research Centre, Working Paper.

Promising Approaches for Enrolment and Retention in Girls' Education at School Level

SAVITA KAUSHAL

Introduction

Education of girls has been a high priority with the Government of India. The National commitment to provide free and compulsory education to all children in the 6-14 years age group is now a Fundamental Right of every child in India after the passing of the Constitution (86th Amendment) Act in December, 2002. Educational attainment has found direct linkage with growth of the individual and nation. However keeping in view the constitutional provisions there is huge task to be accomplished as there is gender gap in education. Women inequality holds back growth of the society, as they form significant proportion of the population. There are 927 women per 1000 male as per 2001 Census in India. It is more than 49 per cent of the total population. For the growth of the country it is essential that women are educated and participate in the economic activities in an equal proportion.

Though the female literacy rate increased from 8.86 per cent in 1951 to 53.67 per cent in 2001, it is still considerably lower than male literacy. As per the 1991 Census, the female literacy rate was 39.42% compared to 63.86% for male. In 2001, the scenario changed. As per the 2001 Census, female literacy rate rose up to 54.16% and male literacy rate to 75.85%. This shows that gap in male-female literacy rate narrowed from 24.4% in 1991 to 21.7% in 2001 (Figure 1).

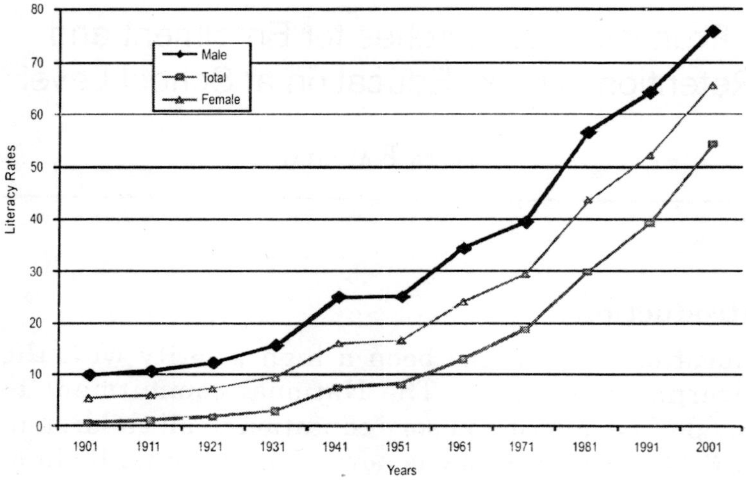

Figure 1: Growth of Literacy : 1901-2001

While India's economy booms, educational opportunities remain out of grasp for large number of rural poor, especially girls. In nine of India's 35 states and territories, illiteracy rates among women are 50 per cent or higher, according to figures from the 2001 Census of India. By contrast, no state or territory has an illiteracy rate of 50 per cent or higher among males. The situation in rural India is even worse (46.1%) against (72.9%) urban India (Census, 2001). Moreover literacy rate for females in rural area is much lower than males (Table I).

This dismal situation has been stated in the various policy documents published by the Government of India from time to time. The Programme of Action (POA), 1992 in the context of National Policy on Education mentioned that "there are significant rural urban disparities among

Table I
Literacy Rates by Sex and Region (1951-2001)

Year	*Literacy Rate* (%)* *Persons*	*Male*	*Female*
1951 (5+ years)			
Rural	12.10	19.02	4.87
Urban	34.59	45.60	22.33
Total	18.33	27.16	8.86
1961 (5+ years)			
Rural	22.50	34.30	10.10
Urban	54.40	66.00	40.50
Total	28.30	40.40	15.35
1971 (5+ years)			
Rural	27.90	48.60	15.50
Urban	60.20	69.80	48.80
Total	34.45	45.96	21.97
1981 (7+ years)			
Rural	36.00	49.60	21.70
Urban	67.20	76.70	56.30
Total	43.57	56.38	29.76
1991 (7+ years)			
Rural	44.70	57.90	30.60
Urban	73.10	81.10	64.00
Total	52.21	64.13	39.29
2001(7+ years)			
Rural	59.40	71.40	46.70
Urban	80.30	86.70	73.20
Total	65.38	75.85	54.16

* (1) Literacy rates for 1951 refer to effective literacy rates and the break up of rural, urban and male-female components are crude literacy rates. (2) Literacy rates for 1981 exclude Assam whereas the 1991 literacy rates exclude Jammu and Kashmir where Census could not be conducted due to disturbed conditions. (3) Literacy rates for Census of 2001 exclude the entire Kuchchh district and three taluka of Rajkot district, one taluka of Jamnagar district in Gujarat and the entire Kinnaur district of Himachal Pradesh where population enumeration of Census of 2001 could not be conducted due to natural calamities.

Source: Government of India (2001a) & Government of India. (2001b) women" .The National Policy on Education (1986) admitted that the school would not reach all the children particularly millions of girls and working children whose participation in the school system was thwarted by socio-economic conditions of life. It is very difficult to achieve Universal Elementary Education (UEE) unless concrete efforts are made to reach out to the girl child. As quoted in POA: "Girls who cannot attend formal schools or have had to dropout will be provided educational opportunities through alternative schooling. The open school, distance education system and other innovative programmes will reach out to girls in rural remote areas and urban slums" (1.7.1.P.O.A.).

Availability of basic facilities in schools not only attracts more girl children to schools but also helps in improving retention rate. An assessment of the scenario in terms of basic facilities that facilitate enrollment and retention of girls also indicates that much more needs to be done. Provision of separate girls' toilet, female teachers facilitates enrollment and retention of the girls. As per the DISE (District Information System for School Education) data about 50 per cent elementary schools in India during the year 2007-08 had separate toilets for girls and nearly about 42 per cent primary schools in India had girls toilet. The figures pertaining to enrollment of girls at primary stage (46.99%) and upper primary stage (47.88%) also indicate that the girls are lagging behind in enrollment. The per centage of female teachers (42.3% in primary schools and 42.72% in all elementary schools) also needs to be improved. The average of 35 states and union territories (UTs) in 2007-08 indicates a gender parity index (GPI) of 0.93 in primary classes and 0.92 in case of all elementary schools taken together. The analysis clearly indicates that boys outnumber girls both at the primary and upper primary level (Table 2).

Table 2
Selected Indicators Pertaining Girls Enrollment as per DISE Data (2007-08)

S. No.	*Indicators*	*Primary Schools*	*All Schools*
1.	% Girls Enrolment (Primary)	46.22	47.88
2.	% Girls Enrolment (Upper Primary)	46.99	47.88
3.	% ST girls to ST Enrolment	48.30	47.64
4.	% SC girls to SC Enrolment	48.10	47.76
5.	% Female Teachers	42.30	42.72
6.	% Schools with Girl's Toilet	41.95	50.55
7.	Gender Parity Index (GPI)	0.93	0.92

**Note:* The term All Schools includes Schools having classes from Class I to VIII

Source: (Mehta, 2008)

Barriers to Girls' Education

Education provides people with knowledge and skill especially in the areas of health, nutrition, water and sanitation, and the environment. The enrolment rates of girls have a positive impact on economic well-being of women, their families and society in the long run. Education empowers women; it increases women's status in the community and leads to greater input into family and community decision-making. Perhaps more importantly, education provides girls with a basic knowledge of their rights as individuals and citizens. Having knowledge and decision-making power can place women on a more equal footing with their male counterparts. Education is a necessary condition to ensure that development efforts will be sustained. Efforts have been made in the last fifty years to identify barriers which restrict full participation of girl child.

There are a number of barriers to education of girl child. Without overcoming these barriers, it is not possible to promote education of girl children. Some of the barriers are as follows:

- Parents are less aware of girl's education, especially in rural areas.
- Parents are unable to understand the utility of girls' education.
- Girl child belonging to labour classes are busy in earning money and sibling care.
- Early childhood marriage trend.
- Long distance between schools and home.
- Lack of separate girls' middle school.
- Lack of separate toilet for girls in schools.
- Negative attitude of community.
- In-sufficient number of female teachers.

The girl child is the worst sufferer. The denial and deprivation affect her self-concept negatively. These girls are not only of rural areas of urban slums and of minority communities which are involved in domestic chores or sibling care, but girls of labourers, homeless street girls etc. They also need appropriate education facilities. These girls are of different age groups. It is a great challenge for society. To meet this demand diversified approaches and strategies were adopted for providing them education under District Primary Education Programme (DPEP) followed by Sarva Shiksha Abhiyan (SSA).

All over the world there is a growing realization that only alternative strategies can meet this challenge. Alternative schooling provides various educational strategies for different groups. Educational programmes of the Government and Non-Governmental Organizations (NGOs) are doing a lot of work for expanding access to schooling. A number of initiatives in this direction were undertaken/promoted under DPEP followed by Sarva Shiksha Abhiyan.

Initiatives for Promoting Girls Education

Some of the initiatives that were taken up by the different states to promote girls education are:

- **Haryana:** Bicycles are given to girls on joining class VI in a Govt. school located outside the village to prevent dropout at the end of class V and help girls to complete 8 years of schooling. 16171 girls in 2004-05 and more than 21000 girls 2005-06 benefitted from the programme
- **Uttar Pradesh**: Intensive campaign for community mobilisation in selected villages; 21 days training of instructors; use of teaching learning material; residential arrangement for girls and instructors; arrangements for sports, cultural programmes, life skills.
- **Uttaranchal**: Provisioning Early Childhood Education (ECE) in a convergent mode with Integrated Child Development Services (ICDS), SSA supporting with additional teaching learning material; capacity building; honorarium; constructing rooms in primary schools for running ECE centres; relocation of Integrated Child Development Services centres in/near primary schools; synchronized timings of ECE and primary school.
- **Orissa – Kalasi Dhara (carrying earthen vessel)** - An initiative to mobilise the community and Mother Teacher Associations to monitor the attendance of teachers and children, cleanliness of the school compound, regularity of classes being held. The designated mothers are also required to bring to school those children found to be absent by motivating their parents etc.
- **Bihar**: Summer Camps for Remedial Teaching, provided to girls.

- **Madhya Pradesh**: Open Learning for many girls who are unable to complete elementary education due to poor access. A tie up with State Open School where there is a 50:50 cost sharing between SSA & State Open School for the girls fees.

Alternative Schooling System

Three kinds of models of Alternative Schooling, especially tailored to the special needs of girls have been launched in the Indian context. The first model provides for adopting flexibility in the existing structure. The second model of alternative schooling stands for face-to-face teaching learning system in a flexible manner through its centres. The third model of alternative schooling is linked with open learning system.

Back to School Model – Bridge Courses and Camp Mode

This strategy supports mainstreaming of children through Bridge Courses of varied duration which bring out-of-school children to appropriate competency levels through condensed courses followed by their admission in appropriate grades in the formal schools. The bridge courses can be residential or non-residential. These are especially beneficial for tribal girls, children living in remote areas and working girls. The M.V. Foundations, Andhra Pradesh is doing a lot of work in this area. M.V. Foundation (MVF) began as an institution and research on issues relating to social change and transformation in 1981. But in 1991 it began focusing on abolition of child labour and making education available to all children. Today, it is a very large NGO in India with presence in 125 districts across the country. A variety of Back-to-School Models in the form of Bridge Courses were evolved in the country under Alternative Schooling Design of the District Primary Education Programme (DPEP) and later on under SSA. According to the needs of learners there is full flexibility of timing, place, content, mode of learning etc. These courses are both motivational and instructional.

Alternative Schooling through Face to Face Mode

Another model envisage alternative innovative centres in different forms viz., Balshalas, Prahar Pathshalas, Maktab Madrasas etc. Community participation and community ownership is an important feature of these models. These kinds of educational delivery systems are playing a vital role towards education of the girl child. Some of the models of alternative schooling initiated under the DPEP and later on under SSA are as follows:

Balshalas

It is very true that girls are doubly disadvantaged by non-availability of educational facilities and by the work they have to do in their homes and outside, such as, care of siblings, household responsibilities including collection of fuel, fodder etc.

The Balshalas were visualized to give a solution to the problem of child care faced by many girls who remain out-of-school as they are required to discharge responsibilities of care of siblings in the family. Balshala targeted two types of groups. One for 3-6 years age-group and the second for 6-11 years age group. The girls in the age-group 6-11 years received primary education in balshalas and those in the age-group 3 to 6 years attended school readiness package. The Balshalas were very popular and successful in so far as the programme relates to education of girls who were required to attend to sibling care and household responsibilities.

Prahar Pathshalas

Prahar Pathshala were evolved out for education of girls of 9+ age who were either non-starters or were drop-outs at an early level. In rural areas, 9+ is the critical age for girls when most parents of first generation learners withdraw their daughters from schools due to several socio-psychological reasons. The Prahar Pathshalas gave opportunities to these girls to receive education. The unique feature of the Prahar Pathshala was provision for training in certain local crafts along with primary education.

The Prahar Pathshalas could be started in any village where 15 or more girls of 9-14 years were willing to attend the centre for primary education and were anxious to learn any local craft which prepared them for self-employment. In Prahar Pathshalas, children were free to learn at their own pace without feeling threatened.

Maktab Madrasas

Muslim Community members do not like to send their girls in formal schools. They educate their children in Maktabs and Madrasas. The Maktab/Madrasas were strengthened as centres of alternative schooling. Selected teachers identified by the community from among the Maktab Madrasas were imparted training in the effective transactions of the formal school curriculum.

These are few examples of Alternative Schooling programmes of the Government of India. Similar schemes have been promoted in may States under Education Guarantee Scheme (EGS). Quality improvement and successful implementation of such schemes in decentralized manner is important to improve education of girls both in terms of access and retention. Several Non-Governmental Organizations (NGOs) are doing a lot of work for the promotion of girls' education.

Kasturba Gandhi Balika Vidyalaya (KGBV)

Kasturba Gandhi Balika Vidyalaya (KGBV) is a scheme launched in July 2004, for setting up residential schools at upper primary level for girls belonging predominantly to the SC, ST, OBC and minority communities. .K.G.B.V. was initiated for the girls of SC, ST, OBC, and minorities to provide elementary level of education. Those girls who drop out because of economic or domestic reasons are able to complete their education with the help of residential school scheme.

The scheme is being implemented in 27 States/UTs *namely*: Assam, Andhra Pradesh, Arunachal Pradesh, Bihar, Chhattisgarh, Dadar & Nagar Haveli, Delhi, Gujarat, Haryana, Himachal Pradesh, Jammu and Kashmir, Jharkhand, Karnataka, Madhya Pradesh, Maharashtra, Manipur, Meghalaya, Mizoram, Nagaland, Orissa, Punjab, Rajasthan, Tamil Nadu, Tripura, Uttar Pradesh, Uttarakhand and West Bengal.

2578 KGBVs were sanctioned by Government of India till date. Of these, 427 KGBVs were sanctioned in Muslim concentration blocks, 612 in ST blocks, 688 in SC blocks. As on 31st January, 2009, 2423 KGBVs were reported to be functional (i.e. 94%) in the States and 1,90,404 girls enrolled in them (50,630 SC girls (27%); 58,682 ST girls (31%); 50,161 OBC girls (26%); 18,206 below poverty line girls (9%); 12,725 Minority girls (7%). Out of the total 2578 sanctioned KGBVs, 547 KGBVs have been constructed and 1262 are in progress and 769 have not been started. The Kasturba Gandhi Balika Vidyalaya scheme was merged with Sarva Shiksha Abhiyan in the XIth Plan with effect from 1st April, 2007. . The objectives of KGBV are to:

- Bring awareness and encouragement of girls' education in the society.
- Ensure elementary education of all girl children.
- Increase literacy rate of the girls of SC, ST, OBC, Minorities and those who are below poverty line.
- Provide schooling facilities for the girls who could not reach to the school because of long distance between home and school.

- Bridge the gender gap of literacy and remove the gender bias at the elementary level.

Kasturaba Gandhi Balika Vidyalaya scheme aims to give benefits to those girls, who are away from education. It gives benefit to:

- The girls of SC, ST, OBC and minorities who could not complete their elementary level education due to social circumstances or some other reasons.
- The girls who could not complete their elementary level education.
- The girls of difficult areas where there is no school near their villages.
- The girls who have school near their village but can not go to school due to forest, river etc.
- Girls who cannot go to school due to bad financial conditions of family.
- Girls who cannot go to schools due to migration of their family.
- Girls who leave the school due to some other problems.
- Overage girls who have not completed their elementary level education.
- Working class girls of families that are below poverty line.
- Girls of suppressed portion of the society.

National Programme for Education of Girls for Elementary Level (NPEGEL)

The NPEGEL, launched in September 2003, is an integral but distinct component of the Sarva Shiksha Abhiyan. It provides additional provisions for enhancing the education of underprivileged/disadvantaged girls at elementary level through more intense community mobilization, the development of model schools in clusters, gender sensitization of teachers, development of gender sensitive learning materials, early child care and education facilities (ECCE) and provision of need-based incentives like escorts, stationery, work books and uniforms etc. for girls. All Educationally Backward Blocks have been included under NPEGEL.

Under NPEGEL, around 40,187 Model cluster schools have been opened, 38,626 ECCE centres are being supported, 38,409 additional classrooms have been constructed, and 2,11,215 teachers have been given training on gender sensitization, remedial teaching to 11,44,370 girls, bridge course covering 89,462 girls, including additional incentives like uniforms etc. to about 1,60,73,048 girls (upto 31st January, 2009).

Model of Open Schooling

Another model for girls/women's education in India is that of the National Institute of Open Schooling (NIOS). In addition to this there are also State Open Schools. Eleven states have already started State Open Schools (Andhra Pradesh, Delhi, Haryana, Jammu and Kashmir, Karnataka, Kerala, Madhya Pradesh, Punjab, Rajasthan, Tamil Nadu, and West Bengal). There are 8 states namely Uttar Pradesh, Orissa, Maharashtra, Gujarat, Assam, Chhattisgarh, Uttaranchal and Bihar which are in the process of setting up state Open Schools.

In order to sustain interest and commitment in continued schooling of girls/women. Open schooling provides a viable and workable solution to the challenges of equity in educational provisions for girls and women. Some salient features of open schooling programmes that addressed to the special needs of the girls/women are:

Policy Regarding Entry to NIOS Courses

The open schooling ensures access to education of the learners as it addresses to the shortcomings of the traditional system in terms of its rigidity. The NIOS expects a minimum level of competence for entry into its courses. Any student who has passed primary school can enroll himself/herself in the Foundation Course or the Secondary Level course. To enroll in the Senior Secondary course, however, a candidate must have a secondary pass certificate, to maintain equivalence with other examining bodies.

In NIOS, there are no rigid entry qualifications and no upper age limit for admission. It has a diverse student profile at the secondary stage having young and old learners ranging from 14 years to 89 years, distributed throughout the country. Flexibilities of the open learning system with regard to entrance policy makes possible for the girls/women engaged in household responsibilities to join NIOS programme at a time suitable to them. They have the freedom as well to discontinue their studies for a while to attend to demanding socio-economic responsibilities. The NIOS provides them an opportunity to rejoin the course at a point where they had left it due to compulsions of life. As such, instead of leaving educational pursuits forever Open Learning and Distance Education delivery system enables girls and women to pursue their educational efforts without disturbing normal rhythm of life and needs of the family. The open learning can benefit the family without taking the wife/mother away from the house to pursue her studies. Thus the open learning enables the mother to stay home with children and study as well. The strength of open learning system is that it allows women a bit of breathing space within a stripling situation of gender inequality.

There is yet another motivating factor that promotes girls/women's education. Many parents cannot afford expenses on private cost of education of their children. Due to this, they have to make a choice as to whom they should send for education; for which they prefer the son, as he is seen as the future earning member of the family. To encourage greater participation of women/girls, the NIOS allows partial fee exemption to them for different levels of courses. In order to promote the involvement of women, the NIOS accepts the name of the mother as parent in the relevant column of the Learner's Admission Form.

Programme Dimensions

NIOS through open learning and distance education delivery process has mounted a number of programmes in this direction. It runs courses for out-of-school girls/ women as well as for other categories of personnel who are desirous of pursuing the secondary/senior secondary level open school courses, general and vocational education courses. Open Basic Education courses cater to the special needs of girls in the age-group 6-14.

It also provides a course for adolescent girls and adult women, named, the Paripurna Mahila (Completely Empowered Women). This course covers vast range areas of women's empowerment, namely, legal literacy, health, nutrition, general awareness and the like. The major objectives of the courses are to equip, especially, rural women with better skills and enhance their knowledge so as to prepare them to face new challenges due to technological development. There are several subject areas in which the NIOS are working for the betterment of the women. Some such subject areas are – Jan swasthaya (public health), early childhood care and education, cutting and tailoring, beauty culture, computer applications and poultry farming.

This basically is a comprehensive programme directed towards gender equity. It is directed at generating awareness among girls/women forging an urge for equity measures. Most often, women lag behind initiating enterprising activities, not because they lack ability, but because they do not posses the required skills and access to relevant information that can enable them to set-up their own enterprise. Components of entrepreneurship are inter-woven in the NIOS programmes. The courses seek to develop requisite attitudes and abilities, enterprise management, growth and sustenance of the creative entrepreneurship. The NIOS is taking steps to identify more such gender sensitive courses that would promote women's empowerment.

The NIOS in collaboration with Bihar Education Project is engaged in implementing the Hunar scheme." Hunar', a scheme has turned out to be a huge success in Bihar with over 13,000 of them joining the programme. The scheme was jointly launched by the Union HRD Ministry and Bihar Government to make women from minority communities' self-dependent. Under the scheme, started on an experimental basis in July 2008,women from minority communities are being trained in computers and other skills like weaving, knitting, and child and maternity care. Altogether 13,884 candidates got enrolled for the programme in the very first year.

Student Support Services

Carefully designed self-instructional printed material, audio-video programmes, teleconferencing and personal contact programmes at the study centres give enormous potential to the open learning system as it provides the women/girl opportunity to learn at their own pace and place. Realizing that the distance education methodology is still new to the learners, the NIOS gives importance to the face-to-face instruction also. The Student Support Services are provided through the study centres. The study centres are located in formal schools, colleges, teacher's training institutions and non-government organizations with the capacity to provide academic support to the learners. Lack of facilities and lack of accessibility are factors that affect education of girls in the rural areas. To send girls to far-off places for education is both economically not viable and socially unacceptable in such a context. The NIOS has been striving to identify good schools and vocational institutes in semi-urban, rural and remote parts of the country to bring more and more girls/women into its fold.

As stated earlier, one of the reasons why less number of rural parents are sending girls to schools is the co-educational status of the schools. Surprisingly this feeling exists as much in the urban areas as in rural areas. NIOS has sought a way out by having study centres for 'girls only' in areas where their strength is significant. These schools usually have lady teachers who can understand the problems of girl students better. It is because of these features that its coverage has been expanding. Classes are held on Saturdays and Sundays or evenings of working days. Some Study Centres have special timings for women. Sometimes classes are held in the afternoon when women are free from their domestic responsibilities. This has widened the scope for gender equity in access to educational opportunities to girls/women.

Learner Evaluation

In pursuance of its policy of learner friendly approach, the National Institute of Open Schooling allows students to take examinations when they are ready. Examinations are held twice a year and students are free to take the examinations according to their convenience. This allows the girls/women to study at their own pace and convenience. In addition to this, the NIOS has introduced a scheme of transfer of credits through which learners who have cleared one or two subjects from the Central Board of Secondary Education (CBSE) but not qualified in the full course can seek admission in the NIOS and take the remaining subjects in the Secondary/Senior

Secondary programmes. The NIOS also introduced On-Demand Examination (ODES). This has opened up wider opportunities for appearing in the examination by parts as per the date and time suitable to the learner.

Conclusion

Education for women's equality is a vital component of overall strategy of securing equality and social justice in education. Through alternative schooling facilitates possibility of improvement in access and retention of female education is immense. In view of the constraints faced by the girls/women, provision of alternative schooling facilities goes a long way in stepping up enrolment and retention. It is important that the emphasis should be on improving conditions for assimilation by enabling women to participate in educational process thereby. It is a positive sign that these strategies have made remarkable efforts to represent the attempt of widening access to education by reaching out to the women in remote rural areas.

One of the purposes of education is to teach us to understand, analyze, question and finally attempt to change all that is wrong. Thus the increased opportunities for education of girl child may result in a better "deal" for women within the family; such influences are usually slow and need to be supplemented by more direct public action if gender equity is to be achieved. But the first step in this direction would be to develop a consciousness among women, regarding their unequal status. And what better agent of consciousness building could there be, than education itself. After all, the benefits of education derive not just from increases in cognitive skills, but to a far greater extent from the increases in non-cognitive abilities as reflected by changes in a person's ideas, perceptions and attitudes.

References

Government of India (1986), National Policy on Education (NPE) – 1986. New Delhi: Department of Education, Ministry of Human Resource Development.

Government of India (2001a), "Literacy in the Context of the Constitution of India". A Consultation Paper of the National Commission to Review the Working of the Constitution. New Delhi.

Government of India (2001b), Census of 2001, Series 1 India: Provisional Population Totals, Paper 1 of 2001. New Delhi: Registrar General and Census Commissioner, India.

Juneja, Nalini (2003), "Constitutional Amendment to Make Education a Fundamental Right: Issues for a Follow-up Legislation". NIEPA Occasional Paper 33. New Delhi: National Institute of Educational Planning and Administration.

Mehta, A. C. (2008), Analytical Report 2007-08, Elementary Education in India: Progress towards UEE, NUEPA, New Delhi.

Wazir, Rekha (ed.) (2000), The Gender Gap in Basic Education: NGOs as Change Agents. New Delhi: Sage.

Literacy Differentials between Scheduled Castes and Non-Scheduled Castes in Haryana

SNEH SANGWAN, ANJU DABAS AND
RANDHIR SINGH SANGWAN

Introduction

Literacy has been treated as an important indicator of social development. For the purpose of the Census, a person who can read and write with understanding in any language is treated as literate. Since the formation of Haryana as a separate state in 1966, a considerable growth has taken place in literacy. As per Census of India (2001), Haryana has literacy rate of 67.91 per cent, which is comparatively higher than the National average (64.80 per cent).

The scheduled caste population, comprising of about one-sixth of the country's total population (16.20 per cent) lags behind the rest of population in terms of literacy and education. The Census of India, 2001 revealed that only 54.69 per cent of the scheduled caste population was literate as against 70.81 per cent of the non-scheduled caste population which was able to read and write. So, there exists a wide gap in the literacy rates of the scheduled caste and the non-scheduled caste population. The present paper deals with this literacy differential.

Objectives

The present study is a modest attempt to describe and interpret the regional aspects of scheduled castes and non-scheduled castes differentials in literacy rates in Haryana. The tahsil has been chosen as a unit of analysis. Two-fold objectives of the present study are as follows:

1. To describe literacy rates of scheduled castes and non-scheduled castes separately as per 2001 Census.
2. To give an in-depth analysis of the spatial patterns of scheduled castes and non-scheduled castes differential in literacy rates in Haryana.

Data Base and Methodology

The study is mainly based on the secondary sources of data. The reference year for the secondary data is 2001, which have been obtained particularly from the secondary sources. Both qualitative and quantitative methods and techniques have been applied for the data analysis. The most important among them are the choropleth method used for the preparation of maps. Sopher's disparity index, as modified by Kundu and Rao, for the computation of scheduled castes and non-scheduled castes differential indices has been applied.

Sopher's disparity index as modified by Kundu and Rao in 1982 is as follows:

$$Ds = \log (X_2/X_1) + \log (200-X_1) / (200-X_2)$$

Where

$X_2 \geq X_1$ and X_1 & X_2 are the literacy rates of *alphas* and *non-alphas* respectively.

Since spatial concept occupies a central position in every geographic enquiry, map is an important tool. On calculating differential index for literacy rates, the tehsils of Haryana were classified into various categories (keeping in view the state average) and represented cartographically so as to present a regional perspective, showing spatial patterns of scheduled castes and non-scheduled castes differential in literacy rates.

Literacy: The Conceptual Analysis

During the earlier censuses up to 1981, it was customary to work out the literacy rate taking into account the total population. Since literacy rate is more meaningful if population in the age group 0-6 is excluded from the total population, it was decided in 1991 to use the term literacy rate for the population aged seven years and above. At the 1991 Census, it was also decided that all children in the age-group 0-6 years will be treated as illiterate by definition and the population aged seven years and above only is to be classified as literate or illiterate. The same criterion has been retained at the Census of India, 2001. The literacy rate taking into account the total population in the denominator has now been termed as *crude literacy rate*. However, the per cent of literates among population aged seven years and above has been referred to as *effective literacy rate* or simply as *literacy rate*.

Literacy Among Scheduled Castes

The Census of India, 2001 reflected that Haryana has only 55.43 per cent of the scheduled caste population as literate. The literacy rate among scheduled caste population was very low in comparison to general population. There are many reasons such as social injustice, lack of awareness, association with menial jobs and economic restriction to participate in economically gainful activities. Mahendragarh district (68.7 per cent) has highest literacy rate among scheduled caste population and Fatehabad district (41.0 per cent) has lowest literacy rate among scheduled caste population. But in case of tehsils, Gurgaon has the highest literacy rate among scheduled castes and Ratia tehsil has the lowest literacy rate (Table 1).

Table 1
Haryana: Literacy Rate of Scheduled Castes by Tehsils, 2001

Sr. No.	*State / Tehsil*	*Literacy Rate (Per cent)*
	Haryana	**55.43**
1.	Gurgaon	69.35
2.	Rewari	69.32
3.	Pataudi	68.03
4.	Kosli	67.69
5.	Kalka	67.30
6.	Bawal	66.51
7.	Sonipat	65.41
8.	Sohna	64.38
9.	Taoru	64.24
10.	Barara	63.92
11.	Mahendragarh	63.89
12.	Jagadhri	63.66
13.	Ambala	63.59
14.	Bahadurgarh	63.55
15.	Narnaul	63.44
16.	Beri	62.40
17.	Jhajjar	61.68
18.	Kharkhoda	61.54
19.	Naraingarh	61.47
20.	Ganaur	61.07
21.	Panchkula	61.00
22.	Rohtak	60.25
23.	Dadri	59.78
24.	Thanesar	59.50
25.	Samalkha	59.43
26.	Ballabgarh	58.80
27.	Faridabad	58.67
28.	Gohana	58.51

table contd.

Sr. No.	State / Tehsil	Literacy Rate (Per cent)
29.	Chhachhrauli	57.50
30.	Bhiwani	57.18
31.	Shahbad	56.97
32.	Maham	56.43
33.	Panipat	55.92
34.	Julana	55.72
35.	Indri	55.61
36.	Palwal	55.50
37.	Israna	55.36
38.	Karnal	55.29
39.	Gharaunda	55.20
40.	Nilokheri	54.86
41.	Loharu	54.29
42.	Jind	53.34
43.	Tosham	53.15
44.	Bawani Khera	52.68
45.	Nuh	52.58
46.	Siwani	52.29
47.	Safidon	52.05
48.	Hathin	51.19
49.	Ferozpur Jhirka	51.12
50.	Hansi	51.07
51.	Pehowa	50.56
52.	Hisar	50.04
53.	Hodal	49.09
54.	Narnaund	47.81
55.	Adampur	46.45
56.	Asandh	46.20
57.	Kaithal	45.76
58.	Punahana	44.40
59.	Tohana	43.34
60.	Fatehabad	43.05
61.	Sirsa	42.23
62.	Narwana	42.03

table contd.

Sr. No.	*State / Tehsil*	*Literacy Rate (Per cent)*
63.	Guhla	41.02
64.	Dabwali	41.01
65.	Ellenabad	40.30
66.	Rania	39.57
67.	Ratia	34.74

Source: Computed from, Census of India (2001), *Primary Census Abstract*, Haryana, Series 7, Tables - A5-A8, Directorate of Census Operations, Haryana.

Note: Tehsils are arranged in descending order of literacy rate, 2001.

On the basis of total literacy rates of scheduled castes, the tehsils of Haryana can be divided into five categories. These are as under:

1. Areas of Very High Schedule Caste Literacy Rate (> 66 per cent)
2. Areas of High Schedule Caste Literacy Rate (58 - 66 per cent)
3. Areas of Moderate Schedule Caste Literacy Rate (50–58 per cent)
4. Areas of Low Schedule Caste Literacy Rate (42–50 per cent)
5. Areas of Very Low Schedule Caste Literacy Rate (< 42 per cent)

1. Areas of Very High Schedule Caste Literacy Rate (66 per cent and above): Six tehsils come under this category and cover only 5.88 per cent area of the state. Gurgaon tehsil has highest scheduled caste literacy rate and is followed by the tehsils of Rewari (69.32 per cent), Pataudi (68.03 per cent), Kosli (67.69 per cent), Kalka (67.30 per cent) and Bawal (66.51 per cent). High scheduled caste literacy rate in these tehsils is mainly because these are highly industrialized, e.g. Maruti Udyog, Stainless Steel industries, etc. are located here. High level of urbanization is also the characteristic of these areas. Their location in the vicinity of National and State capitals like New Delhi and Chandigarh is also responsible for the high literacy rates.

2. Areas of High Schedule Caste Literacy Rate (58-66 per cent): Twenty two tehsils come under this category and cover 32.89 per cent area of the state. Out of these, the tehsils of Barara, Jagadhri, Naraingarh, Ambala, Panchkula, Thanesar are located in the north; Samalkha, Ganaur, Sonipat, Kharkhoda, Gohana, Rohtak, Beri, Jhajjar, and Bahadurgarh in the east; Sohna, Taoru, Faridabad, and Ballabgarh in the south-east; Mahendragarh, Narnaul, and Dadri in the south of state. These tehsils are economically sound and industrially developed. In these tehsils, many industries such as sugar industry, cotton industries, milk powder industry, Atlas cycle industry, Cattle Feed Plant, etc. are located. The educational, transportation and communication facilities are also well developed in these tehsils.

3. Areas of Moderate Schedule Caste Literacy Rate (50-58 per cent): In this category, twenty four tehsils come and cover 33.5 per cent area of the state. These are Chhachhrauli, Shahbad, Pehowa in northern side; Maham, Panipat, Indri, Israna, Karnal, Gharaunda, Nilokheri in eastern side; Palwal, Nuh, Hathin, Ferozepur Jhirka in south-eastern side; Bhiwani, Loharu, Tosham, Bawani Khera, Siwani in south-western side; Hansi, Hisar in western parts; Julana, Jind, Safidon are located in central parts of the state. In these tehsils, scheduled caste population is mainly engaged in primary activities. Moreover, these tehsils have low educational development.

4. Areas of Low Schedule Caste Literacy Rate (42-50 per cent): Only ten tehsils come under this category and cover 20.48 per cent area of the state. These are Narnaund, Adampur, Tohana, Fatehabad, Sirsa in western side; Hodal, Punahana in south-eastern; Assand,

Kaithal and Narwana in north-western parts of the state. Most of the scheduled caste people in these tehsils are agricultural labourers. They have no work most of the year. These low literacy rates were attributed to large concentration of scheduled caste population and their backwardness both socially and economically.

5. Areas of Very Low Schedule Caste Literacy Rate (42 per cent and below): Only five tehsils come under this category and cover only 7.25 per cent area of the state. These are Guhla in north-western; Dabwali, Ellenabad, Rania and Ratia in western parts of the State. Ratia has the lowest scheduled caste literacy rate i.e. 34.74 per cent. These tehsils are backward in agriculture as well as in industry. The reasons of low scheduled caste literacy are lack of awareness, association with menial jobs and economic restrictions to participate in economic gainful activities.

Literacy Among Non-scheduled Castes

In 2001 Census, 70.81 per cent of the non-scheduled caste population was able to read and write (Table 2). Here both the male and female literacy rates were higher in comparison to scheduled caste population.

Table 2
Haryana: Literacy Rate of Non-Scheduled Castes by Tehsils, 2001

Sr. No.	*State / Tehsil*	*Literacy Rate (Per cent)*
	Haryana	**70.81**
1.	Gurgaon	82.50
2.	Ambala	82.22
3.	Faridabad	80.01
4.	Sonipat	78.62
5.	Rohtak	78.45
6.	Rewari	78.21
7.	Bahadurgarh	77.37

table contd.

Sr. No.	*State / Tehsil*	*Literacy Rate (Per cent)*
8.	Barara	76.81
9.	Jagadhri	76.77
10.	Shahbad	76.48
11.	Panchkula	76.44
12.	Pataudi	76.30
13.	Kosli	75.70
14.	Karnal	74.45
15.	Panipat	74.43
16.	Kalka	74.34
17.	Thanesar	74.25
18.	Beri	74.22
19.	Bhiwani	73.23
20.	Gohana	72.04
21.	Sohna	71.96
22.	Pehowa	71.95
23.	Mahendragarh	71.88
24.	Jhajjar	71.70
25.	Ganaur	71.50
26.	Kharkhoda	71.26
27.	Dadri	71.24
28.	Hisar	70.80
29.	Indri	70.77
30.	Nilokheri	70.77
31.	Maham	70.46
32.	Narnaul	70.45
33.	Bawal	70.36
34.	Naraingarh	70.26
35.	Jind	69.60
36.	Julana	68.96
37.	Sirsa	68.78
38.	Palwal	68.13
39.	Hansi	68.02
40.	Bawani Khera	67.88
41.	Israna	67.50
42.	Ballabgarh	66.94
43.	Safidon	66.87
44.	Gharaunda	66.68
45.	Asandh	66.31
46.	Loharu	66.06

table contd.

Sr. No.	*State / Tehsil*	*Literacy Rate (Per cent)*
47.	Dabwali	66.01
48.	Samalkha	65.99
49.	Rania	65.82
50.	Hodal	65.52
51.	Tohana	65.39
52.	Tosham	65.12
53.	Siwani	64.11
54.	Adampur	63.96
55.	Fatehabad	63.66
56.	Ratia	63.19
57.	Kaithal	62.94
58.	Narnaund	62.67
59.	Ellenabad	62.47
60.	Guhla	62.32
61.	Chhachhrauli	60.47
62.	Narwana	59.12
63.	Taoru	56.16
64.	Hathin	45.01
65.	Nuh	44.33
66.	Ferozpur Jhirka	39.20
67.	Punahana	37.00

Source: Computed from, Census of India (2001), *Primary Census Abstract*, Haryana, Series 7, Tables-A5-A8, Directorate of Census Operations, Haryana.

Note: Tehsils are arranged in descending order of literacy rate, 2001

On the basis of total literacy rates of Non-Scheduled Castes, tehsils of state can be divided into five categories. These are:

1. Areas of Very High Non Scheduled Caste Literacy Rate (82 per cent and above).
2. Areas of High Non Scheduled Caste Literacy Rate (71–82 per cent).
3. Areas of Moderate Non Scheduled Caste Literacy Rate (60–71 per cent).
4. Areas of Low Non Scheduled Caste Literacy Rate (49–60 per cent).

5. Areas of Very Low Non Scheduled Caste Literacy Rate (49 per cent and below).

1. **Areas of Very High Non Scheduled Caste Literacy Rate (82 per cent and above):** Only two tehsils come under this category. Gurgaon (82.50 per cent) and Ambala (82.22 per cent) cover 3.27 per cent of state area. Both of these tehsils are highly urbanized, industrialized and more educational institutions are located here.

2. **Areas of High Non Scheduled Caste Literacy Rate (71-82 per cent):** Twenty five thesils come under this category and cover 36.61 per cent area of the state. These are Barara, Jagadhri, Shahbad, Panchkula, Kalka, Thanesar, Pehowa in northern parts; Sonipat, Rohtak, Bahadurgarh, Karnal, Panipat, Beri, Gohana, Jhajjar, Ganaur, Kharkhoda in eastern parts; Faridabad, Pataudi, Sohna in south-eastern; Rewari, Kosli, Mahendragarh, Dadri in southern parts and Bhiwani in western parts of state. Development in agriculture, industries, transportation and communication system is the main reason for high literacy in these tehsils.

3. **Areas of Moderate Non Scheduled Caste Literacy Rate (60-71 per cent):** Thirty four tehsils come under this category and cover 53.25 per cent area of the state. These are Hisar, Sirsa, Hansi, Bawani Khera, Dabwali, Rania, Tohana, Tosham, Siwani, Adampur, Fatehabad, Ratia, Narnaund, Ellenabad in western parts; Nilokheri, Indri, Maham, Israna, Gharaunda Samalkha in eastern parts; Naraingarh, Chhachhrauli in northern parts; Narnaul, Bawal, Loharu, in southern parts; Palwal, Ballabgarh, Hodal in south-eastern parts; Jind, Julana, Safidon, Assandh, Kaithal, Guhla are in central parts of the state. In all these tehsils, population is engaged in primary activities. Here, more prejudices against the females' mobility and education, low degree of urbanization are also found.

4. **Areas of Low Non Scheduled Caste Literacy Rate (49-61 per cent):** Only two tehsils of Narwana and Taoru come under this category and cover only 3.11 per cent of area of the state. Both tehsils have also low general literacy rate and people are mainly engaged in primary activities and are characterized with low status of woman in society.

5. **Areas of Very Low Non Scheduled Caste Literacy Rate (49 per cent and Below):** Only four tehsils of Hathin (45.01), Nuh (44.33), Ferozepur Jhirkha (39.20) and Punahana (37.00 per cent) are characterized with very low literacy rate among the non scheduled castes population. These tehsils cover 3.76 per cent area of the state. All these tehsils are located in the south- eastern part of the state. Very low literacy rates in these tehsils are mainly because of the dominance of Meo-Muslim population and are also characterized with low levels of urbanization and industrialization alongwith poor development of educational institutions.

Literacy Differential Among Scheduled Castes And Non Scheduled Castes

In 2001, the index of literacy differential between Non-Scheduled caste and scheduled caste population is 0.1550. It is highest in Ratia tehsil (0.3419) and negative in Ferozepur Jhirka tehsil (-0.1488). The highest index in Ratia is due to low rates of general literacy, widespread poverty, prevalent socio-economc segregation, lack of employment opportunities and strong feeling of insecurity particularly among rural females. The negative index of Ferozepur Jhirka was associated with high concentration of scheduled caste population.

Spatial Pattern

There are noticed wide regional variation in the index of literacy differential between non-scheduled caste and scheduled caste population.

On the basis of state average i.e. 0.1550, five types of areas can be classified (Table 3 & Figure 1):

1. Areas of Very High Literacy Differential Index (0.2501 and above).
2. Areas of High Literacy Differential Index (0.1701 – 0.2500).
3. Areas of Moderate Literacy Differential Index (0.0901–0.1700).
4. Areas of Low Literacy differential Index (0.0100–0.0900).
5. Areas of Negative Literacy differential where the literacy among non-scheduled castes was relatively lower than that among scheduled castes.

1. **Areas of Very High Literacy Differential Index (0.2501 and Above):** Five tehsils of state have very high literacy differential. These are Ratia, Rania, Sirsa, Dabwali and Ellenabad, all are situated in the western part of the state. The main reason of very high differential index is because the proportion of Scheduled Caste population is high to total population. This social group lags much behind in the race of socio-economic development. Very high literacy differential is also related to poverty, less developed agriculture, low standard of living, low per capital income, less educational facilities, less developed transportation and industrialization, rigid caste system.

2. **Areas of High Literacy Differential (0.1701 – 0.2500):** Sixteen tehsils come under high literacy differential. These are Tohana, Fatehabad, Hisar, Adampur, Hansi in western part; Guhla, Assandh, Narwana, Kaithal in north-central part; Shahbad, Karnal, Panipat, Rohtak, Pehowa in eastern part, Faridabad in south-eastern and Ambala in northern parts of the state.

Table 3
Haryana: Literacy Differential between Non-Scheduled Castes and Scheduled Castes Population by Tehsils, 2001

Sr. No.	*State / Tehsil*	*Differential Index*
	Haryana	**0.1550**
1.	Ratia	0.3419
2.	Rania	0.2985
3.	Sirsa	0.2919
4.	Dabwali	0.2811
5.	Ellenabad	0.2552
6.	Tohana	0.2475
7.	Guhla	0.2442
8.	Fatehabad	0.2310
9.	Asandh	0.2177
10.	Hisar	0.2154
11.	Faridabad	0.2058
12.	Narwana	0.1979
13.	Shahbad	0.1916
14.	Adampur	0.1916
15.	Karnal	0.1909
16.	Kaithal	0.1896
17.	Panipat	0.1840
18.	Hansi	0.1769
19.	Ambala	0.1754
20.	Rohtak	0.1752
21.	Pehowa	0.1750
22.	Jind	0.1666
23.	Narnaund	0.1621
24.	Nilokheri	0.1610
25.	Bhiwani	0.1592
26.	Bawani Khera	0.1574
27.	Safidon	0.1547
28.	Indri	0.1529
29.	Panchkula	0.1492
30.	Hodal	0.1455
31.	Thanesar	0.1444
32.	Maham	0.1411
33.	Julana	0.1344

table contd.

Sr. No.	*State / Tahsil*	*Differential Index*
34.	Gohana	0.1340
35.	Bahadurgarh	0.1319
36.	Palwal	0.1288
37.	Jagadhri	0.1253
38.	Tosham	0.1251
39.	Sonipat	0.1248
40.	Siwani	0.1248
41.	Israna	0.1241
42.	Barara	0.1230
43.	Loharu	0.1217
44.	Gurgaon	0.1215
45.	Gharaunda	0.1179
46.	Beri	0.1143
47.	Dadri	0.1133
48.	Ganaur	0.1024
49.	Jhajjar	0.0980
50.	Kharkhoda	0.0953
51.	Naraingarh	0.0866
52.	Rewari	0.0830
53.	Ballabgarh	0.0821
54.	Pataudi	0.0779
55.	Mahendragarh	0.0774
56.	Kosli	0.0757
57.	Sohna	0.0733
58.	Narnaul	0.0685
59.	Kalka	0.0669
60.	Samalkha	0.0663
61.	Bawal	0.0372
62.	Chhachhrauli	0.0310
63.	Hathin	-0.0736
64.	Taoru	-0.0835
65.	Nuh	-0.0978
66.	Punahana	-0.0994
67.	Ferozpur Jhirka	-0.1488

Source: Computed from, Census of India (2001), *Primary Census Abstract*, Haryana, Series 7, Tables-A5-A8, Directorate of Census Operations, Haryana.

Note: Tahsils are arranged in descending order of differential index, 2001.

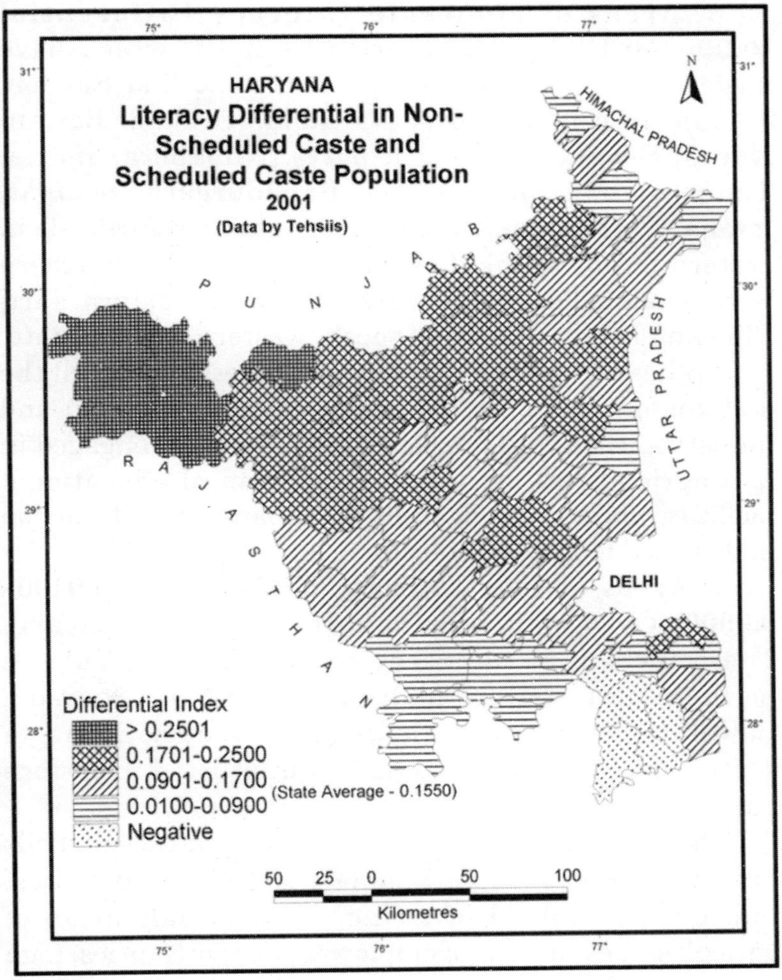

Figure 1

These areas of high literacy differential were associated with having low degree of urbanization and industrialization, having feudal landlord system, high proportion of scheduled caste population, deep roots of caste system either by natural growth rate or by immigration.

3. **Areas of Moderate Literacy Differential (0.0901–0.1700):** These are 29 tehsils, which have moderate differential index. These include Jind, Safidon, Julana, in central parts, Narnund, Tosham, Bawani Khera, Siwani in western parts, Nilokhera, Indria, Thanesar, Meham, Gohana, Bahadurgarh, Sonipat, Israna, Gharaunda, Beri, Ganaur, Jhajjar, Kharkhoda in eastern part; Panchkula, Jagadhri, Barara in northern part; Hodal, Palwal, Gurgaon in south-eastern part; Bhiwani, Loharu, Dadri in south- western parts of state.

In these tehsils, increasing awareness among all the sections of society, development of transportation and industrialization, high proportion of workers engaged in non agricultural activities, expansion of educational facilities, and increasing degree of urban-rural interaction had caused moderate literacy.

4. **Areas of Low Literacy Differential (0.0100–0.0900):** Only twelve tehsils have low literacy differential. These include Naraingarh, Kalka, Chhachhrauli, in northern part, Rewari, Mahendragarh, Kosli, Narnaul, Bawal in southern part, Ballabgarh, Pataudi, Sohna, in south-eastern part and Samalkha is situated in eastern part of state.

The main reasons of low differential in these tehsils are: high literacy rate among the scheduled castes, improved standard of living, commercialization of agriculture, expansion of employment opportunities, high mobility and awareness among people in rural areas and developed transportation.

5. **Areas of Negative Literacy Differential:** Only five tehsils of Hathin (-0.0736), Taoru (-0.0835), Nuh (-0.0978), Punahana (-0.0994), and Ferozpur Jhirka (-0.1488) fall in this category of negative literacy differential. The negative differential value varied from -0.0736 in Hathin tehsil to -0.1488 in Ferozepur Jhirka tehsil. In all these tehsils, the scheduled caste literacy

was high than non-scheduled caste literacy because of the dominance of Muslims in non-schedules castes population.

Concluding Remarks

The literacy rate among scheduled caste population was very low in comparison to general population. There are many reasons such as social injustice, lack of awareness, association with menial jobs and economic restriction to participate in economically gainful activities. Six tehsils are characterized with very high Schedule Castes literacy rate (66 per cent and above). Amongst these, Gurgaon tehsil has highest scheduled caste literacy rate (69.35 per cent) and is followed by the tehsils of Rewari, Pataudi, Kosli, Kalka and Bawal. High scheduled castes literacy rate in these tehsils is mainly because they are highly industrialized, because of the establishment of important large scale industries like Maruti Udyog, etc. Urbanization is also high. Moreover, these are also located in the vicinity of the National and State capitals like New Delhi and Chandigarh.

On the other hand, only five tehsils experienced very low Scheduled Castes literacy rates (42 per cent and below). Amongst these, Ratia has the lowest scheduled castes literacy rate i.e. 34.74 per cent and is followed by the tehsils of Rania. Ellenabad, Dabwali, and Guhla. These tehsils are backward in agriculture as well as in industry. The reasons of low scheduled caste literacy are lack of awareness, association with menial jobs and economic restrictions to participate in economic gainful activities. By comparision, in 2001 Census, 70.81 per cent of the non-scheduled caste population was able to read and write. Only two tehsils are characterized with very high Non-Schedules Castes literacy rate (82 per cent and above). Amongst these, Gurgaon tehsil has the highest non-scheduled castes literacy rate (82.50 per cent) and is followed by the tehsil of Ambala (82.22 per cent). Both the tehsils are highly urbanized, industrialized and educationally developed. On the other hand, only four tehsils of Hathin, Nuh, Ferozepur Jhirkha and Punahana are characterized with very low literacy rate among the Non-Scheduled castes population. All these tehsils are located in the south- eastern part of the state. Very low literacy rates in these tehsils are mainly because of the dominance of Meo-Muslim population and are also characterized with low levels of urbanization and industrialization alongwith poor development of educational institutions.

Spatially, there are wide regional variations in the index of literacy differential between non-scheduled castes and scheduled castes population. Five tehsils of state have very high literacy differential (0.2501 and above). Amongst these, Ratia tehsil has the highest literacy differential between non-scheduled castes and scheduled castes population, and is followed by the tehsils of Rania, Sirsa, Dabwali and Ellenabad, which are situated in the western part of the state. The main reason of very high differential index is because the proportion of Scheduled Caste population is high to total population. This social group lags much behind in the race of socio-economic development. Very high literacy differential is also related to poverty, less developed agriculture, low standard of living, low per capita income, less educational facilities, less developed transportation and industrialization, rigid caste system, subsistence farm economy, low degree of urbanization, involvement of large proportion of scheduled caste population as agricultural labour, in construction activity and informal service sector. On the other hand, five tehsils of Hathin, Taoru, Nuh, Punahana, and Ferozpur Jhirka have experienced negative literacy differential ranging from highest in Hathin tehsil to lowest in Ferozepur Jhirka tehsil. In all these tehsils, scheduled castes literacy was higher than non-scheduled castes literacy because of the dominance of Muslims in non-scheduled castes population.

References

Bhardwaj, P. D. (1999), "Literacy in Himachal Pradesh", *Population Geography*, Vol. 21, Nos. 1 & 2, pp. 31-42.

Khan, A. Q. (2004), "Regional Imbalances and Changing Pattern of Literacy in Jammu & Kashmir State (1961-2001)", *Population Geography*, Vol. 26, Nos. 1&2, pp..49-60.

Krishan, G. and Chandna, R.C. (1974), "Patterns of Literacy in Haryana", *Punjab University Research Bulletin (Arts)*, 5, 2, pp. 95-104.

Krishan, G. and Shyam, M. (1978), "Regional Aspects of Urban-Rural Differentials in Literacy in India: 1971", *The Journal of Developing Areas*, 13, 1, pp. 11-21.

Kundu, A. and Rao, J. M. (1982), "Inequality in Educational Development: Issues in Measurement Changing Structure and its Socio-Economic Correlates with Special Reference to India", Paper Presented at Regional Workshop on *Long-Term Educational Planning*, NIEPA, New Delhi.

Samanta, G. (2003), "Gender Disparity in Literacy in West Bengal", *The Deccan Geographer,* Vol. 31, No. 1, pp. 31-42.

Sangwan, Sneh and Sangwan, R.S. (2003), *Rural-Urban Divide: Changing Spatial Pattern of Social Variables*, Concept Publishing Co., New Delhi, pp. 126-154.

Sangwan, Sneh and Sangwan, R. S. (2010), "Haryana: Urban-Rural Differentials in Literacy", in Dubey, R. N. (ed.), *Urbanisation and Urban Planning in India: Vision and Reality*, Shree Nataraj Prakashan, New Delhi, pp. 320-333.

Sopher, D. E. (1974), "A Measure of Disparity", *The Professional Geographer*, 26, 4, pp. 389-392.

Sopher, D. E. (1980), "Sex Disparity in Indian Literacy", in Sopher D. E. (ed.), *An Exploration of India Geographical Perspectives on Society and Culture*, Cornell University Press, Ithaca, New York, pp. 130-188.

Usha Rao, N. J. (1982), "Inter-State Disparities in Literacy Rates in Karnataka with Special Reference to Scheduled Castes", in Bhat, L.S. (ed.), *Regional Disparities in India*, Society for the Study of Regional Disparities, New Delhi.

Women's Participation in Rural Environment: A Case Study of Rural Delhi

MONICA AHLAWAT

Introduction

The direct and critical relationship between women and natural resources draws its strength not from biology—that is, not because women are born female—but from gender, and the socially created roles and responsibilities that continue to fall to women in households, communities and ecosystems throughout the world. Women have primary responsibility for rearing children, and for ensuring sufficient resources to meet children's needs for nutrition, health care and schooling. In the rural areas, they are also the main managers of essential household resources like clean water, fuel for cooking and heating and fodder for domestic animals. Women grow vegetables, fruit and grain for home consumption and also for sale. The term environment means whole sum of the surrounding external conditions within which an organism, a community or an object exists. All the components of the environment are interrelated and mutually interdependent. The environment includes components of land, water, air, climate, soil, forests and fauna and biogeochemical and gaseous cycles (Hussain, 1996). More people are using more resources with more intensity than at any point in human history. Utilization of natural resources is the interaction between two systems – the natural and human. In course of the utilization, sometimes the very resource base is eroded leading to devaluation or diminution of environment, or some sort of disturbance is triggered in the ecosystem functioning and structure causing ecological imbalance. Environmental degradation is qualitative and quantitative decline/ impoverishment/ decrease/ devaluation/ degeneration/ diminishment in the ecosystem potentiality/ productivity affecting habitat of man, plant and animal (Hussain, 1996). Smith (1972) has called 'degraded environments' as 'endangered category, for life of the biotic community and existence of abiotic elements are threatened. Lately man has realized that loss or decline in environment is his loss as well. That's why; expression of extreme concern for environment by man is stemmed from his sense of fear about the deteriorating health of environment, because its further decline is a threat to his very survival itself. Women have primary responsibility for rearing children, and for ensuring sufficient resources to meet children's needs for nutrition, health care and schooling (Figure 1). In the rural areas they are also the main managers of essential household resources like clean water, fuel for cooking and heating and fodder for domestic animals. Women grow vegetables, fruit and grain for home consumption and also for sale.

"Environmental problems begin with people as a cause and end with people as victim. They are usually born of ignorance and apathy. It is the people who create a bad environment and a bad environment brings out the worst in people. Man and nature need each other and by hurting one we would wound the other. There is much that needs to be done to halt the destruction of our world environment, so many prejudices and so much self-interest to be overcome"—**Sir Edmund Hillary (As quoted in Sapru and Bhardwaj, 1990)**

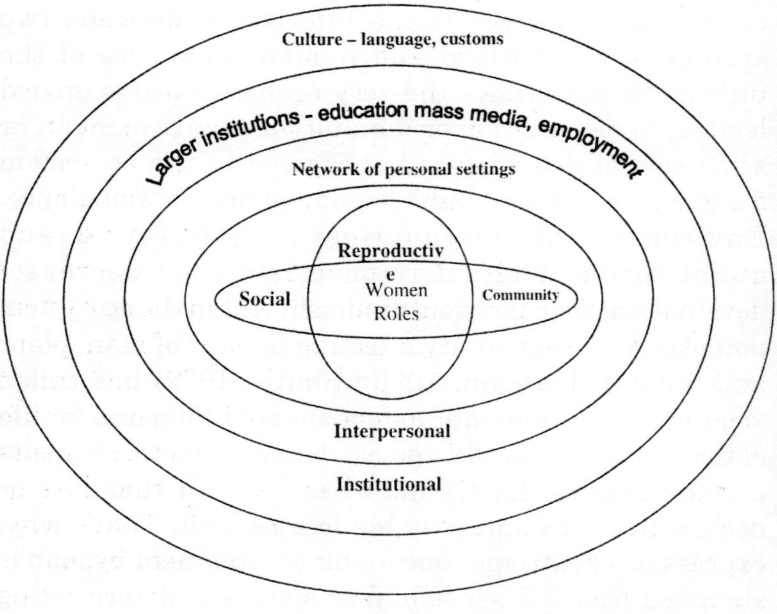

Figure 1: Women and their Social Relationships, Source: Millican and Smithells, 1992

It is important to study women's work because:

- A substantial part of women's time and energy is spent in working.
- Women's work is most of the time invisible and unrecognized, primarily because in the accepted capitalist paradigm any 'economic' activity is synonymous with 'market' activity and much of women's work is done outside the realm of the market. It's also usually undervalued since the unpaid element is not taken into consideration while computing national income.
- A number of issues confronting the women's movement today, particularly in developing countries such as in India, are related to different aspect of women's work.

- In the contest of globalization women's work is probably one of the most important areas that are being affected by the changing situation.

Study Area

In 2001 Census, the total population of N.C.T. of Delhi was 138, 50,507. The rural population of N.C.T. of Delhi has grown from 17, 44,072 with 304 inhabited villages in 1951 to 138, 50,507 with 165 inhabited villages in 2001. The research work was done in rural Delhi based on old administrative structure. It is spread over an area of 783.23 square kilometers and is administrated by 2 Tehsils – the Delhi Tehsil and Mehrauli Tehsil. Clubbed together these tehsils, encompass 209 villages which are further divided among 5 community Blocks. These are: Alipur Block, Kanjhawala Block, Najafgarh Block, Mehrauli Block, Shahadara Block. Two major community Blocks: Alipur and Najafgarh were selected for the primary survey. In Alipur Block 4 villages were surveyed, namely, Bankauli, Hiranki, Qadi Pur and Sanoth. Alipur block is located between 28° 42' N to 28° 52' N latitudes, while its longitudinal extent is 77° 2' 30" E to 77° 13' 50" E. It is located in the north east part of National Capital Territory of Delhi. The Yamuna demarcates its eastern boundary. A total of 131 households were surveyed in Alipur Block. Overall, there has been an increase in the population of the villages. In Alipur block, Bankauli and Hiranki village being nearer to Alipur town are more developed as compared to the other two villages i.e. Sanoth village and Qadi Pur village. Alipur Block has fertile soils, especially near the river, and the quality of groundwater also makes the area quite suitable for cultivation. The main road running towards Amritsar (GT Road) separates the block into an eastern and a western part. The eastern part has relatively more land under intensive cultivation and yields better harvests, including horticultural crops.

Sanoth is a cosmopolitan village having a mixture of all castes and religious i.e. Jats, Brahmins and lower castes along with Muslims. This is the only village with a Muslim habitation. These are engaged either in carpentry or are masons.

While in Najafgarh Block the 4 villages were surveyed: Dhansa, Dichaun Kalan, Kair and Ujwa. Najafgarh is located at the outskirts of New Delhi, India. In the southwestern part of Delhi, it is a mixture of Jat, Yadav and Brahmin populations. Najafgarh has the distinction of being the most populous electorial region in the National Capital Region of India. In Najafgarh block Dichaon kalan village is a sample of urban land use change. It is the largest village in terms of area as well as in population. Dichaon Kalan has 1853 main workers which is the highest among the surveyed villages. It also has maximum number of cultivators. Najafgarh Block occupies 1st position in literacy among all the rural blocks, with 84.1 per cent of males and 59.6 per cent of female's literate. Distance is one of the chief reasons for the lagging behind of Dhansa in the race for urbanization. Dhansa is a village undergoing agricultural land use change. In Najafgarh Block, Kair is the only village, which has multiple sources of irrigation. The availability of educational facilities is available in Kair village. Recently a woman's college was opened in Kair for the rural girls. Many young girls from Dichaon Kalan village, Najafgarh towns study in Kair College. There are very few villages in Delhi's fringe which are purely rural in character. But what is essentially missing from a few villages is the daily interaction with the city. Lack of proper bus service is one of the main reasons for the lesser development and neglect of the village from urban forces. Being away from the main road Ujwa is one of the classic examples of a rural village in urban vicinity (Figure 2).

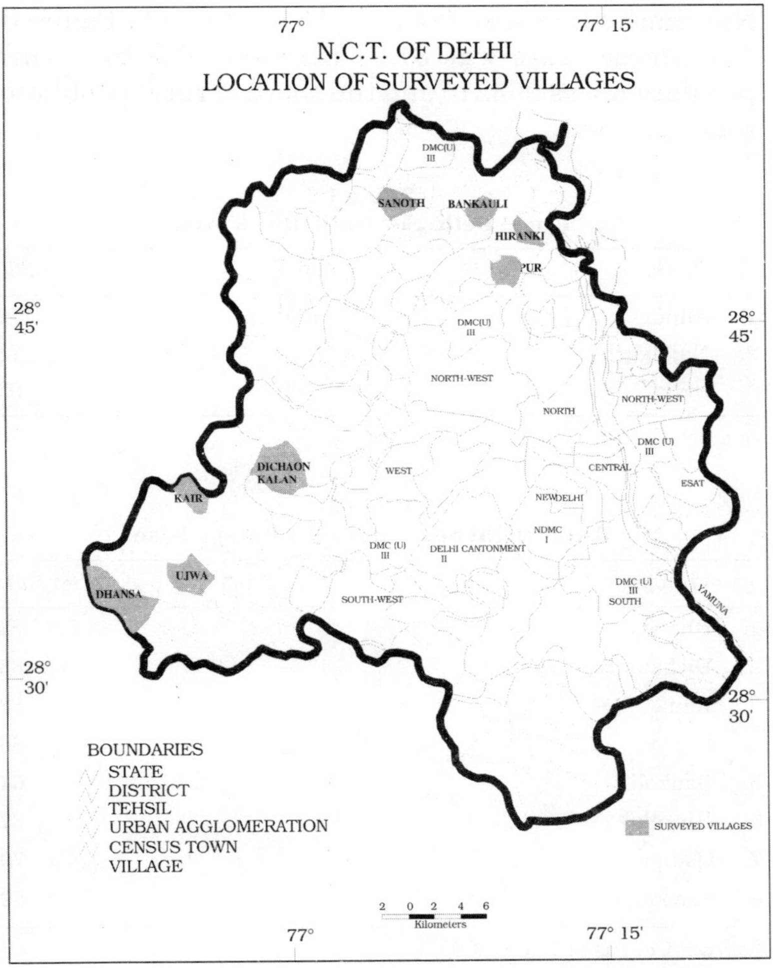

Figure 2: Surveyed Villages of Rural Delhi

Sex Ratio

In 2001 Census, the total population of N.C.T. of Delhi was 138, 50,507. The rural population of N.C.T. of Delhi has grown from 17, 44,072 with 304 inhabited villages in 1951 to 138, 50,507 with 165 inhabited villages in 2001. The sex ratio of N.C.T. of Delhi was 821 females per 1000 males. But the sex ratio of Alipur block was 799 and

Najafgarh block was 784 only (Table 1 and 2, Figure 3). The literacy rate was 81.97 per cent. The total rural population was 963215 and the share of rural population was 7 per cent.

Table 1
Rural Delhi Sex Ratio Block Wise

	Blocks	*1991*	*2001*
1.	Alipur	785	799
2.	Najafgarh	820	784
	India	924	933

Source: Census of India, 1991, 2001

Table 2
Rural Delhi Sex Ratio 2001 Village Wise

	Villages	*Sex Ratio*
1.	Dhansa	875
2.	Dichaun Kalan	830
3.	Kair	839
4.	Ujjwa	863
5.	Bankoli	648
6.	Hiranki	879
7.	Qadipur	795
8.	Sanooth	898

Source: Census of India, 2001

Women and Land

In rural economy, agriculture occupies a place of pride. Almost all the persons living in the village are in one way or the other connected with agriculture. It can be considered to be the pivot or focus of rural economy (Sharma, 1979). Of the total Indian population of about 1028 million (Census of India, 2001), nearly 827 million live in villages. Women constitute 49 per cent of the rural

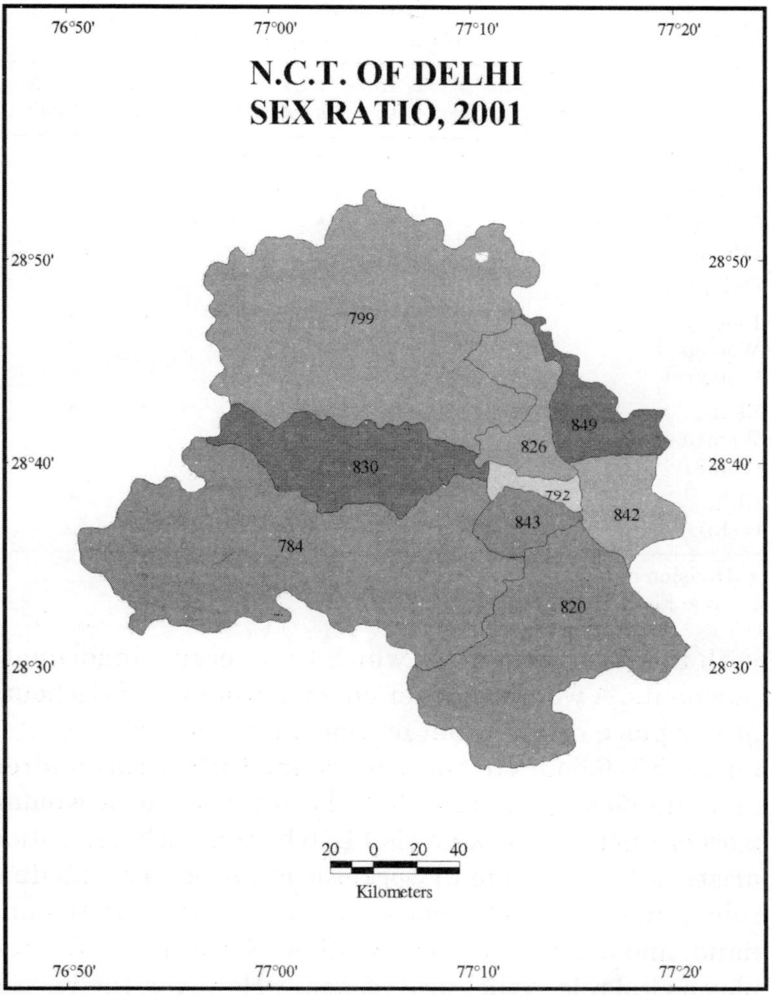

Figure 3: Sex Ratio of Delhi

population. About 75 per cent of the rural women belong to the families of small and marginal farmers (Table 3).

The average hour's daily work for the agriculture is about 12. On an average a women devotes between 11-12 hours a day to her work in fields, in addition to her normal household activities. Farm labourers have often to work

Table 3
Division of Labour in Agriculture

	Ploug-hing	*Sow-ing*	*Weed-ing*	*Har-vest*	*Thres-hing*	*Winno-wing*	*Stor-age*	*Milling, Cooking*
Elderly Men (>50 yrs)	•	•		•	•			
Young Men (15-50)	•	•		•	•	•	•	
Child (<15)	•				•			
Elderly Women (>50 yrs)					•		•	•
Young Women (15-50)			•	•	•	•	•	•
Child (< 15)						•		•

• Division of labour
Source: Field Survey

with the farm owners for which they receive inadequate payments. A women worker not only works for 8-10 hours at her place of work, but on reaching home, she devotes another 5- 6 hours to house work and the care of children and the sick. It means that the duration of a woman worker's normal work-day is 14-16 hours. Such a situation arises because a woman worker is required to play a dual role namely, that of a housewife and mother on the one hand, and of a worker, on the other (Sethi, 1982).Women play a vital role in the rural economy. However, their socio-economic status remains marginal. Agriculture is the major economic activity for women (Figure 4).

The landless households have no land to guarantee subsistence and the men's income is not sufficient to meet the family's needs therefore, it becomes necessary for the women to supplement the income and become agricultural labourers. Some households have sufficient land for their own subsistence. They do not hire labour but the women

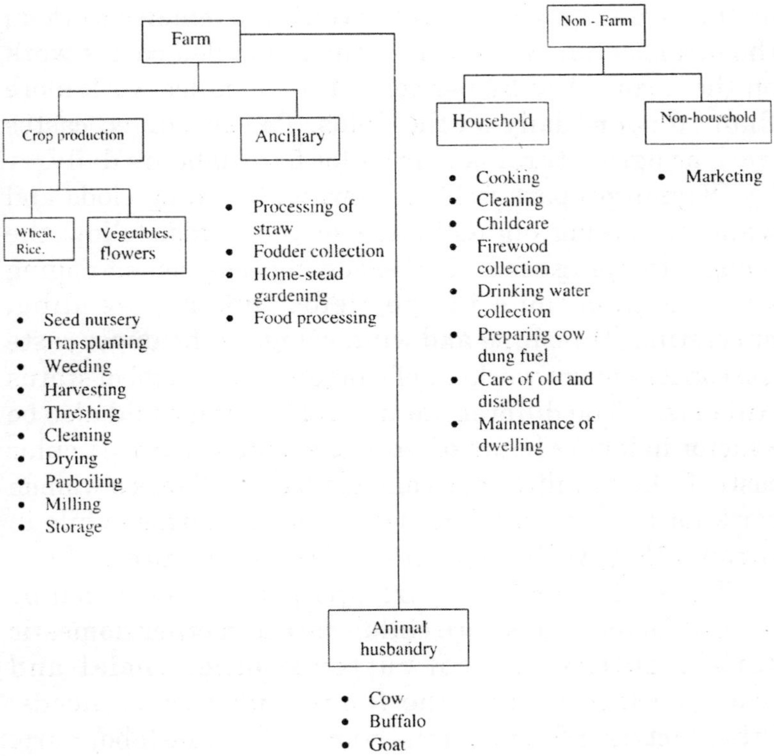

Figure 4: Facets of Farm Women's Work, Source: Mitra, B. *et al.* 1990

are engaged in agricultural work on their own farms. Large cultivators have more land than they need for their own subsistence and so they can afford labour force from outside. In upper castes the concept of not working of women as a symbol of status still exists.

The main occupation of the females is agriculture. They work on the farms as cultivators or as agricultural labourers. Woman work in the farms mainly during the sowing and harvesting season. In Alipur block 52 per cent of the females do agricultural work. And 44 per cent of the females work 5 to 10 hours daily on the farms. Among the different communities 19 per cent Jat females work on the farms, 11 per cent of the Brahmin females work on the farms, while 8 per cent of the scheduled castes work on the farms. And 22 per cent of the females work more than 10 hours daily on the fields. The scheduled castes work as agricultural labourers for 5 to 10 hours daily.

Woman prepare field for sowing, breaking clods and wallets, making embankment of the fields, follow the plough dropping seeds during sowing seasons, managing water applications in the fields, hoeing, weeding, harvesting , threshing and winnowing. In the upper caste the concept of not working of women as a symbol of status still exists. The differential in land holding could also be a factor in longer hours of work the female of a particular caste. In Bankauli, Sanoth and Qadi Pur villages a women work for more than 8 hours daily on the farms, while in Hiranki they work between 4 to 8 hours (Figure 4).

Female labour force participation is conditioned by several factors; these are child - care and other domestic duties, nutrition, local customs, other social and economic factors, and the family subsistence needs. Other factors influencing the supply of female labour are: the extent of mobility allowed to women, the availability of suitable jobs, the desire on the part of women to avail job opportunities, educational attainments and income levels and the husband's occupational status. Interplay of these factors determines the overall participation rates for the community as a whole and for the women specifically.

Rural women are primarily involved in land based activities - agriculture and forestry. They were the early practitioners of agro forestry. They worked with the hoe, axe to turn the earth and prepare the nurseries for the paddy seedlings. They carried out transplantation; weeding, harvesting, and irrigation were organized by women. They were skilled in land and water management. Mixing cow dung with leaf mould in specially constructed ditches produced organic manure. The women applied the manure when the fields were prepared for planting. Crop rotation was known and practiced and contemporary authors listed a large variety of crops, including several varieties of paddy.

Female participation rates in the farm work are high as they perform every kind of field labour except ploughing and carting. The main occupation of the females in Alipur block is agriculture, service and agricultural labourers. A total of 65 per cent of the females are engaged in agricultural work as cultivators on their own land. And 21 per cent of the Jat females are involved doing agricultural activity or animal husbandry or both the activities. Only 10 per cent of the Brahmins are involved in agricultural activities. The scheduled caste females were employed as agricultural labourers. They worked 8 to 10 hours daily on the farms of Jats or Brahmins and were paid Rs. 25 to Rs. 40 per day. The major crops grown here are - *Wheat, Bajara, Jowar, and Mustard*. A total of 30 per cent of the females do household chores and most of these females belong to the scheduled caste. The lowest percentages of women are confined in the service class.

In larger land holdings Grain Farming is important which has mechanized, so role of women in farming is less, like in Hiranki. Whereas in Qadi Pur which is located very near to urban Delhi, 60 per cent of the women respondents were engaged in farming and animal husbandry activities. There is more Floriculture in Qadi Pur so more women are engaged in farming with 27 per cent of women working more than 8 hours on farm. In Hiranki which is located nearby river Yamuna, 80 per cent of the women were engaged in farming and animal husbandry activities. Here more than 40 per cent women were landless as the population of Scheduled Caste was more. The Jat community had only 2-4 acres of land with 30 per cent respondents. And about 45 per cent of women worked more than 4 hours per day on farms. In Bankauli near Alipur 64 per cent of women were employed in farming and animal husbandry of which 60 per cent were landless as Bankauli has Scheduled Caste and Other Backward Class families who were employed as agricultural labourers. In Sanoth 65 per cent of the women were engaged in farming and animal husbandry activities with 46 per cent landless with Scheduled Caste and Other Backward Class families as agricultural labourer. About 30 per cent had less than 2 acres of land (Plate 1).

In the Najafgarh block agriculture and animal husbandry is main occupation of the female population. A total of 75 per cent of the females are engaged in agriculture, animal husbandry or both or as agricultural labourers. Within the Jat community 42 per cent women are doing either agricultural activity and animal husbandry or both. And in the Brahmin community only 12 per cent is engaged in agricultural activity or animal husbandry or both the activities. The scheduled castes have 9 per cent females doing agricultural activity. The scheduled castes female agricultural labourers work on the farms of the Jats and Brahmin family. They work during the sowing and harvesting season and are paid between Rs. 25 - Rs. 50 per day depending on land owner. Wheat, Jowar, Bajara, Mustard are the main crops grown here. Only 2 per cent of the OBC's are engaged as potters and some as agricultural Labourers. The female potters make 10 to 15 pots a day and sell them at a price of Rs. 10 to Rs. 14 per pot. Mostly the villagers bought these pots and that too in summers only. During the winter season when no pots were made or purchased these potters work on the farms of Jats and Brahmin. Nearly 20 per cent of the females were doing household work. Agriculture and animal husbandry is the main occupation of females in Dhansa, Dichaun Kalan, Kair and Ujwa. Animal husbandry is dominant in Dhansa village only (Plate 2).

In Najafgarh block, Dhansa village which is located near Haryana border has more than 60 per cent of women engaged in agriculture and animal husbandry activities. The size of land holdings was small i.e. 2-4 acres only with more than 50 per cent landless. The percentage of female respondents working on farms was also low with only 25 per cent . Dichaun Kalan village which is located near Najafgarh has more than 80 per cent of women engaged in agriculture and animal husbandry activities. About 25 per cent have 2-4 acres of land with more than 70 per cent women working on farms. In Ujwa more than 75 per cent of women are engaged in agriculture and animal husbandry activities. Here more than 25 per cent have more than 4 acres of land and 50 per cent women worked more than 4 hours per day on farms. In Kair village located near Haryana border 65 per cent of women are engaged in agriculture and animal husbandry activities with more than 40 per cent women working 4-8 hours daily on farms. About 40 per cent women were landless (Figure 5 and 6).

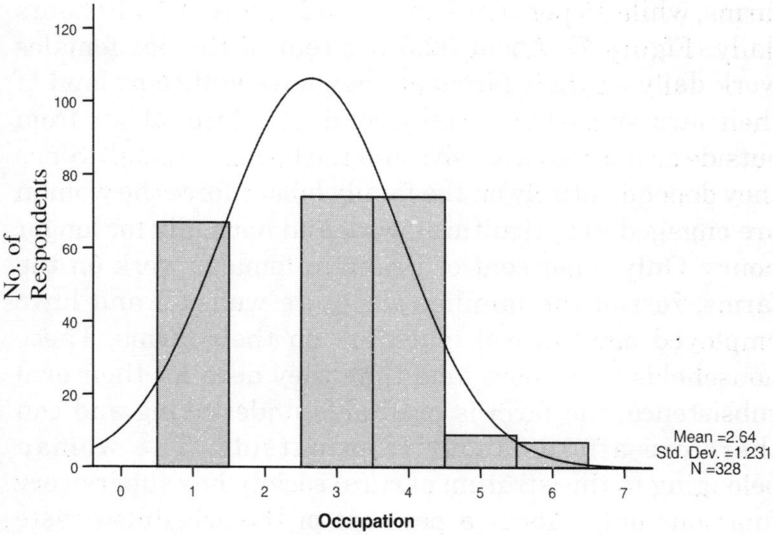

Figure 5: Women's Occupational Structure

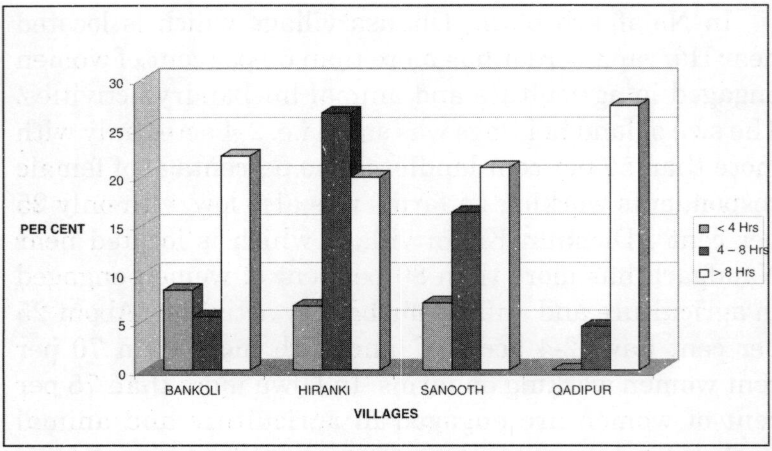

Figure 6: Women's Working Hours on Farm (ALIPUR BLOCK)

In Najafgarh block a female works 8 to 10 hours per day in the fields. Nearly 53 per cent of the females work on the farms from morning till the evening. And 38 per cent of the females work nearly 5 to 10 hours daily on the farms, while 15 per cent females work more than 10 hours daily (Figure 7). About 32.5 per cent of the Jat females work daily on their farms as they have sufficient land of their own subsistence and they do not hire labour from outside as it is too expensive and the farms too small. Since they depend entirely on the family labour force the woman are engaged in agricultural work and normally for longer hours. Only 8 per cent of Brahmin females work on the farms, rest of the families are quite well- off and have employed agricultural labourers on their farms. These households have more land than they need for their own subsistence, the farm is profitable undertaking and can therefore afford labour from outside. The woman belonging to this stratum of rural society has supervisory functions only. About 5 per cent of the scheduled caste and the OBC woman females work as agricultural labourers on the farms 5 to 10 hours daily. Such houses are landless households and have no land to guarantee subsistence and the men's income is not sufficient to meet the family's need and therefore, it becomes necessary for the woman to supplement the income and become agricultural labourers.

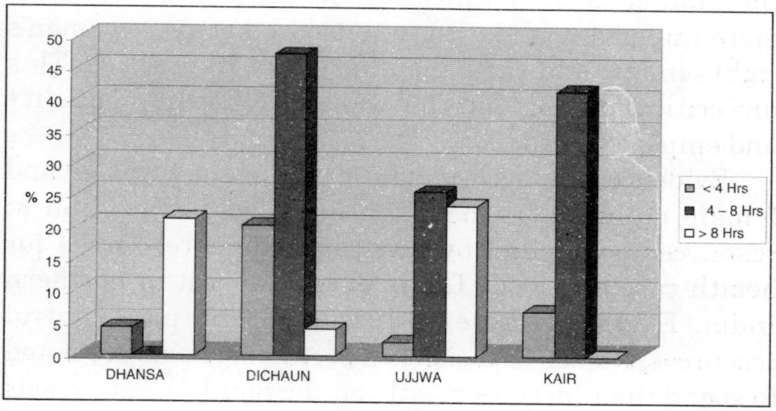

Figure 7: Women's Working Hours on far (NAJAFGARH BLOCK)

Land Rights: The single most important economic factor affecting women's situation is the gender gap in command over property, an issue virtually neglected in research, policy and grassroots action. In rural Delhi, the most significant form of property is arable land, a critical determinant of economic well-being, social status, and political power. Yet few women possess a field of their own. Despite the symbolic association between women and land and the widespread cultural perception of earth as mother, women hardly own any land. Women produce majority of the food, but they share limited control over, ownership of, and access to land (Sachs, 1996). Rights are defined here as claims that are legally and socially recognized and enforceable by an external legitimized authority be it a village-level institution or some higher-level judicial or executive body of the State. Rights in land can take the form of ownership. Today most arable land is privatized; in India, about 86 per cent is private and 89 per cent of rural households own some land.. Inheritance laws vary by religion and region, and although they give women today much greater rights than did custom, significant inequalities remain. Even more important is the enormous gap between women's rights in law and their disinheritance in practice. This has critical implications for welfare, efficiency, equality and empowerment.

Evidence shows a systematic bias against women and female children in the intra-household allocation of resources controlled by men, including resources for health care and food. The bias is strongest in northern India. Further, where both women and men control resources, women especially in poor households are noted to spend their incomes mostly on the family's basic needs and men in greater part on Field needs. Direct access to economic resources would thus reduce both women's own and the children's risk of poverty. Indeed, even women of rich households can face poverty following marital breakdown. For Female -headed households with little or no male support (estimated to be about 20 per cent of households in India and Bangladesh), the link between direct resource access and physical well-being needs no emphasis.

Among female-headed rural households a lack of land titles reduces access to credit, inputs, and information on new technologies. Land titles could motivate and enable women farmers to increase output by adopting improved technique and practices. Titles for women become especially important given a growing feminization of agriculture, as more men shift to non-farm employment. Supporting women farmers would also enlarge the information base of farming in many regions. Women know more than men about indigenous seed selection and cultivation methods. Apart from gender equality being a measure of a just society, more equal property rights would enhance women's bargaining power within the family and community. Women owning even small plots, in contrast to landless women, feel more respected by family members and by villagers, are better able to challenge social inequities, and have a greater say in public decision-making bodies. In other words, equality in land rights is linked with women's empowerment, defined here as a process that enhances the ability of disadvantaged ('powerless') individuals or groups to challenge and change (in their favour) existing power relationship that place them in subordinate economic, social and political positions.

> "Advancing gender equality, through reversing the various social and economic handicaps that make women voiceless and powerless, may also be one of the best ways of saving the environment, and countering the dangers of overcrowding and other adversities associated with population pressure. The voice of women is critically important for the world's future—not just for women's future."
>
> **—Amartya Sen**

Women and Water

"Water, water everywhere, yet not a drop to drink". Samuel Taylor Coleridge (As quoted in Rogers, and Feiss, 1998)

The Ancient Mariners immortal lines have a nightmarish reality in today's world. Water as is well known is one of the five basic elements of Nature held sacred by our own scriptures. While all environmental concerns are important, there is a growing realization that water is the most important common element in environment. In a very real sense water is life (Figure 8).

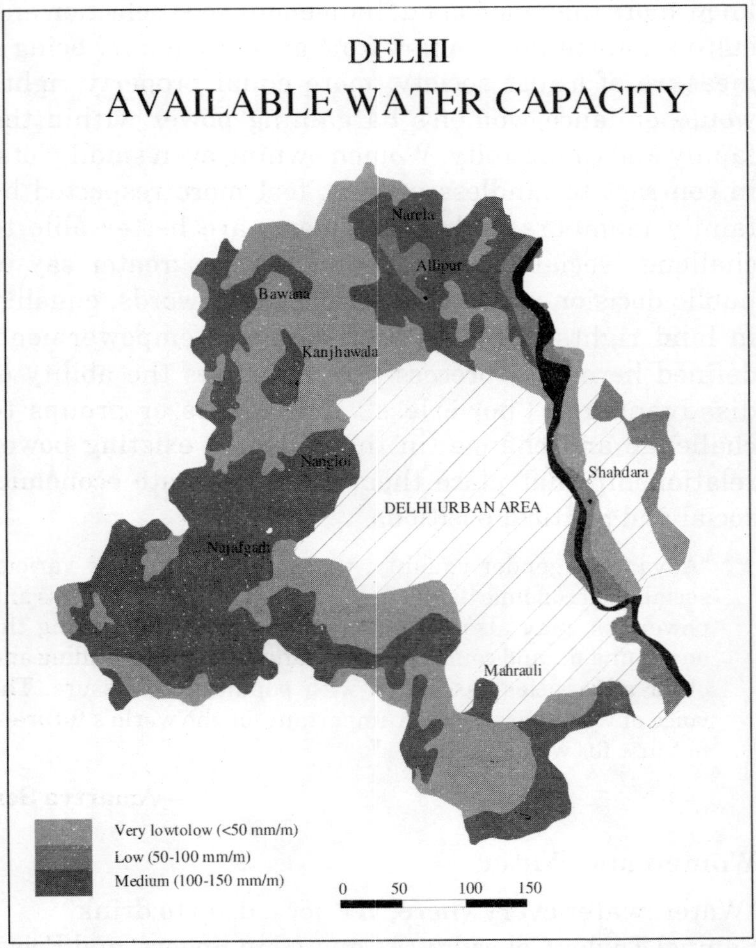

Figure 8: Available Water Capacity in Delhi

Water Collection Women are the main carriers of water. It takes up much of the women's energy intake, as they have been reported to carry as much as 5 to 10 Kg weight of water in a single trip. The women of the household collect the water for household consumption i.e. for cooking, washing and bathing only. Tap within the house is the main source of water. In addition, hand pump is also the source of water for household consumption. But the villages generally do face an acute shortage of water (Table 4). The respondents reported that some times supply of water was suspended for as long as three days and the females had to travel 2 -4 kms. daily to collect water for household consumption. In one of the villages, to meet the water shortage, the Jat community had ordered a water tanker separately for their community only. There were frequent fights over water. Majority of the women i.e. 88 Per cent in Alipur Block collected water daily from their household taps or hand pumps. But 12 Per cent of the female respondents had to travel up to 3 km. (Figure 9) daily to fetch water for household consumption. In the summer season the region faces acute water shortage when the taps of the households go dry. At this time the higher class communities like the Jats, Brahmins and Rajputs arrange for Water Tankers of the Delhi Jal Board. It is a typical sight to watch village women, children carrying buckets, cans, jars etc. to gather around the tank hustling to collect water. Sometimes there are little disputes also over who will collect water first and how much. But the poor communities like the Scheduled Castes, Scheduled Tribes and Other Backward Classes have to face acute drinking water shortage as they cannot afford to call water tankers (Plate 3 & 4).

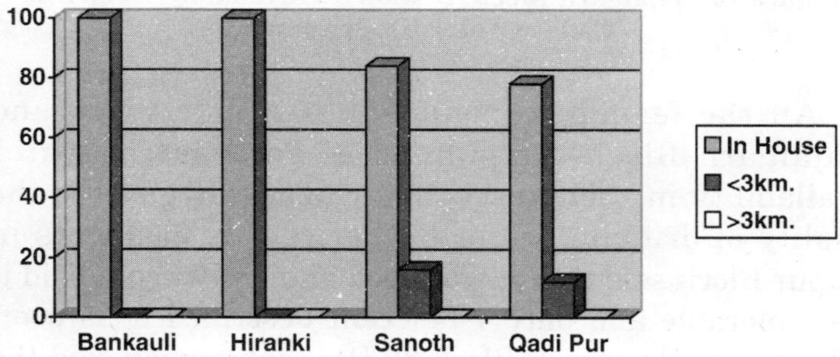

Figure 9: Alipur Block: Distance Traveled by women to Collect Water (in Percent)

Women play an important role in domestic water provisioning, but a gender analyst is also interested in relations between women and other women and here we find that relation of power between women in households frequently gives rise to patterns of delegation of water work within the households (Figure 10). Older women characteristically shift the more strenuous tasks to younger women, daughters before their marriage and young wives of sons, who are expected in many cultures to obey their mother-in-law (Merry, and Bainskar, 1998,). In Najafgarh Block 77 Per cent of women collected water from a source within their house itself, but 23 Per cent of them had to travel 3- 5 km. every day to find and collect water for the household.

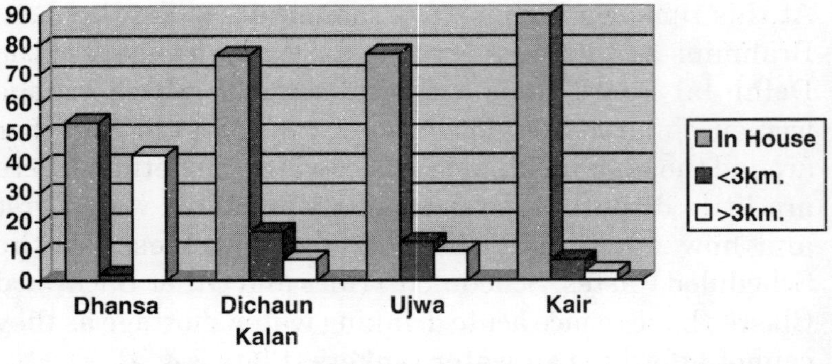

Figure 10: Najafgarh Block: Distance Traveled by Women to Collect Water (in Per cent)

All the female respondents of Alipur Block and Najafgarh Block complained of decrease in water availability in their respective villages. Regarding the quality of drinking water 63 Per cent respondents in Alipur Block said that it was good and 34 Per cent said it was tolerable and only 4 Per cent described it as poor. Whereas in Najafgarh Block 30 Per cent women said the quality of water was good, 22 Per cent as tolerable, a substantial 48 Per cent responded that the quality of water was poor (Figure 11 and 12).

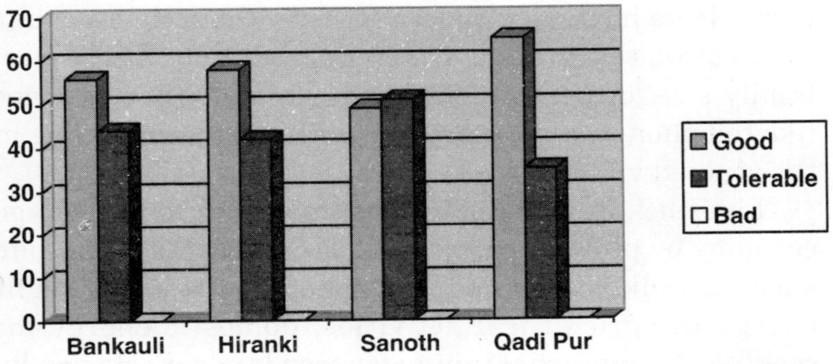

Figure 11: Alipur Block: Quality of Water (in Per cent)

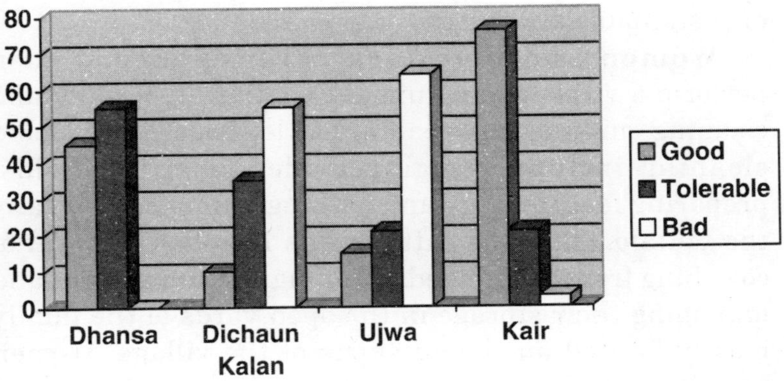

Figure 12: Najafgarh Block: Quality of Water (in Per cent)

The impact of water availability can be gauged through the lower female literacy rate and almost an absence of working rural women. The awareness levels of women from both Alipur and Najafgarh were found to be high because of the accessibility of Cable T.V. The women of Alipur had a higher literacy rate, because of easy availability of water which left them with enough time to engage in education and other vocations. In Delhi's fringe area, it was observed that there is a direct relationship between the water availability and income levels of the farmers as well as their general social development. Four parameters have been chosen to study the socio-economic implications of ground water degradation. These are: family size, ownership of tube wells and other luxuries like telephones, cars, etc. along with the income levels in the studied villages.

Livestock in India plays multiple roles in the village economy by providing products for consumption and sale such as milk, milk products, meat and wool, draught energy in agricultural activities, domestic energy for cooking and manurial inputs for crop lands. Ecologically, the cow has been central to Indian civilization. The integration of livestock with farming has been the secret of sustainable agriculture (Shiva, 1988).

Women and Livestock Maintenance Women perform a vital role in animal care that is, feeding and cleaning, milking and cleaning cattle sheds. Feeding and cleaning include bringing fodder from the fields, preparing feed, feeding and bathing animals, especially the buffalo. Cleaning cattle sheds included removal of cow dung from cattle sheds, making cow dung cakes and managing their storage in the open yards of the family mostly located on the outskirts of the village. Women have been the Field experts in animal husbandry as well as the food processors in the traditional dairy industry, making curd, butter, ghee and butter milk. But also with the dwindling population of milch cattle that expertise is not fully utilized now (Shiva, 1988).

Women's contribution to the economy of the village through their work with cattle has been very significant. In Alipur block 60 per cent of females spend some time during the day looking after their livestock. Only 16 per cent of the female respondents spend less than 4 hours daily, 34 per cent females 4-8 hours daily and 11 per cent spend more than 8 hours daily looking after their livestock's. While 40 per cent of female respondents spend no time as they don't have livestock's. And 8 per cent of the Jat females spend more than 8 hours daily looking after the livestock, while 7 per cent of the Scheduled Caste females spend less than 8 hrs looking after livestock, because they usually have small animals like goats as compared to Jats or Brahmins who have bigger animals like cows or buffaloes that need more time in maintenance. Some of the female respondents who owned livestock looked after the maintenance of the livestock entirely by themselves. Usually a household had 2- 4 Buffaloes and a female spent nearly the whole day in this activity (Plate 5 & 6).

Table 4
Women's Role in Livestock Management

Activity	*Gender*	*Time*
1. Clearling of Shed	Female	Daily
2. Bathing Animals	Female	Daily
3. Feeding Fodder	Female	Daily
4. Milking	Female	Daily
5. Grazing	Male / Female	Daily
6. Storage of Milk	Female	Daily
7. Taking Animals in & out of Shed	Female	Daily
8. Giving Water to Animals	Female	Daily
9. Care of Lactating Animals	Female	Daily
10. Care of Sick Animals	Female	Daily
11. Selling Milk	Male / Female	Daily
12. Making Cow dung Cakes	Female	Daily
13. Storage of Cow dung Cakes	Female	Daily

Source: Guhathakurta and Sarin, (1991)

Women possess good knowledge of various aspects of livestock production management and particularly of feed resources. They know each animal's production characteristics, temperament and feeding behaviour. Most women were aware of the need for good quality feed to achieve better production but felt that feeding a non-productive animal with quality feed was not necessary. In any case such cattle make up the deficiency on their own while they were herded in the jungle. Women perform a vital role in animal care that is, feeding and cleaning, milking and cleaning cattle sheds. Though women play a significant role in livestock management and production, women's control over livestock and its products is negligible. They need to have more control over sale and purchase of cattle and their milk and milk products. Around 25–30 per cent of the milk is sold collectively both by men and women. And 30 to 40 per cent of cattle is sold and bought collectively (Plate 7).

In Alipur block 54 per cent of the females collected fodder for their animals. Among the Jat community 21 per cent of the female respondents collected fodder. A total of 15 per cent of Scheduled Caste women collected fodder. In the Najafgarh block 70 per cent of the female respondents collected fodder, while only 2 per cent males of the house is involved in this activity. Only 2 per cent of the servants in the households collected fodder. Among the Jats 43 per cent of females went for collection of fodder and only a per cent of servants were involved in this activity. In the Brahmin community 11 per cent of the females of the household collected fodder, only one household had a servant. The Scheduled Caste and OBCs, all had females of the households collecting fodder for their animals. The purpose of raising cattle in Alipur Block was mainly domestic with 29 per cent of respondents, while in Najafgarh Block it was 45 per cent households (Figure 13 and 14).

Livestock in India plays multiple roles in the village economy by providing products for consumption and sale such as milk, milk products, meat and wool, draught

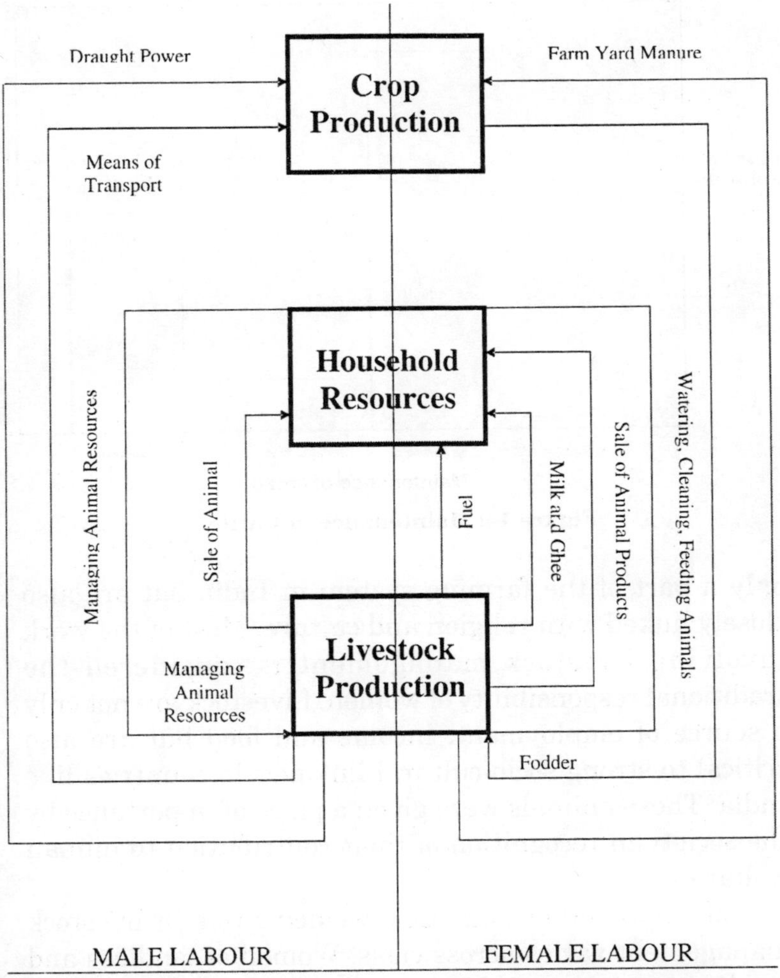

Figure 13: Environment, Livestock Rearing and Women, Gender Division of Labour within the Farm Livestock Interface

Source: Freedman and Wai, 1988

energy in agricultural activities, domestic energy for cooking and manurial inputs for crop lands. Ecologically, the cow has been central to Indian civilization. The integration of livestock with farming has been the secret of sustainable agriculture (Shiva, 1988). Livestock are not

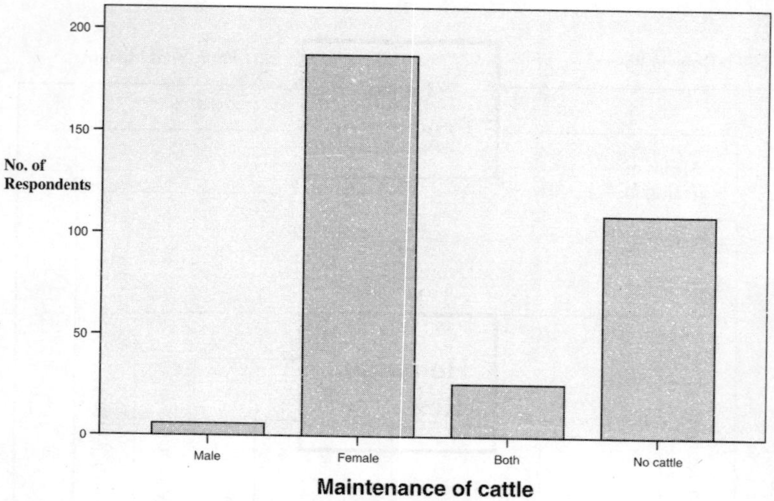

Figure 14: Maintenance of Cattle

only a part of the farming system in India but are also closely linked with religion and culture. Most of the work involving livestock management is considered the traditional responsibility of women. Livestock are not only a source of employment, income and food but are also critical to strong socio-cultural linkages in countries like India. These animals were given a place of importance by the society in recognition of their contribution to human welfare.

Like most other activities, women's role in livestock management varies across class. Women of landless and small-peasant households perform all the activities listed in above table, whereas women of medium farmer households perform tasks confined to the house, such as feeding and milking. Women of rich farmer households, on the other hand, perform mainly supervisory functions (Venkateswaran, 1995).

In the Najafgarh block 77 per cent of the females spent 4-8 hours daily looking after their livestock. Almost 43 per cent of the females of the Jat community spend time maintaining their livestock. While 5 per cent spend more than 8 hours daily. But only 13 per cent of the females of the Brahmin community spend time looking after their livestock (Figure 15).

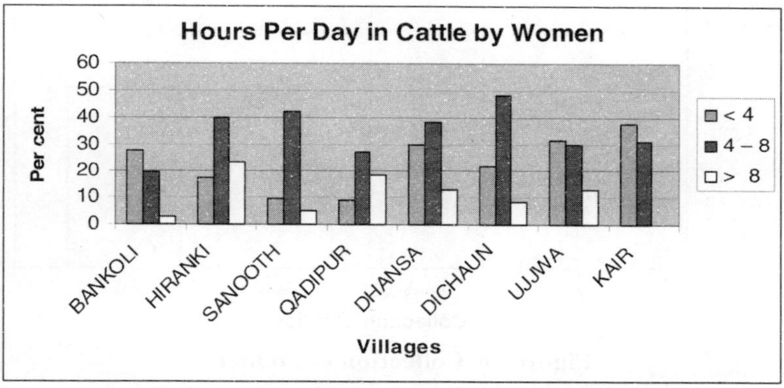

Figure 15: A Women's' Time Spent in Cattle Care (hours per day)

Conservation of feed and fodder is usually discussed by most technicians as a new technology (to be transferred to the farmers). However, while studying the traditional practices of feeding animals it was noticed that conserving supplementary feeds for use during the dry season is a traditional practice. Through their experience, women have identified supplementary feeds beneficial to animal productivity. Most of these are available seasonally and are generally dried and stored for use in the dry season. Collection of fodder for animals is primarily the activity of a women only in the rural households (Figure 16).

Utilization and Management of Domestic Energy by Women Self-reliance in energy supply is a vital factor in the development of any nation. Energy from biomass holds a promising scope under Indian conditions because this sector encourages self-reliance through efficient use of indigenous resources and employment of simpler technologies consistent with minimum pollution hazards.

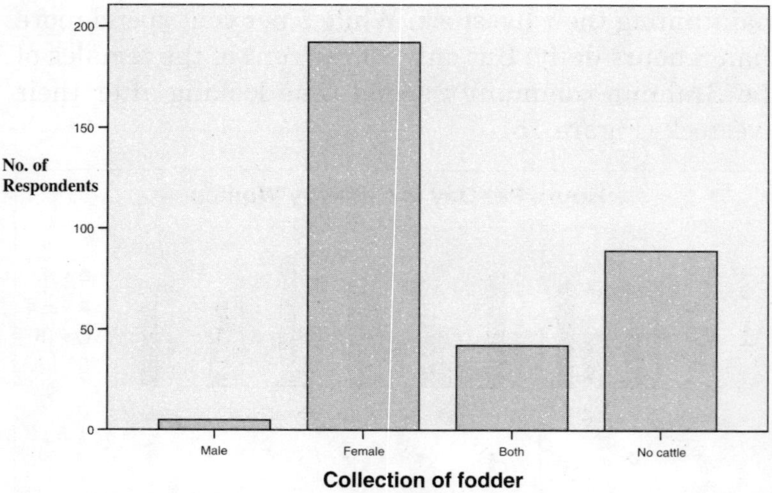

Figure 16: Collection of Fodder

Energy consumption is an index of a country's economic development. In rural areas, firewood constitutes to be the major source of fuel, followed by animal dung (Table 5).

Table 5
Patterns of Energy Use in Rural Delhi

	Activities	*Non-Commercial Energy*
1.	Domestic (cooking, housework)	Firewood, cow dung, agricultural waste
2.	Agricultural	Human / Animal Energy, cow dung manure
3.	Non-agricultural work (Pottery, Brick making)	Firewood, Charcoal, Human Energy

Source: Field Survey

The disposal rate of dung collected in the household varies from season to season. In rainy season, 85 per cent of the dung collected is diverted into manure pits and only 13 per cent is converted into dung cakes vis-à-vis 58 per cent diverted into manure pits and 40 per cent converted into dung cakes during winter (Table 6).

Table 6
Utilization of Dung Collected in Rural Households having Bovines (Season Wise)

Seasons	*Percentage of Total Dung (Collected)*		
	Thrown into manure pits	*Converted into dung cakes*	*Used for other purposes*
Summer	66	32	2
Rainy	85	13	2
Winter	58	40	2
Overall	69	29	2

This is because in winter, the requirement of domestic fuel is high and the cultivators find it a cheap fuel. In rainy season, it becomes difficult to prepare and dry the dung cakes. During summer, about 1/3rd of the total dung is converted into cakes to meet the fuel requirements (Plate 8).

Cooking Mode in the Households Domestic energy for a vast majority of rural homes derives primarily from biomass fuels as by products of forestry, livestock and agriculture. Biomass fuels comprise of wood, crop residue and dung cakes. A substantial proportion of firewood is collected from local forest. Rural households commonly used twigs and small branches for fuel. Women labourers often take part of their wages in agricultural residues that can be used as fuel (Venkateswaran, 1995). The major source of collection of firewood is the own farms of the respondents with 65 per cent of households, and forests with only 10 per cent. Source of wood fuels consists of stems, branches, twigs, saw dust and other residues from logging and wood processing activities (Table 7).

Table 7
Major Sources of Collection of Fire wood in Alipur and Najafgarh Blocks

	Sources	*Per cent of Households*
1.	Own Farms	65
2.	Roadside bushes, trees	20
3.	Forests	10
4.	Others	5

Source: Field Survey

Apart from wood, agricultural land produces biomass residues, part of which is available as fuel on an environmentally sustainable basis. Villagers rely predominantly on twigs, roots and thin branches, gathered mainly by women and children. Firewood is a biomass resource, but it is being extracted in a non- renewable manner. The origin of the problem lies in the fact that the bulk of the firewood is consumed in the poorest households for cooking and heating (Reddy, 1983). Traditionally women have gathered products from the trees and other plants, products which have provided them with the basic three 'Fs' of fuel, food and fodder, and for variety of other uses (Figure 17). The source of most of domestic energy used in rural households is provided by the burning of biomass and it is the women who are mainly responsible for its collection. To do so, the women walk long distances and carry heavy loads. The time spent will depend on the availability of the supply (Rodda, 1994).

Women cooking in rural kitchen with biomass fuels are exposed to pollutants found in biomass smoke, such as carbon monoxide, nitrogen oxide and suspended particulates. Prolonged exposure to these pollutants has adverse impact on the health of women, who spend between 3 to 5 hours daily cooking with biomass fuels (Table 8). The burning of the cooking fuel envelops the indoor environment with heavy smoke. As women have

Table 8
Health Effects of Biomass Fuels

	Biomass Fuels	*Health Effects*
1.	Effects of Smoke (Acute)	Conjunctivitis, Upper Respiratory Irrigation/ Inflation, A cute Respiratory Infection.
2.	Effects of Toxic Gases (CO)	Acute Poisoning
3.	Effects of Smoke (Chronic)	Chronic Bronchitis, Lung Cancer, Adverse Reproductive Outcomes
4.	Effects of Heat : Acute	Burns
	Chronic	Cataract
5.	Ergonomic effects of Crouching Over Stove	Arthritis

Source: WHO, 1991.

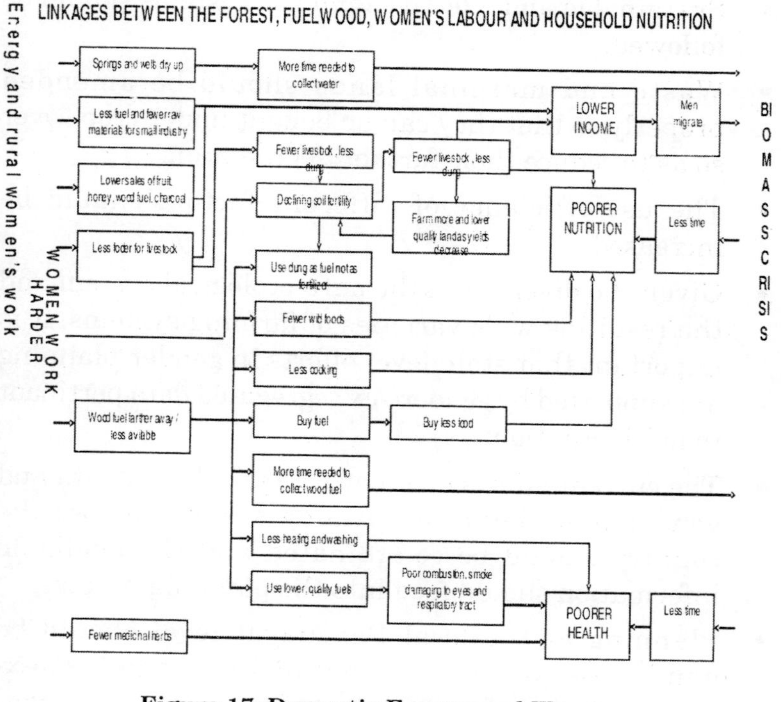

Figure 17: Domestic Energy and Women

Source: Cecelski, 1987

to do all the cooking daily the effects could include eye problems, respiratory problems, chronic bronchitis, and lung cancer (Sugumar, 1990). Many studies conducted by World Health Organisation (WHO) have revealed that the concentration of pollutants inside the Indian kitchen is much higher than the prescribed standards. Most of the women keep their children in the kitchen/cooling area while they are busy in cooking, thereby exposing the children to the prevailing pollutants there (Table 8).

Conclusion

- Proper records of land productivity status should be prepared with the help of soil scientists along with the latest technologies like remote sensing;
- Sustainable and effective land use systems should be followed;
- Waste and marginal lands should be amended properly so that they can be bought under crop cover so as to reduce their further degradation;
- The use efficiency of different factors should be increased;
- Given the diversity of the agro-ecological system and the resulting wide varieties of farming systems, it is important that state level efforts in gender planning are supported by gender/sex segregated data pertinent to local situation;
- The current efforts on creating agricultural data and gender information in farming systems across the country should be co-ordinated and the available information should feed into the planning process;
- Planning at the local Panchayat level should be gender-sensitive and should have gender/sex-segregated information to support for local development efforts;

- Agricultural education institutions and training centers should develop regular curricula to integrate a gender approach in all technical areas of agricultural sector;
- Existing policies related to targeting women and gender-equity should be widely communicated to all involved development work including women clientele;
- Review of the land right policies and implementation of land distribution for gender equity and women's land rights;
- Farmer's right initiatives should explicitly address concerns of women farmers;
- Develop programmes to train Panchayat leaders in gender-integrative participatory approaches and local planning;
- Develop programmes for rural women to build leadership skills for managing agriculture community-based development activities;
- Develop economic and political literacy programmes for rural women to utilize development inputs effectively;
- Support agriculture extension in-service training programmes and other relevant agricultural curricular activities to support gender-sensitive extension agents;
- Put in place appropriate institutional mechanisms for information exchange and cooperation between the agencies that focus on women's affair and agriculture-related technical activities;
- Support women-managed rural production and marketing ventures in horticulture, floriculture and post-harvest processing in commodities;
- Provide technology training and input support to women to take advantage of emerging high-value agribusiness sector including bio-technology and forest products.

- Roof top rainwater harvesting, Uproot concrete surfaces, and re-lay them with more permeable material
- Create percolating tanks, Revive dead tanks, Recycling waste water
- Rural strategies, Rotation of crops, Reduce dependence on chemicals/fertilizers
- Promote sensible water conservation/ management

References

Aggarwal, Bina (1994), A Field of One's Own: Gender and Land Rights in South Asia, Cambridge University Press, Cambridge.

Cecelski, E. (1987), Energy and Rural Women Works: Crisis, Response and Policy Alternatives, International Labour Review, 126: 1, Jan. Feb.

Census of India (1991), Delhi District Census Handbook, Town and Village Directory, Primary Census Abstract, Delhi.

Census of India (2001), Delhi District Census Handbook, Town and Village Directory, Primary Census Abstract, Delhi.

Centre for Science and Environment (1999), State of India's Environment, The Citizens' Fifth Report, Part II: Statistical Data Base, New Delhi.

Desai, Neera and Usha Thakkar (2001), Women in Indian Society, National Book Trust, New Delhi.

Druijven, P. C. J. and R. B. Singh (1996), Environmental Degradation and its Impact on Livelihood Strategies in the Urban Fringe of Delhi-Some Theoretical Reflections, in Disasters, Environment and Development, Ed. R. B. Singh, Oxford and IBH Publication, New Delhi.

Freedman, J. and Wai, I. (1988), Gender and Development in Barani Areas of Pakistan, Agriculture Canada Report.

Guhathakurta, Prabir and Sarin (1991), Gender and Forestry Case study in Himachal Pradesh, World Bank, New Delhi.

Hussain, Zahid (1996), Environmental Degradation and Conservation in North East India, Omsons Publications, New Delhi.

Krishnaraj Maithreyi (Ed.) (2007), Gender, Food Security and Rural Livelihoods, Stree an Imprint of Bhatkal and Sen, Kolkata.

Merchant, Carolyn (1995), Earth Care: Women and the Environment, Routledge Press, New York.

Merry, Douglas and Bainskar, S. (eds.), (1998), Gender Analysis and Reform of Irrigation Management: Concepts, Cases and Gaps in Knowledge, International Water Management Institute, Srilanka.

Mies, Maria (1986), Indian Women in Subsistence and Agricultural Labour, International Labour Office, Geneva.

Millican, J. and Smithells, J. (1992), Women Working for Women's Development: Barriers and Strategies in P. R. Reddy and P. Sumangala, (eds.) Women in Development, Perspectives from Selected States of India, Vol. 2, B. R. Publishing Corporation, Delhi.

Mitchell, B. (1997), Resource and Environmental Management, Longman Ltd., England.

Mitra, B. et al. (1990), Women in Rice Based Farming System. A Case Study in C. Prasad and Shri Ram (eds.), Women in Agriculture Technological Perspective, International Federation for Women in Agriculture, Delhi.

Nangia, Sudesh (2006), Women's Empowerment and Gender Equity, Presidential Address, Annals of the Association of Geographers, India, Volume XXVI, Number 1, June 2006.

Prabhakar, Vani (2004), Women in Rural India, Dominant, New Delhi.

Prasad, Kiran (2006), Women in Rural Development: Contemporary Social Policy and Practice, edited by, The Women Press, New Delhi.

Rao, N. and R'urup, L. (1997), A Just Right: Women's Ownership of Natural Resources and Livelihood Security, Friedrich Ebert Stiftung, New Delhi.

Rodda, Annabel (1994), Women and Environment, Zed Books, London.

Rogers, John J. W. and Feiss, P. Geoffrey (1998), People and the Earth, Basic Issues in the Sustainability of Resources and Environment, Cambridge University Press, U.K.

Sachs, Carolyn E. (1996), Gendered Fields: Rural Women, Agriculture and Environment, Westview Press, Colorado.

Sachs, Carolyn E. (1997), Women Working in the Environment, Taylor and Francis, Washington.

Sen, Amartya (1999), Beyond the Crisis: Development Strategies in Asia, Capital Publishers, New Delhi.

Sethi, Raj Mohini (1982), Female Labour in Agriculture: A Case of Punjab, Punjab University, Chandigarh.

Sharma, Ram Nath (1979), Indian Rural Sociology, Munshi Ram Manohar Lal Publishers, Pvt. Ltd. Delhi.

Shiva, Vandana (1988), Staying Alive: Women, Ecology and Survival, Kali for Women, New Delhi.

Sontheiwer, Sally (1991), Women and Environment, Earthscan Publication, London.

Sugumar, V. Ragi (1990), Smoke Exposure and Health Hazards, Yojana, Vol. 34, No. 3, Feb.

Swarankar, G. P. (1988), Women's Participation in Rural Environment, Chugh Publications, Allahabad.

Tripathy, S. N. (Ed.) (2006), Women and Rural Development, Sonali, New Delhi.

Venkateswaran, S. (1995), Environment, Development and the Gender Gap, Sage Publication, New Delhi.

WHO (1991), Epidemiological Social and Technical Aspects of Indoor Pollution from Biomass Fuel, WHO Report, Geneva.

Plate 1: Rural Women Cutting Wheat in Dichaon Kalan Village (Najafgarh Block)

Plate 2: Agricultural Labourers Working on the Farms in Bankauli Village (Alipur Block)

Plate 3: Women Going to Fetch Water in Najafgarh Block

Plate 4: Women Fetching Water from Water Tankers

Plate 5: Women taking Buffaloes to the Jhor

Plate 6: Women Carrying Fodder for Animals

Plate 7: Rural Women Cutting Fodder for the Animals

Plate 8: Women making Dung Cakes

Economic Growth in India: Structure and Trend

NITIN PUNIT

Introduction

Economic growth indicates the increasing ability of a society to produce goods and services to satisfy its wants. The most frequently used measure of economic growth is the rate of growth of real Gross Domestic Product (GDP) per person per year on the basis of national income accounts. It includes in itself all relative parameters of economic development.

The economic growth of a country is the manifestation of different productive sector-primary, secondary and tertiary. The primary sector is in direct contact with nature as it includes activities like, agriculture, forestry, fishing, mining, etc. Primary sector is the base of economic structure because it shows the natural resources stock of a region. Besides, it also provides the raw material to the manufacturing and construction activities to boost the secondary sector. In a true sense, secondary sector is largely dependent on the primary sector. For instance, the cotton textile industry needs a good crop of cotton to run their machines. Finally, the tertiary sector as service sector works to facilitate the activities of primary and secondary sector. It includes transportation, communication, insurance etc, and may have lesser direct impact on depletion of natural resources in comparison to primary and secondary sector. However, the transport sector is one of the important sources of pollution.

National income and its industrial origin mirror the macro flows of goods and services, and the economic structure of the nation. It is the money value of final goods and services produced in an economy over some period of time, usually a year. Final goods and services are those purchased for their own sake; they are what the economy finally wanted and obtained. Around the concept of national income, there are various concept with different contents.

Gross Product: Gross Domestic Product (GDP) measures the total value of final output of goods and services produced within the country's domestic economy by residents and non-residents. Gross National Product (GNP) is comprised of the value of GDP plus factor-income check to residents from abroad, less the income earned in the domestic economy accruing to persons abroad. In other words, GNP measures the total domestic and foreign output claimed by residents (or nationals) of a country.

Net Product: Net Domestic Product (NDP) is the GDP minus depreciation (reduction in the value of assets through wear and tear).

All these estimates are made at market prices. The market prices are made up of costs of production, indirect taxes and subsidies. If we deduct indirect taxes and add subsidies to the various estimates, the resulting value will equal the cost of producing goods and services. Thus, we have two types of estimates: at market prices, and at factor cost. NNP, when calculated at factor cost, is called National Income.

Per Capita Product/Income: Per capita income is another much-used concept. It refers to the average income of a country. It is arrived at by dividing the national income of a country by its population. An increase in the per capita income indicates a rise in the availability of goods and services and for this reason is often used as an index of welfare of the people.

Economic Growth in India: 1951-2001

For a low-income economy like India, a very important indicator of its economic performance is the increase in the national income and per capita income over a period of time. There are two notable features that mark the growth-profile of the economy since the early 1950's. *One* is that there has been an uptrend in the growth rate of the economy, with a faster growth in the recent years. *Two*, associated with it there is some increase in the per capita income.

Gross Domestic Product: The rate of growth of GDP since 1950 to 1980 has been a little over four per cent. Within this long-term trend, there are some significant sub-trends. In the first three decades (1950-80), the growth rate hovered around a low of 3.5 per cent per annum (often described as the Hindu Growth-Rate for its stubbornness to remain stuck at the low level for long). The break from this trend came in the 1980's with the growth-rate going above 5 per cent (during 1980-81 to 1991-92). A further acceleration of the growth-rate has taken place in the 1990's. The growth-rate moved over to 6 per cent from 1992-93 to 2000-2001 (Table 1).

In absolute terms, the GDP (at constant 1993-94 prices or in real terms) rose somewhat slowly in the three decades since 1950 from, over Rs. 100,000 million in 1950-51 to over Rs. 400,000 million in 1980-81. Thereafter the rise was rapid and large, with the GDP rising to about Rs. 700,000 million in 1990-91, and further to nearly Rs. 1200,000 million in 2000-01.

As for the sectoral contributions to the GDP is concerned all the three sectors have added to the final outcome. There are, however, differences in the contributions of different sectors over time. In the early low growth phase of three decades since 1950, agriculture was an important contributor, alongside industry and services. In the recent high-growth phase since 1980, the service sector, with the fastest growth, emerged as the largest contributor. This is followed by industry and agriculture (Table.1).

Table 1
Real Growth Rates (at Factor Cost)

Year	*GDP*	*Agriculture and Allied Activities*	*Industry*	*Services*
1980-81 to1991-92	5.4	3.9	6.3	6.4
1992-93 to2000-01	6.4	3.3	6.5	8.2
1993-94	5.9	4.1	5.2	7.7
1994-95	7.3	5	10.2	7.15
1995-96	7.3	0.9	11.6	10.5
1996-97	7.8	9.6	7.1	7.2
1997-98	4.8	2.4	4.3	9.8
1998-99	6.5	6.2	3.7	8.5
1999-2000 (P)	6.1	1.3	4.9	9.5
2000-2001 (Q)	4	9.2	6.3	4.8
2001-02 (A)	5.4	5.7	3.3	6.5

A = Advance estimates Q = Quick estimates P = Provisional

Source: Central Statistical Organization/Economic Survey 2000-01.

Per Capita Income: The growth-rate of per capita income since 1950 has been of the order of about 2 per cent per annum. This is because the difference between the growth-rate of national income and that of population has been of this size. Within this long-term trend, there are some important sub-trends. In the three decades since 1950, the growth-rate has been dismally low because of the slow growth of national income and a rapid increase in population. For almost half the number of years, it has been less than one per cent, with negative growth for as many as ten years. In the 1980's, the growth-rate has been much higher at an average of over 3 per cent. It increased further in the 1990's and moved on an average of about 4 per cent.

Growth Performance of Major States: Inter State Comparisons

The growth performance of the fourteen major states in the pre and post reform period can be studied on the basis of the available data on the National State Domestic Product (NSDP) for each state. A comment on data problems is appropriate at this stage. Ideally, the NSDP data series for individual states should be fully consistent with the national account estimates of Gross Domestic Product, so that the desegregated picture of economic performance at the state level corresponds with the picture for the country as a whole that emerges from the national accounts. This type of consistency is not possible at present.

Though there are differences in the growth rates as between gross SDP and net SDP – total as well as per capita, these differences are not so significant as to alter the sequential order of states arranged in descending order of growth rates except for a very few cases. Therefore, the discussion being carried on here is based on net SDP, though some references are made to the growth rates based on gross SDP (Figure 1 to Figure 3).

All measures of SDP growth suggest that there has been an overall acceleration between 1980 and 1990. The number of states whose annual growth rates have been higher than the average growth for all states together, has risen somewhat in gross as well as net total SDP, but such rise has been much sharper in terms of per capita SDP, which combined with the acceleration in overall growth, portrays a situation of greater inequality in the 1990s. This difference in the picture as between total SDP

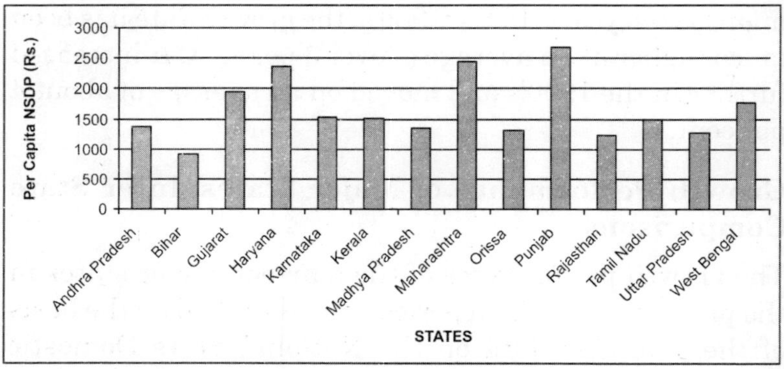

Figure 1: Per Capita Net State Domestic Product in Selected Indian States (1980-81)

growth and per capita SDP growth arises from the relative differences in the population growth of states. For instances, Kerala's total SDP growth at compound rate is about 5.0 per cent has been below the all-states growth during the 1990s, but in per capita SDP, Kerala's annual growth in the same decade at over 4.10 per cent has upstaged the average per capita growth for all states which has worked out to about 3.80 per cent per annum. On the other hand, Maharashtra, which has experienced 6.5 per cent annual growth in total NSDP during the 1990s and remained above the all-states growth, has slipped below in rank and comes closer to the average growth in terms of per capita SDP, though fractionally, that is, at 3.82 per cent against the all-states average growth of 3.78 per cent. This is understandable, for Kerala had, for instance, a population growth of 9.42 per cent only between the 1991 and 2001 censuses, while Maharashtra had a population expansion of 22.57 per cent, that is, slightly higher than even the nation's overall population increase of 21.35 per cent during the decade (Ahluwalia, 2002).

In terms of both SDP and per capita SDP growth, all states have shown accelerated growth during the 1980s and many major states like Punjab, Haryana, Gujarat, West Bengal, and Kerala have moved up the ladder above or near the all-states average SDP growth. In this respect, amongst the major states that have shown outstanding performance in the 1990s are Karnataka, West Bengal, Rajasthan, Tamil Nadu and Gujarat which belong to different regions in the country, but at the same time, deceleration in the growth of high-income states of Punjab, Haryana and Maharashtra stands out. It should be noted in parenthesis here that Punjab has remained as the state with the highest per capita NSDP throughout the past two decades but has persistently experienced reduced growth rates.

At the other end, the low-income and poorly-performing major states of Uttar Pradesh, Madhya Pradesh, Bihar and Orissa, have not only persisted with their low growth syndrome but have also experienced further deceleration in growth rates in the 1990s (except Bihar and Orissa which have shown some differing trends as between per capita GSDP and per capita NSDP.

With the general phenomenon of relatively higher population growth in low-income and low-growth states, the growth of per capita income in such states is found to be relatively lower than the growth of total state incomes. However, the per capita income growth of high-income and middle-income states is comparatively higher than the growth in their total incomes.

As explained earlier, the measure of annual average of incomes seems to become sensitive to the years covered in the period of the study.

The inclusion of the first three years in the decade of the 1990s by extending backward the 1993-94 series through splicing as above alters the relative growth rates of various states; this gives a clue regarding the possible differences in the growth rates of state SDP and per capita SDP as between the initial years of the decade and its

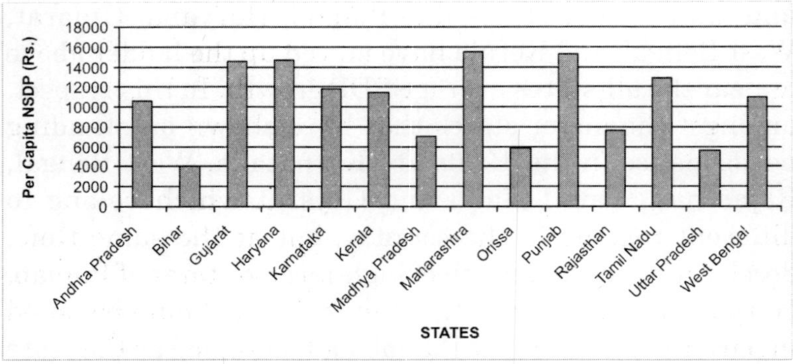

Figure 2: Per Capita Net State Domestic Product in Selected Indian States (2002-03)

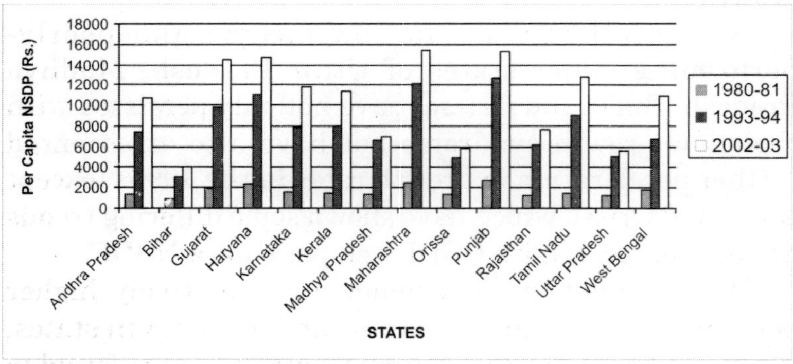

Figure 3: Per Capita Net State Domestic Product in Selected Indian States (1980-81 to 2002-03)

latter years. Overall, the difference in growth has been negligible; while all states total GSDP shows a fractional deceleration in growth from about 5.60 per cent in the second half, all-states per capita NSDP sustains a growth of about 3.60 per cent per annum throughout the two decades. But, growth differences at individual states levels are significant. While a number of major states like Karnataka, West Bengal, Punjab, Haryana, Andhra Pradesh and Rajasthan have shown acceleration in growth in the second half, some crucial ones, on the other hand namely, Maharashtra, Gujarat, Madhya Pradesh and

Orissa have experienced deceleration. The loss in the growth momentum of Maharashtra as between the two five year periods is indeed steep – from 6.5 per cent to 4.2 per cent in the case of Maharashtra in NSDP. Though Kerala's growth in total NSDP shows a decline in the second half of the decade, its performance nevertheless remains above the national average in terms of per capita income growth.

Structural Changes in State Economies

The mass of SDP data presented by industry of origin enables us to understand the nature of structural changes taking place in state economies. To be able to do so, the sectoral compositions of GSDP /NSDP, expressed in percentages, have been worked out and presented. Apart from depicting the nature of changes taking place in the composition of state incomes as between agriculture, industry and services, the structural ratios also facilitate an assessment of the changes taking place in the composition of state incomes as between agriculture, industry and services, the structural ratios also facilitate in assessment of the changes in the relative prices experienced by various sectors (Figure 4 to Figure 17).

Table 2 exhibits that amongst the major states, the share of agriculture in total NSDP remained at one-half or above until the early 1980s in respect of a large number of important states, namely, Bihar, Uttar Pradesh, Madhya Pradesh, Rajasthan, Orissa, Punjab and Haryana. That was the period when the national share of agriculture in Gross Domestic Product (GDP) was around 40 per cent. Thus, the high agricultural share was observed both in agriculturally advanced states with the highest per capita income in the country (Punjab and Haryana) and in many poorer states reflecting general backwardness. Towards the end of the 1990s, Punjab and Bihar are the only two states which have retained a 40

Table 2
Share of Primary, Secondary and Tertiary Sector in Total NSDP (1980-81)

States	*Primary*	*Secondary*	*Tertiary*
Andhra Pradesh	45.59	17.64	36.77
Bihar	51.06	21.51	27.43
Gujarat	39.93	28.12	31.95
Haryana	54.23	19.35	26.42
Karnataka	45.41	21.41	33.18
Kerala	39.11	24.49	36.4
Madhya Pradesh	52.12	21.87	26.01
Maharashtra	27.75	35.45	36.8
Orissa	53.14	17.64	29.22
Punjab	49.48	18.49	32.02
Rajasthan	50.34	19.95	29.71
Tamil Nadu	25.43	33.97	40.6
Uttar Pradesh	52.01	15.63	32.36
West Bengal	31.63	30.26	38.12

*Source:*Domestic Product of States of India: 1960-61 to 2000-01

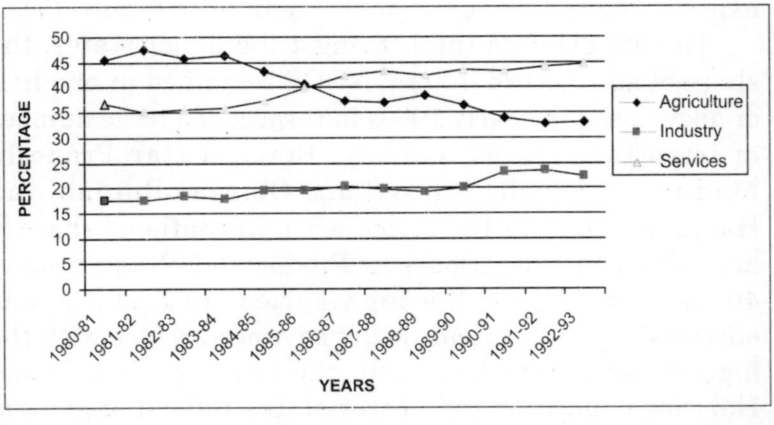

Figure 4: Share of Agriculture, Industry and Service Sectors in Total NSDP – Andhra Pradesh (1980-81 to 1992-93)

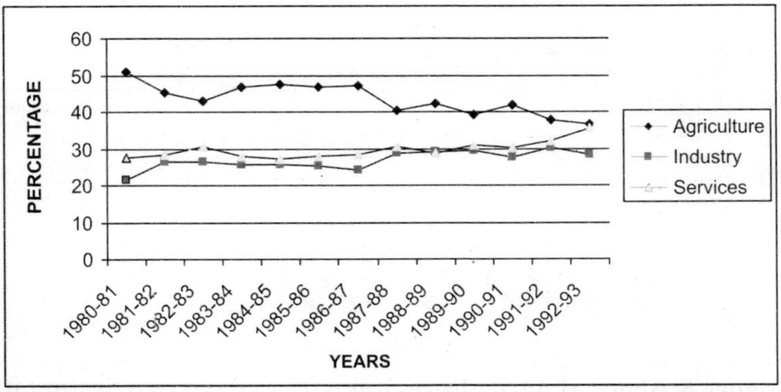

Figure 5: Share of Agriculture, Industry and Service Sectors in Total NSDP – Bihar (1980-81 to 1992-93)

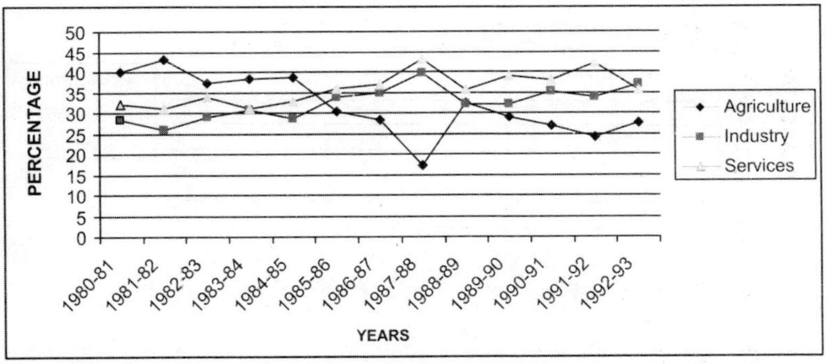

Figure 6: Share of Agriculture, Industry and Service Sectors in Total NSDP – Gujarat (1980-81 to 1992-93)

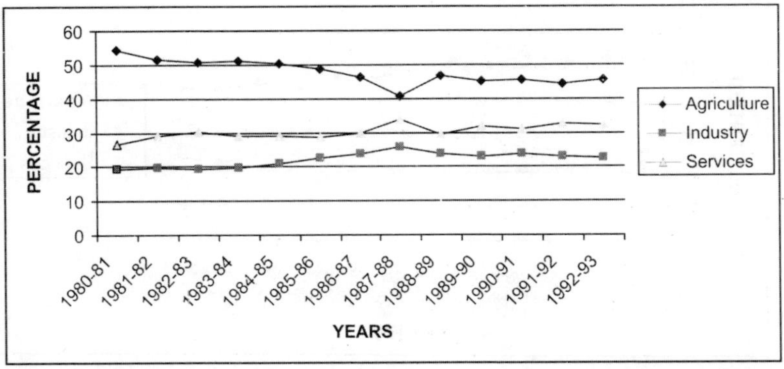

Figure 7: Share of Agriculture, Industry and Service Sectors in Total NSDP – Haryana (1980-81 to 1992-93)

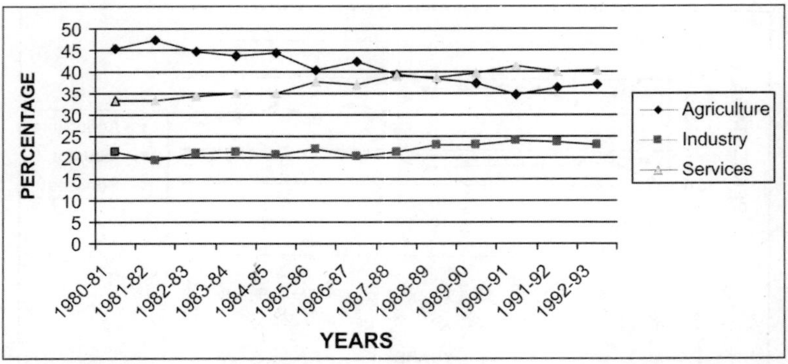

Figure 8: Share of Agriculture, Industry and Service Sectors in Total NSDP – Karnataka (1980-81 to 1992-93)

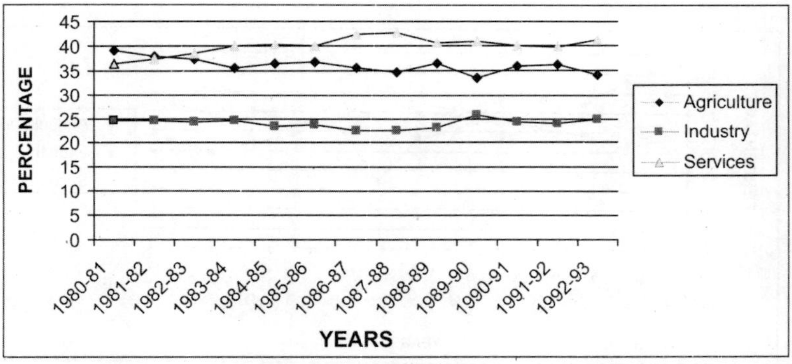

Figure 9: Share of Agriculture, Industry and Service Sectors in Total NSDP – Kerala (1980-81 to 1992-93)

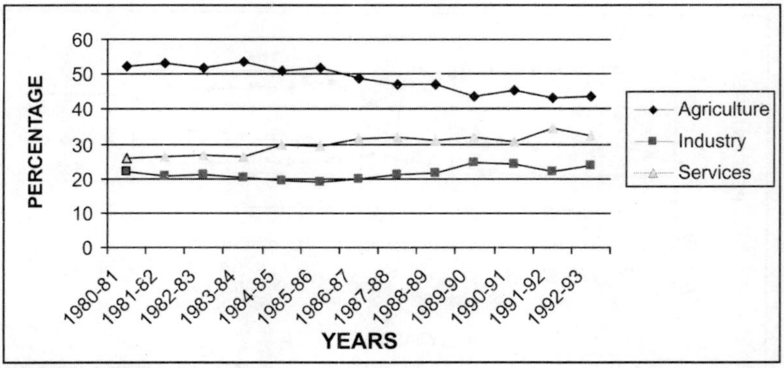

Figure 10: Share of Agriculture, Industry and Service Sectors in Total NSDP – Madhya Pradesh (1980-81 to 1992-93)

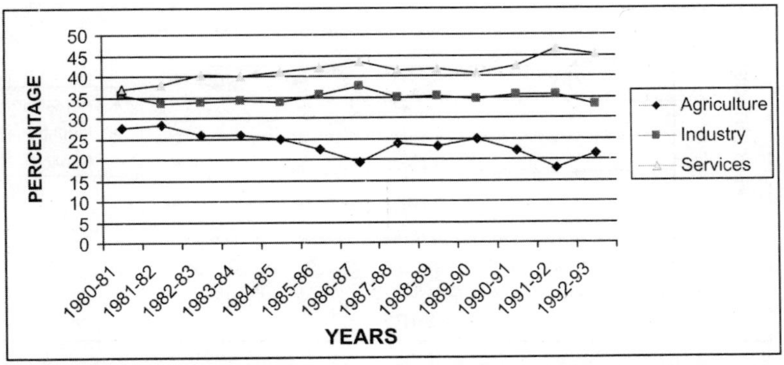

Figure 11: Share of Agriculture, Industry and Service Sectors in Total NSDP – Maharashtra (1980-81 to 1992-93)

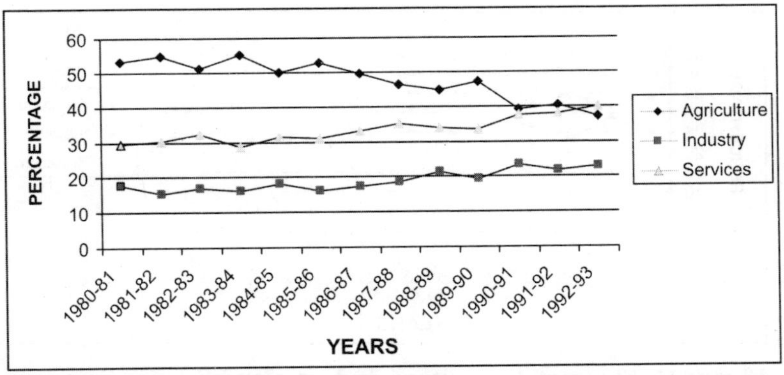

Figure 12: Share of Agriculture, Industry and Service Sectors in Total NSDP – Orissa (1980-81 to 1992-93)

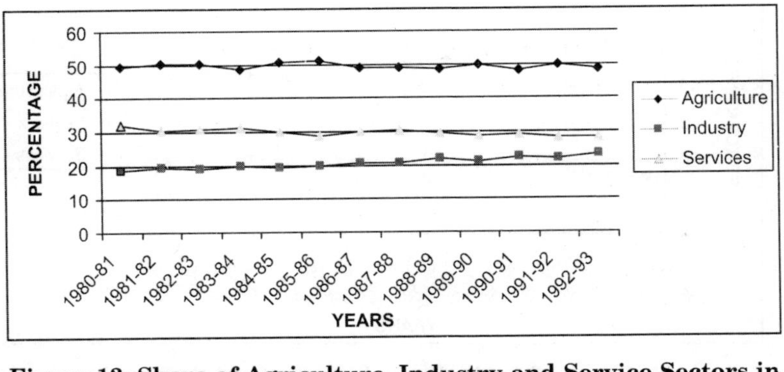

Figure 13: Share of Agriculture, Industry and Service Sectors in Total NSDP – Punjab (1980-81 to 1992-93)

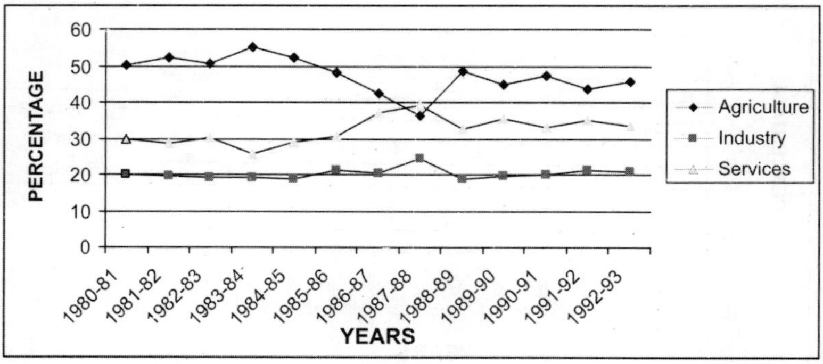

Figure 14: Share of Agriculture, Industry and Service Sectors in Total NSDP – Rajasthan (1980-81 to 1992-93)

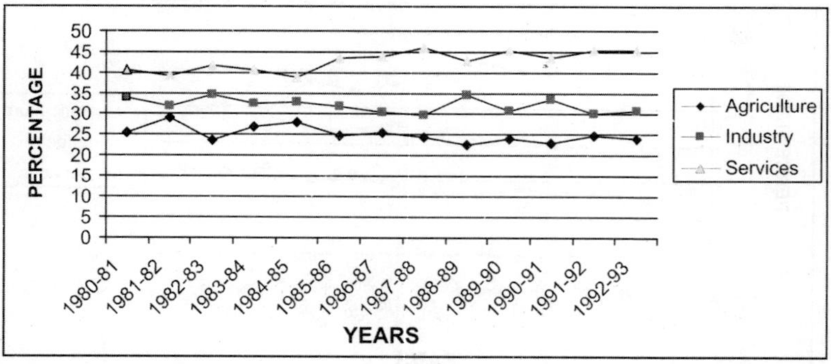

Figure 15: Share of Agriculture, Industry and Service Sectors in Total NSDP – Tamil Nadu (1980-81 to 1992-93)

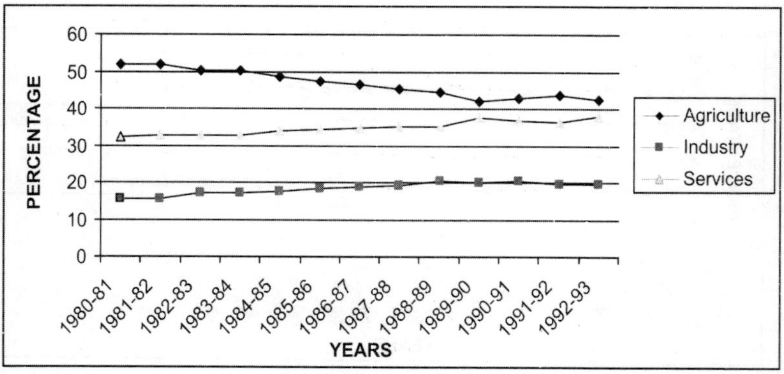

Figure 16: Share of Agriculture, Industry and Service Sectors in Total NSDP – Uttar Pradesh (1980-81 to 1992-93)

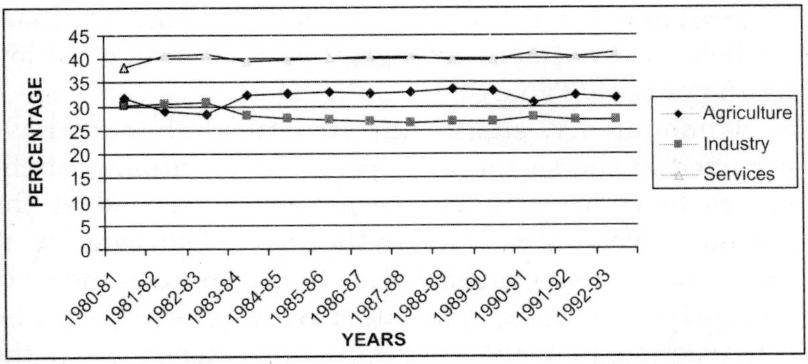

Figure 17: Share of Agriculture, Industry and Service Sectors in Total NSDP – West Bengal (1980-81 to 1992-93)

Table 3
Share of Primary, Secondary and Tertiary Sector in Total NSDP (2002-03)

States	*Primary*	*Secondary*	*Tertiary*
Andhra Pradesh	30.92	15.03	54.05
Bihar	45.46	3.73	50.81
Gujarat	16.45	30.11	53.44
Haryana	32.79	18.68	48.53
Karnataka	31.61	14.61	53.78
Kerala	20.14	13.23	66.63
Madhya Pradesh	28.25	21.33	50.42
Maharashtra	17.00	22.15	60.85
Orissa	31.39	16.89	51.72
Punjab	42.02	16.03	41.95
Rajasthan	26.30	17.37	56.33
Tamil Nadu	18.71	22.09	59.20
Uttar Pradesh	38.46	13.51	48.03
West Bengal	27.73	16.93	55.33

Source: Handbook of Statistics on the Indian Economy, 2004-05

per cent or over as its share of agriculture in NSDP; all other states mentioned above have agricultural share which is around 35 per cent or less. The relatively low national average is because there are a few states relatively industrially advanced like Maharashtra,

Gujarat and Tamil Nadu, which have agricultural share far below the national average; they hover around as low as 15 per cent (Table 3).

Undoubtedly, significant structural changes have occurred in the sectoral compositions of incomes at the states level broadly on the pattern observed at the national level: a steady fall in the share of agriculture (to about 25 per cent), a gentle rise in the share of industry (comprising mining, manufacturing, electricity, and construction) to near 27 per cent, and a steady rise in the share of services to above 48 per cent. At one end, even Punjab's share in agricultural share has fallen from over 50 percent in the mid-1980s to around 40 per cent by end of the 1990s; this relatively moderate fall is accompanied by very moderate rise in the state's shares of NSDP originating in industry and services sectors. In regard to non-agricultural sector shares, what stands out is this niggardly growth of industry share accompanied by a relatively more rapid rise in the share of the services sector. The only exception in this respect has been the state of Gujarat which has experienced a distinct improvement in the share of industry both in the 1980s and in the 1990s; this share went up from 28.12 per cent in 1980-81 series to 30.90 per cent in 1993-94 as per the 1993-94 series (both at constant prices) Gujarat thus support the highest share of industry in the state NSDP. Interestingly, Maharashtra, Tamil Nadu and West Bengal, all of which have had higher industry shares in the early 1980s, have experienced a little setback. Maharashtra and Tamil Nadu improved their industry share in the 1980s, but have suffered a decline in the 1990s, whereas West Bengal is yet to recover from the declining trend in industry share that the state began to experience in the early 1980s. As explained earlier, the share of 'services' sector has risen in all the states, but amongst the major states, the most rapid expansion has taken place in Kerala and Maharashtra, both of which now possess the highest services share of 67and 61 percent each in their respective NSDP at constant prices; both of them had begun the decade of the 1980s with the same ratio of around 36/37 per cent. Except Punjab, Haryana, Uttar Pradesh and Bihar all remaining states had more than 50 percent share in respect of services (Table 4 and Figure 18 to 32).

There are the inter-state differences in sectoral shares as between the current and constant price series, which have implications for trends in inter-sectoral terms of trade. While analyzing the sectoral composition of GDP at the national level, it was said: "agricultural sector, for instance, enjoyed relatively better terms of trade between the mid-1960s and mid-1970s but suffered adverse terms during the 1980s, which has been only partially corrected since the beginning of the 1990s. The manufacturing sector has experienced relatively adverse terms of trade in the 1990s as reflected in the fact that the proportion of GDP originating in the sector at current prices has steadily declined from 16.1 per cent in 1993-94 to 15.3 per cent in 2000-01, while it has stayed put at constant prices – a feature which has significant implication for the process of industrialization'. Obviously, this composite national scenario was the sum total of divergent state-level experiences. For instance, the agriculturally advanced state of Punjab has not benefited from improved terms of trade in agriculture in the 1990s, not even Andhra Pradesh, and Madhya Pradesh, whereas in some other states like Haryana, Bihar, Uttar Pradesh, Karnataka and Tamil Nadu, agriculture has enjoyed some edge in matters of the terms of trade. In the industrially advanced states of Maharashtra, Gujarat, West Bengal and Tamil Nadu, the terms of trade have moved in favour of industry, as seen in a relative rise in the sector's share in GSDP at current prices as compared with that at constant prices.

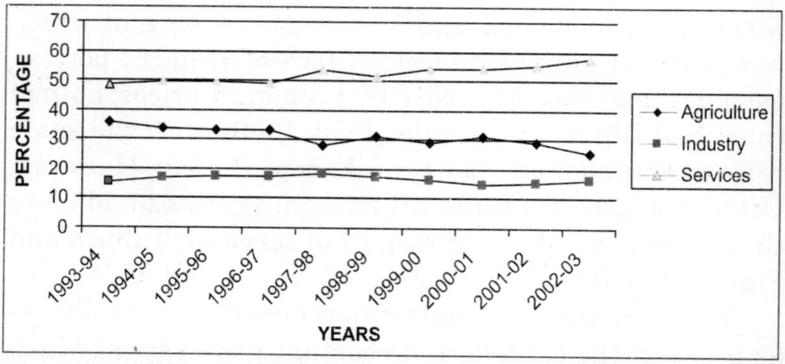

Figure 18: Share of Agriculture, Industry and Service Sectors in Total NSDP – Andhra Pradesh (1993-94 to 2002-03)

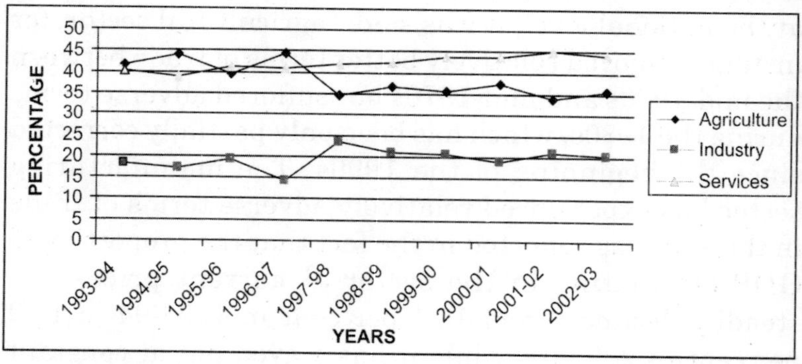

Figure 19: Share of Agriculture, Industry and Service Sectors in Total NSDP – Bihar (1993-94 to 2002-03)

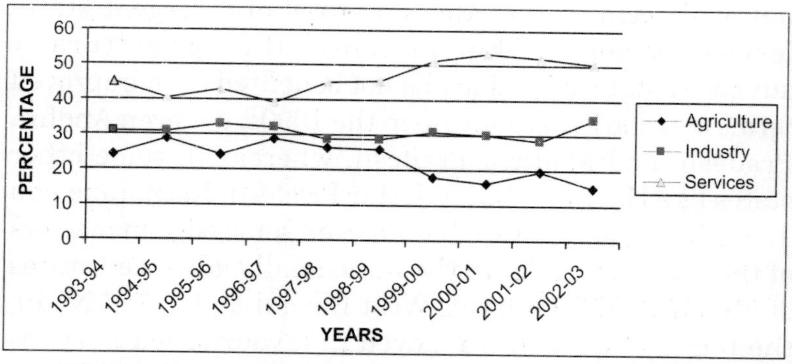

Figure 20: Share of Agriculture, Industry and Service Sectors in Total NSDP – Gujarat (1993-94 to 2002-03)

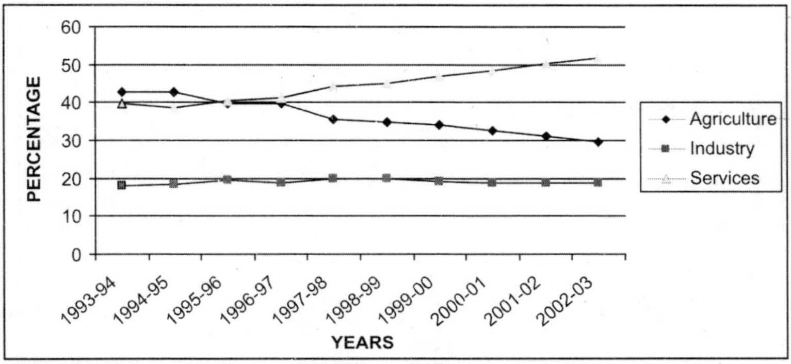

Figure 21: Share of Agriculture, Industry and Service Sectors in Total NSDP – Haryana (1993-94 to 2002-03)

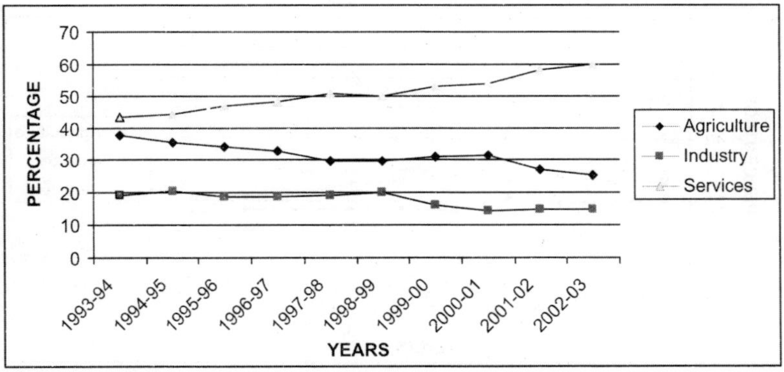

Figure 22: Share of Agriculture, Industry and Service Sectors in Total NSDP – Karnataka (1993-94 to 2002-03)

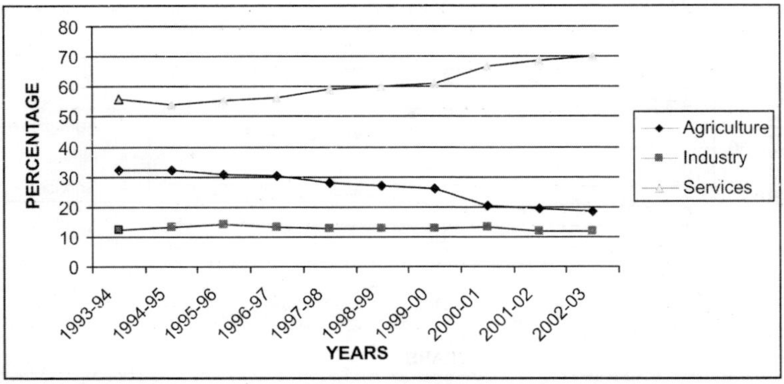

Figure 23: Share of Agriculture, Industry and Service Sectors in Total NSDP – Kerala (1993-94 to 2002-03)

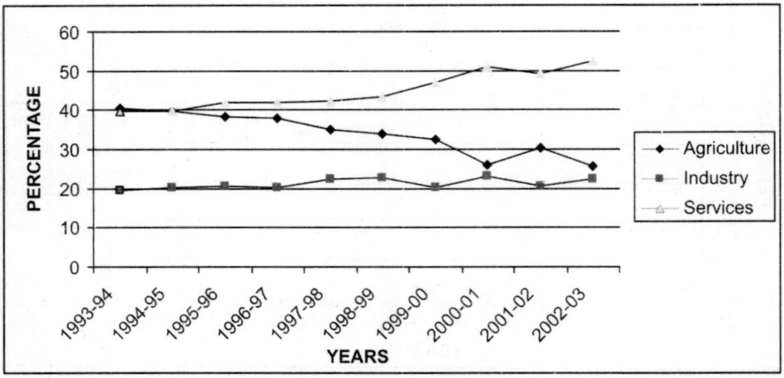

Figure 24: Share of Agriculture, Industry and Service Sectors in Total NSDP - Madhya Pradesh (1993-94 to 2002-03)

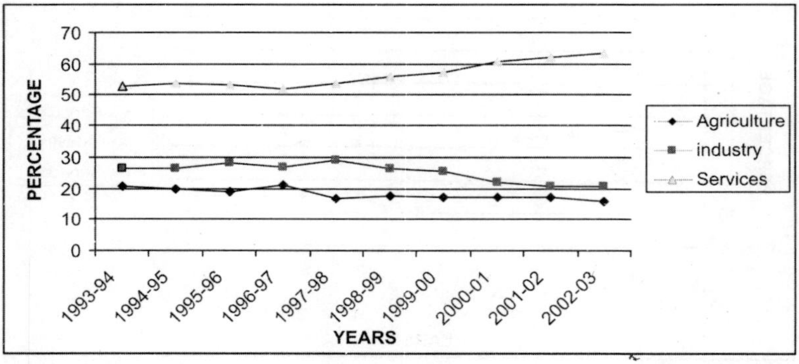

Figure 25: Share of Agriculture, Industry and Service Sectors in Total NSDP - Maharashtra (1993-94 to 2002-03)

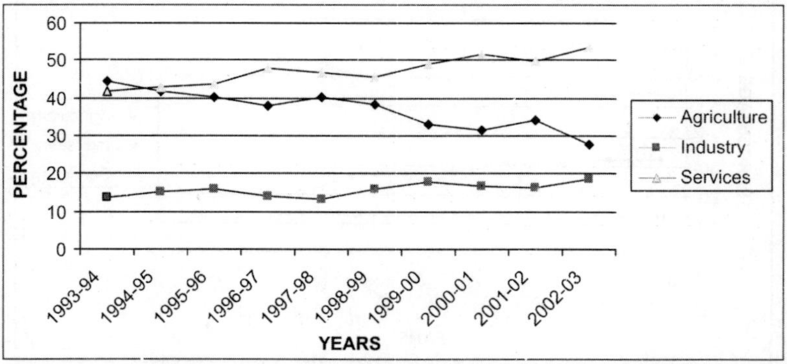

Figure 26: Share of Agriculture, Industry and Service Sectors in Total NSDP - Orissa (1993-94 to 2002-03)

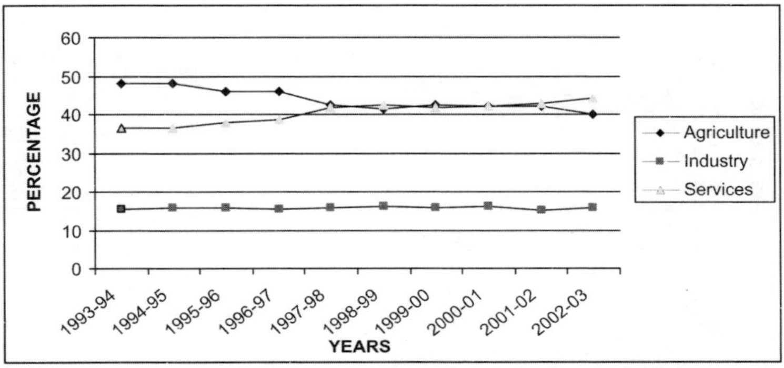

Figure 27: Share of Agriculture, Industry and Service Sectors in Total NSDP - Punjab (1993-94 to 2002-03)

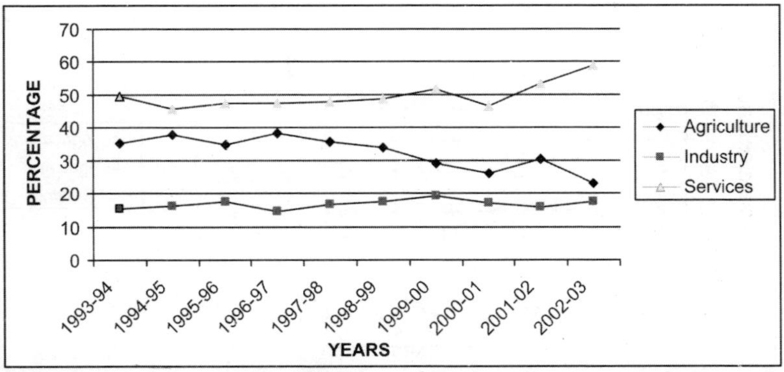

Figure 28: Share of Agriculture, Industry and Service Sectors in Total NSDP - Rajasthan (1993-94 to 2002-03)

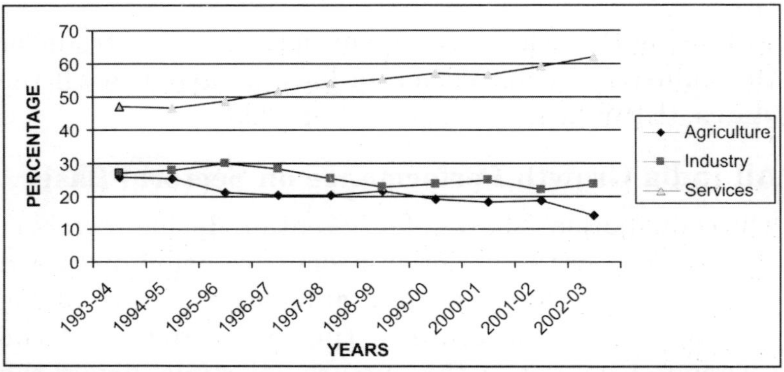

Figure 29: Share of Agriculture, Industry and Service Sectors in Total NSDP - Tamil Nadu (1993-94 to 2002-03)

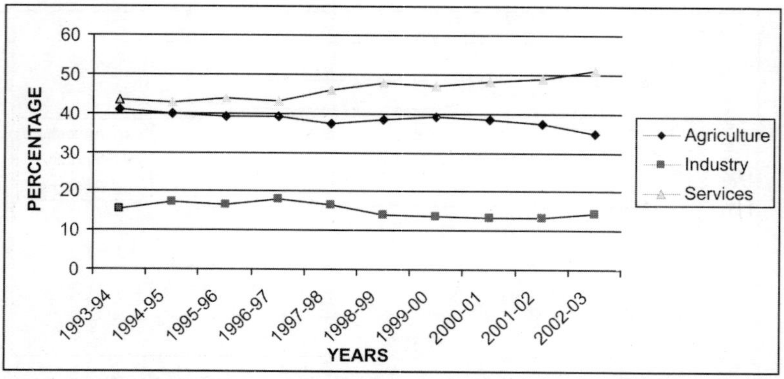

Figure 30: Share of Agriculture, Industry and Service Sectors in Total NSDP – Uttar Pradesh (1993-94 to 2002-03)

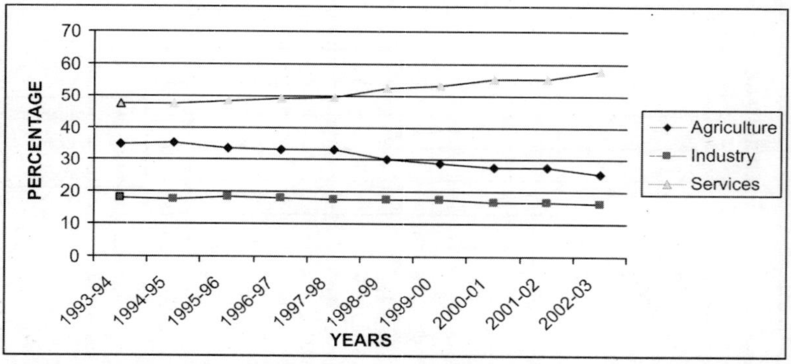

Figure 31: Share of Agriculture, Industry and Service Sectors in Total NSDP – West Bengal (1993-94 to 2002-03)

Interestingly, such improvement in the terms of trade for the industrial sector seems to have occurred in all the states. (EPW Research Foundation, 2002).

All India Growth Performance on Sectoral Basis

The contributions/shares of different producing sectors of the economy of the national income are being discussed in this section. Such a study provides a picture of the composition or structure of the total product of the economy. Changes in this picture, overtime, depict the changes in the standing of the producing sectors in the economy. It enables to know whether the economy is moving forward or backward.

Agriculture and Allied Activities

A low growth-rate in this sector has pulled down its share in the domestic product. However, it needs to be stressed that a low growth-rate of this sector is not desirable. It is its share in the domestic product that will, continue to decline vis-a-vis the other two sectors as the economy develops.

Table 4
Sectoral Distribution Of Gross Domestic Product at 1993-94 Prices

Sector	*1950-51*	*1960-61*	*1970-71*	*1980-81*	*1990-91*	*2000-01*
Agricultural and Allied activities	59.20	52.6	48.12	39.8	34.93	26.55
Manufacturing construction, electricity, gas & water supply	13.29	19.6	19.91	24.3	24.49	25.00
Services	27.51	27.8	31.97	38.9	40.58	48.45
GDP (at factor cost)	100.00	100.00	100.00	100.00	100.00	100.00

Source: Government of India, Economic Survey, 2001-2002.

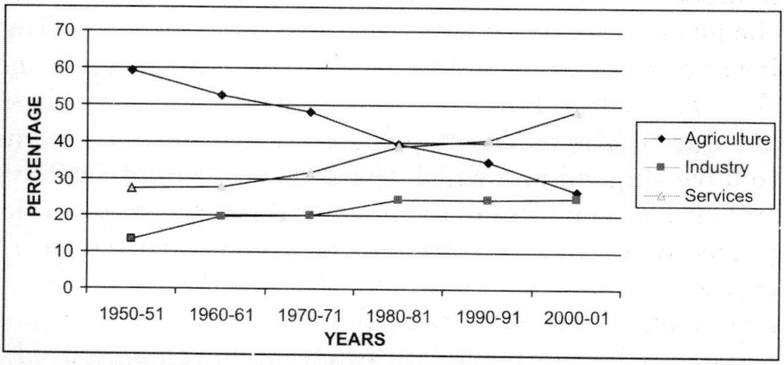

Figure 32: Sectoral Distribution of Gross Domestic Product in India (1950-51 to 2000-01)

Industry

The growth-rate of industrial production, though fair, has not been sufficient to raise its share in the domestic product significantly. It is because of several reasons. The investment in this sector, in particular in the infrastructure related to industry, has been very low indeed. Again, for many years, till the nineties of the 20th century, the strategy of development has been focused on the heavy and capital-goods industries. As these industries have a high capital-output ratio and are long-gestating, the output flows have been small and slow in coming. These are unlike the consumer goods industries, being emphasized in the present, that have a low capital-output ratio and are short-gestating, and therefore cause more output to flow and over a short period.

Services

The growth-rate has been significant enough to raise its share in the domestic product-to more than that of the commodity sectors put together. An important aspect of this change is that, when compared to the present developed countries, it has taken place in a relatively short period of time. As such India's experience in this respect is an altogether different one. It marks a big jump over or leap for the economy from a lower stage to a higher stage, the former associated with the Agriculture sector-making lower order of value additions to the national income, and the latter with the service sector contributing higher level of value additions. The stage in between, that India seems to have skipped and which the developed countries have traversed, is one where technological improvements take place to further the process of steady growth of an economy. India has made use of the already developed technologies, an advantage that curiously enough, has accrued to her for being late starter in development. The much larger share of the service sector in the domestic product will have several implications, mostly beneficial, for the economy.

Comparative Assessment of Major States: Pre and Post Reform Period

The growth rate of the combined NSDP of all the fourteen states taken together increased from 4.8 per cent in the pre reform period to 5.4 percent in the post reform period. This acceleration in the combined NSDP is similar to the picture that emerges from the national accounts, except that the post reform acceleration of GDP in the national accounts is much sharper. GDP grew at 5.5 per cent per year in the first period, which really faster than the 4.8 per cent growth recorded by the combined NSDP of the fourteen states. However, GDP growth accelerated to 6.5 percent in the second period, which was much faster than the 5.4 per cent growth in the combined NSDP. The faster growth recorded in the national accounts GDP series introduced from 1993-94 onward. It is possible that if the NSDP data were revised similarly, the growth rates of NSDP of the different states in the second period would be correspondingly higher. We note that if such a re-vision were to affect some states more than others, it could also alter our assessment of the relative performance of states, but in the absence of specific information that might have guided us on this issue, we assume that the adjustment would adjust growth rates upward across all states, leaving relative performance unaffected.

There is variation in growth performance across states in both periods, with some states growing faster than others, but the degree of dispersion in growth rates increased very significantly in the 1990s. The range of variation in the first period was from a low of 3.0 percent per year for Bihar to a high of 6.5 per cent in Maharashtra. In the second period, the range increased from a low of 3.3 percent per year for Orissa to a high of 7.7 percent for Karnataka (Table 5).

Table 5
Compound Annual Growth Rate of Net State Domestic Product

States	*1980-81 to 1993-94*	*1993-94 to 2002-03*
Andhra Pradesh	5.9	5.8
Bihar	3.0	6.2
Gujarat	5.2	5.6
Haryana	5.5	5.8
Karnataka	5.5	7.7
Kerala	3.9	5.0
Madhya Pradesh	4.3	3.4
Maharashtra	6.5	4.2
Orissa	3.1	3.3
Punjab	5.1	4.4
Rajasthan	5.4	6.5
Tamil Nadu	5.5	6.5
Uttar Pradesh	4.2	3.7
West Bengal	4.6	7.1

Source: Handbook of Statistics on the Indian Economy, 2004-05

The increased variation in growth performance in the 1990s reflects very different behavior at different ends of the spectrum of per capita NSDP. Growth accelerated sharply for two states at the upper end of the spectrum, namely Gujarat and Maharashtra, but it actually decelerated in Bihar, Uttar Pradesh, Gujarat and Orissa, all three of which were not only at the lower end of the capita NSDP spectrum but also had relatively low rates of growth to begin with. The growth pattern in the 1990s therefore increased regional inequality.

Only four states achieved relatively strong growth with growth rates of NSDP in the 1990s above 6.0 percent in the second period, with Karnataka (7.7 per cent) , West Bengal (7.1 per cent), Tamil Nadu (6.5 per cent) and Rajasthan (6.5 per cent). In addition, Andhra Pradesh and

Haryana grew at 5.8 per cent, with Gujarat at 5.6 percent, which is almost at the 6 per cent level. It is very likely that if the NSDP series were revised to reflect the changes made in the national accounts, the growth rates of all seven states would exceed 6 per cent. Except for Rajasthan, all these states also show an acceleration in growth compared with the pre reform period.

An interesting feature of the performance in the 1990s is that the popular characterization of the so-called BIMARU states (Bihar, Madhya Pradesh, Rajasthan and Uttar Pradesh) as a homogeneous group of poor performers, a grouping originally proposed in the context of observed commonalties in demographic behaviour, does not hold as far as economic performance in the post reform period is concerned. Orissa and Madhya Pradesh performed very poorly, growing much more slowly than the average, but the state of Madhya Pradesh, has performed reasonably well. Rajasthan shows a acceleration in growth of NSDP compared with the 1980s, therefore it remained a good performer in the 1990s, growing at about the average for all states. Madhya Pradesh, on the other hand, which had grown more slowly than the average in the decade of 1980, has shown deceleration in the decade of 1990.

Simplistic perception about the role of geography in determining performance, such as the view that only the coastal states or the southern states have done well in the period of liberalization, are also not universally valid. Orissa is a coastal state, but its growth performance is very poor, while Rajasthan is a heartland state and has performed reasonably well. Only two of the southern states, Tamil Nadu and Karnataka, made it to the top six in terms of growth of NSDP in the 1990s. The southern states as a group have done well, but they were by no means the only beneficiaries of the growth acceleration witnessed in the 1990s.

The remarkable performance of Karnataka and West Bengal, grew at over 7 percent per annum in the decade of 1990, a rate normally associated with "miracle growth" economies, deserves careful study. These states clearly benefited the most in the post reform period, but it is important to note that their superior performance was not the result of any conscious policy that limited the benefits of liberalization to these states. Their superior performance must be attributed primarily to the ability of these two states to provide an environment most conducive to benefiting from the new policies. Their experience, together with the experience of the other strong performers, should provide the basis for identifying the critical ingredients of success in accelerating growth, which should be follow by others.

The deceleration of growth in the poorer states witnessed in the 1990s has important implications for regional balance. Regional differences in per capita income level have long been a matter of concern in India, and for good reason. The per capita NSDP of Punjab, the richest state, is five times that of Bihar at the other end of the spectrum. Balanced regional development has always been stated as an objective in India's plans, and although this objective has never been quantified in terms of rates of convergence of per capita NSDP, or a reduction in regional inequality to some specified target in terms of one of the inequality measures, the objective surely implies that regional differences in per capita incomes should narrow with development, and in any case not widen.

Conclusion

- The growth-performance, when viewed in overall terms, appears fair enough in one sense that it exhibits a persistent rise both in the national income and the per capita income. There are, no doubt, years of poor showings with the rate of growth going down or even becoming negative. Yet the over riding fact is that the rise, often modest and sometimes sizeable, continues, though the up curve is not a smooth one. This uptrend is particularly remarkable in the recent years of the 1990s'. At over 6 per cent, it is much above the 3.5 per cent growth which prevailed in the first three decades since the early 1950's. It is also a more noteworthy achievement than the 5 per cent growth of the 1980's. The growth rate is of great importance also for reasons of welfare. As it has been higher than the population growth, there has been a rise in the per capita income, sometimes quite large as in the 1990's. Although there have been years when the per capita income has not risen much or has fallen, yet for most of the time, it has been on the rise.

Another positive aspect of the recent growth-scenario is its developmental character. This is obvious from the fact that, there is a significant change in the contributions of the three sectors. The service sector has now become the leading sector in the composition of the gross domestic product because of its faster growth vis-a-vis industry and agriculture. This is different from what prevailed before 1980's, when agriculture, followed by industry dominated the scene. As a result the structure of the economy has also undergone a big change. The share of agriculture in the national product, though still very significant, has declined considerably from about 57 per cent in 1950 to about 27 per cent in 2000. These are features of a developing economy. Along with this the share of the non-agricultural sectors industry and services has gone up from 43 per cent to over 73 per cent. More significant, however, is the sea change that has marked the services sector. Its share in the national product at 30 per cent in 1950 rose sharply to 48 per cent. This change, by itself, is evidence enough to prove that the Indian economy is moving the way the present developed countries had moved in the past. These structural changes also indicate that the growth has been a broad based one. At the same time, it proves that the growth is no longer dependent on agriculture. On the contrary, it is now being led by the two dynamic sectors, namely, industry and services.

- It is not true that all the richer states got richer relative to the poorer states. Maharashtra, Punjab and Haryana were richer states in 1990-91, but their growth rates of compound NSDP not only were lower in the 1990s than in the 1980s, but in all cases actually fell below the national average. Except Madhya Pradesh, Uttar Pradesh and Orissa, which grew even more slowly, all other states narrowed the distance between themselves and these richer states.
- Karnataka, West Bengal and Gujarat, which are in the group of high growth rate of NSDP and were ranked just below three richer states of Maharshtra, Punjab and Haryana at the start of the 1990s, accelerated very significantly and achieved the fastest rates of growth in compound NSDP. These states clearly pulled ahead of all other states.
- Three of the poorest states, Madhya Pradesh, Uttar Pradesh, and Orissa, which together account for about one-third of the population of the fourteen states, fared very poorly in the 1990s. It is also important to note that not all the poorer states performed badly. Rajasthan, for example, experienced fairly good growth in compound NSDP, more than double that of the other poor states, though still below the national average.
- Performance of four middle-income states of 1980s (West Bengal, Tamil Nadu, Kerala and Karnataka)

was above the average, with Karnataka and West Bengal showing very strong growth. These states not only improved their position relative to the average, but also grew faster in terms of NSDP than they did in the 1980's.

References

Aggarwal, M. et al. (1998), *"Indian Economy in Transition-Environmental and Development"*, Her Anand Publications Ltd. New Delhi, India

Ahluwalia, I. J. and Little, I. M. D. (1998), "India's Economic Reforms and Development- Essays for Manmohan Singh, Oxford University Press, New Delhi, India.

Ahluwalia, I. J. and Williamson, J. ed. (2003), 'The South Asian Experience with Growth', Oxford University Press, New Delhi, India.

Ahluwalia, M. S. (2002), *"State-Level Performance Under Economic Reforms in India"*, In Economic Policy Reforms and the Indian Economy, (eds) Anne O. Krueger, Chicago and London, The University of Chicago Press.

Ahmed, N., (2004), *"India: State of Economy"*, Kilaso Books, New Delhi, India Alberini, A. and Kahn, J.R. ed., (2006), *"Handbook on Contingent Valuation"*, Edward Elgar Publishing Ltd, UK & USA.

Alternative Economic Survey 2001-2002. "Economic 'Reforms': Development Denied", Alternative Survey Group, Rainbow Publishers, India.

Anonymous, (1982), *"Indian Minerals Year Book"*, Indian Bureau of Mines, Nagpur.

Anonymous, (1985), *"Statistical Abstract: India-Vol. 28"*, Central Statistical Organization, Department of Statistics, Ministry of Planning, Govt. of India.

Anonymous, (1996), *"Statistical Abstract: India"*, Central Statistical Organization, Department of Statistics, Ministry of Planning, Govt. of India.

Anonymous, (2003), "Compendium of Environment Statistics: India", Central Statistical Organization, Ministry of Statistics and Programme Implementation, Govt. of India.

Anonymous, (2003), *"Domestic Product of States of India: 1960-61 to 2000-01"*, Mumbai, EPW Research Foundation.

Anonymous (2004), *"Statistical Abstract: India"*, Central Statistical Organization, Ministry of Statistics and Programme Implementation. Govt. of India.

Anonymous (2004-05), *"Handbook of Statistics on the Indian Economy"*, Reserve Bank of India.

Anonymous, (Nil Dated), *"Industrial Databook, 2000-01"*, Centre for Industrial and Economic Research.

Arnold, F. S. (1995), *"Economic Analysis of Environmental Policy and Regulation"*, John Wiley & Sons, Inc. USA.

Byres, T. J. ed. (1998), *"The Indian Economy: Major Debates Since Independence"*, Oxford University Press, Delhi, India.

Dhawan, S. K. (2006), *"A Handbook of Indian Economy: Vol. I and II"*, Max Books, New Delhi, India.

Mozoomdar, A. (2002), *"The Indian Economy: A Different View"*, Har Anand Publications Pvt. Ltd., New Delhi, India.

Perdikis, N. ed. (2000), *"The Indian Economy: Contemporary Issues"*. Ashgate Publishing Ltd., England.

Sagoff, M. (1988), *"The Economy of the Earth"*, Cambridge University Press, Cambridge.

Singh, R. and Mishra, D., ed (1996), *"Development and Environmental Change in India"*, New Delhi: APH Publishing Corporation.

Sury, M. M. (2004), *"Indian Economy in the 21st Century: Prospects and Challenge"*, New Century Publications, New Delhi, India

Swanson, T.M., ed. (1996), *"The Economic of Environmental Degradation: Tragedy for the Commons? "*, *Edward* Elgar Publishing Ltd. UK and US.

Tisdell, C. (1993), *"Environmental Economics: Policies for Environmental Management and Sustainable Development"*, Edward Elgar Publishing Ltd., UK & USA.

Index

K

L

M

N

O

P

R